Didactic Strategies and Resources for Innovative Geography Teaching

Alfonso García de la Vega
Universidad Autónoma de Madrid, Spain

IGI Global
PUBLISHER of TIMELY KNOWLEDGE

A volume in the Advances in Educational Technologies and Instructional Design (AETID) Book Series

Published in the United States of America by
IGI Global
Information Science Reference (an imprint of IGI Global)
701 E. Chocolate Avenue
Hershey PA, USA 17033
Tel: 717-533-8845
Fax: 717-533-8661
E-mail: cust@igi-global.com
Web site: http://www.igi-global.com

Library of Congress Cataloging-in-Publication Data

Names: García de la Vega, Alfonso, editor.
Title: Didactic strategies and resources for innovative geography teaching / Alfonso Garcia de la Vega, Editor.
Description: Hershey, PA : Information Science Reference, [2022] | Includes bibliographical references and index. | Summary: "This reference book provides various new approaches to teaching geography, to include such issues as sustainability and climate change by offering didactic strategies related to geographical principles"-- Provided by publisher.
Identifiers: LCCN 2022012071 (print) | LCCN 2022012072 (ebook) | ISBN 9781799895985 (hardcover) | ISBN 9781799895992 (paperback) | ISBN 9781799896005 (ebook)
Subjects: LCSH: Geography--Study and teaching. | Climatic changes--Study and teaching. | Sustainability--Study and teaching. | Educational innovations.
Classification: LCC G73 .D533 2022 (print) | LCC G73 (ebook) | DDC 910.71--dc23/eng20220611
LC record available at https://lccn.loc.gov/2022012071
LC ebook record available at https://lccn.loc.gov/2022012072

This book is published in the IGI Global book series Advances in Educational Technologies and Instructional Design (AETID) (ISSN: 2326-8905; eISSN: 2326-8913)

British Cataloguing in Publication Data
A Cataloguing in Publication record for this book is available from the British Library.

Advances in Educational Technologies and Instructional Design (AETID) Book Series

ISSN:2326-8905
EISSN:2326-8913

Editor-in-Chief: Lawrence A. Tomei Robert Morris University, USA

MISSION

Education has undergone, and continues to undergo, immense changes in the way it is enacted and distributed to both child and adult learners. In modern education, the traditional classroom learning experience has evolved to include technological resources and to provide online classroom opportunities to students of all ages regardless of their geographical locations. From distance education, Massive-Open-Online-Courses (MOOCs), and electronic tablets in the classroom, technology is now an integral part of learning and is also affecting the way educators communicate information to students.

The **Advances in Educational Technologies & Instructional Design (AETID) Book Series** explores new research and theories for facilitating learning and improving educational performance utilizing technological processes and resources. The series examines technologies that can be integrated into K-12 classrooms to improve skills and learning abilities in all subjects including STEM education and language learning. Additionally, it studies the emergence of fully online classrooms for young and adult learners alike, and the communication and accountability challenges that can arise. Trending topics that are covered include adaptive learning, game-based learning, virtual school environments, and social media effects. School administrators, educators, academicians, researchers, and students will find this series to be an excellent resource for the effective design and implementation of learning technologies in their classes.

COVERAGE

- Instructional Design Models
- Social Media Effects on Education
- Web 2.0 and Education
- Virtual School Environments
- Online Media in Classrooms
- Collaboration Tools
- Bring-Your-Own-Device
- Hybrid Learning
- E-Learning
- Educational Telecommunications

IGI Global is currently accepting manuscripts for publication within this series. To submit a proposal for a volume in this series, please contact our Acquisition Editors at Acquisitions@igi-global.com or visit: http://www.igi-global.com/publish/.

The Advances in Educational Technologies and Instructional Design (AETID) Book Series (ISSN 2326-8905) is published by IGI Global, 701 E. Chocolate Avenue, Hershey, PA 17033-1240, USA, www.igi-global.com. This series is composed of titles available for purchase individually; each title is edited to be contextually exclusive from any other title within the series. For pricing and ordering information please visit http://www.igi-global.com/book-series/advances-educational-technologies-instructional-design/73678. Postmaster: Send all address changes to above address.

Titles in this Series

For a list of additional titles in this series, please visit:
www.igi-global.com/book-series/advances-educational-technologies-instructional-design/73678

Guide to Integrating Problem-Based Learning Programs in Higher Education Classrooms Design, Implementation, and Evaluation
Pam Epler (Youngstown State University, USA) and Jodee Jacobs (Youngstown State University, USA)
Information Science Reference • © 2022 • 315pp • H/C (ISBN: 9781799881773) • US $205.00

EdTech Economy and the Transformation of Education
Sara Fazzin (Xenophon College London, UK)
Information Science Reference • © 2022 • 294pp • H/C (ISBN: 9781799889045) • US $215.00

Methodologies and Use Cases on Extended Reality for Training and Education
Anacleto Correia (Naval Academy, Portugal) and Vitor Viegas (Naval Academy, Portugal)
Information Science Reference • © 2022 • 300pp • H/C (ISBN: 9781668433980) • US $215.00

English as a Foreign Language in a New-Found Post-Pandemic World
Walaa M. El-Henawy (Port Said University, Egypt) and Maria del Mar Suárez (Universitat de Barcelona, Spain)
Information Science Reference • © 2022 • 335pp • H/C (ISBN: 9781668442050) • US $215.00

Handbook of Research on Learner-Centered Approaches to Teaching in an Age of Transformational Change
Billi L. Bromer (Brenau University, USA) and Caroline M. Crawford (University of Houston-Clear Lake, USA)
Information Science Reference • © 2022 • 385pp • H/C (ISBN: 9781668442401) • US $270.00

For an entire list of titles in this series, please visit:
www.igi-global.com/book-series/advances-educational-technologies-instructional-design/73678

IGI Global
PUBLISHER of TIMELY KNOWLEDGE

701 East Chocolate Avenue, Hershey, PA 17033, USA
Tel: 717-533-8845 x100 • Fax: 717-533-8661
E-Mail: cust@igi-global.com • www.igi-global.com

To Esther

Omaha, oh, somewhere in middle America
If you're right to the heart that matters, oh
It´s the heart that matters more

Adam Duritz et al.

As long as I gaze on
Waterloo sunset
I am in paradise

Ray D. Davies

Table of Contents

Preface xv

Acknowledgment xxi

Section 1
Teaching Approaches and Strategies

Chapter 1
Using Inquiry-Based Learning Approaches to Teach Undergraduate Geography. 1
Rachel A. Spronken-Smith, University of Otago, New Zealand
Simon Kingham, University of Canterbury, New Zealand
Ralf A. Ohlemüller, University of Otago, New Zealand

Chapter 2
Application of Inquiry-Based Science Education, Anticipatory Learning Strategy, and Project-Based Learning Strategies in the Context of Environmental Problems of the Contemporary World 23
Iwona Piotrowska, Adam Mickiewicz University in Poznań, Poland
Małgorzata Cichoń, Adam Mickiewicz University in Poznań, Poland
Jakub Sypniewski, Adam Mickiewicz University in Poznań, Poland
Dawid Abramowicz, Adam Mickiewicz University in Poznań, Poland

Chapter 3
The GeoScenario: A Methodological Resource From Problem-Based Learning (PBL) in Landscape Teaching 51
Alfonso García de la Vega, Universidad Autónoma de Madrid, Spain

Chapter 4
Importance and Educational Possibilities of the Place-Based Education Through Virtual Fieldtrips 79
Okkyong Yoon, Cheongju National University of Education, South Korea

Chapter 5
Future-Proofing Geography Teaching and Learning Using Remote Learning and Co-Creation Approaches 102
Theresa G. Mercer, University of Lincoln, UK
Andrew P. Kythreotis, University of Lincoln, UK
Hannah Kelly, University of Lincoln, UK
Zoe P. Robinson, Keele University, UK
Sharon M. George, Keele University, UK
David Sands, Independent Researcher, UK

Section 2
Resources in Geography Teaching

Chapter 6
Landscape and Heritage: Didactic Resources for Sustainability Education 117
María Casas Jericó, University of Navarra, Spain
Luis Erneta-Altarriba, University of Navarra, Spain
Ana Villarroya Ballarín, University of Navarra, Spain

Chapter 7
Identifying Challenges in Openstreetmap That Potentially Prevent New Users From Contributing: Implications for Addressing OSM in Education 144
Robert Vogler, Paris-Lodron-University of Salzburg, Austria
Sabine Hennig, Paris-Lodron-University of Salzburg, Austria

Chapter 8
A Didactical Approach of Rhumb Line vs. Great Circle in Web Mercator Projection for 6th Grade Pupils 168
Nikolaos Lambrinos, Aristotle University of Thessaloniki, Greece
Ioanna Repanidou, Aristotle University of Thessaloniki, Greece
Georgia Intzidou, Mandoulides Schools, Greece

Chapter 9
Reading and Spatial Analysis Through Mental Maps in School Geography..... 203
Denis Richter, Universidade Federal de Goiás, Brazil

Chapter 10
River Landscape: A Didactic Proposal for Primary Education 228
Elena María Muñoz Espinosa, Universidad Autónoma de Madrid, Spain
Celia Laguna Mora, University of Castilla-La Mancha, Spain

Chapter 11
The Spatial Experience for Teaching Geography in the Anthropocene............259
Sylvie Joublot Ferré, Haute École Pédagogique de Lausanne, Switzerland

Compilation of References 282

About the Contributors 318

Index 324

Detailed Table of Contents

Preface xv

Acknowledgment xxi

Section 1
Teaching Approaches and Strategies

Chapter 1
Using Inquiry-Based Learning Approaches to Teach Undergraduate Geography. 1
Rachel A. Spronken-Smith, University of Otago, New Zealand
Simon Kingham, University of Canterbury, New Zealand
Ralf A. Ohlemüller, University of Otago, New Zealand

This chapter discusses the use of inquiry-based approaches in teaching geography. Following an introduction of definitions for inquiry approaches and how they might be used in teaching, two case studies are presented. The cases involve similar final-year research methods courses in which students work in groups to address real-world issues or problems. Both courses are highly rated by students, and there is evidence of students attaining high-level learning outcomes. The benefits and challenges of teaching using inquiry approaches are discussed, and articles for the use of inquiry-based approaches in other geography contexts are provided. Finally, recommendations are provided for teachers wishing to use such approaches, whether they wish to incorporate small inquiry activities in their courses through to the use of inquiry assignments or the design of an inquiry course.

Chapter 2
Application of Inquiry-Based Science Education, Anticipatory Learning Strategy, and Project-Based Learning Strategies in the Context of Environmental Problems of the Contemporary World 23
Iwona Piotrowska, Adam Mickiewicz University in Poznań, Poland
Małgorzata Cichoń, Adam Mickiewicz University in Poznań, Poland
Jakub Sypniewski, Adam Mickiewicz University in Poznań, Poland
Dawid Abramowicz, Adam Mickiewicz University in Poznań, Poland

The global challenges of melting ice caps, rising global ocean levels, increased methane release from permafrost, and restrictions on access to water resources are primarily associated with climate change. Geographical education is of particular importance in this regard. At all levels of education, it should therefore provide knowledge about the changes taking place, shape practical skills, and foster the acquisition of appropriate attitudes in caring for the state of individual components of the geographical environment. Thus, the mentioned problems of the modern world require the selection of appropriate educational strategies. The chapter presents three educational strategies: teaching through scientific inquiry (IBSE), anticipatory learning strategy (ALS), and teaching through project-based learning (PBL). A general characterization was made, and then a proposal for their application to geography education classes was prepared. Content contexts were proposed for the respective strategies: IBSE (global warming), ALS (ice sheet changes), and PBL (water resources and retention).

Chapter 3
The GeoScenario: A Methodological Resource From Problem-Based Learning (PBL) in Landscape Teaching 51
Alfonso García de la Vega, Universidad Autónoma de Madrid, Spain

The geographical scenario (geoscenario) is considered a methodological resource in geography teaching. The geoscenario may be defined as the minimal significant landscape framework unit in landscape teaching that embodies a synthetic representation of a real-world acute issue involving a particular geographical problem. The application of geographical abilities and skills and the selection of real global issues in the geographical scenarios promotes lifelong learning. A broadly-based planning is needed for an integrated and interdisciplinary curriculum in geography. PBL method is aligned with the fundamentals of the constructivist approach and promotes collaborative and cooperative learning in small group work. BPL implementation involves six steps and three progression levels for problem-solving in relation to cognition processes, classroom dynamics, and the identification degree of features and events in the geoscenario. Experiential knowledge and analogical thinking are applied to problem formulation and solving in the geoscenario.

Chapter 4
Importance and Educational Possibilities of the Place-Based Education Through Virtual Fieldtrips 79
Okkyong Yoon, Cheongju National University of Education, South Korea

This chapter deals with the possibilities of place-based education through virtual field trips and strategies to realize them. These are based on the environmental educational value of the experience in a certain place, the geography of the field trip, and the merits of an environmental education method. In video content production, the selection of a place and the composition of the story is very important. In addition to the story of the place, the narration of the content can supplement the actual field trip by including the fieldwork skills of the outdoors. Video content has the effect of combining various types of media so it can be effectively used along with field trips. Virtual field trips using video content have inherent limitations compared to outdoor field trips, but they can be used for place-based education by preparing a plan and support system for more effective blended learning.

Chapter 5
Future-Proofing Geography Teaching and Learning Using Remote Learning and Co-Creation Approaches 102
Theresa G. Mercer, University of Lincoln, UK
Andrew P. Kythreotis, University of Lincoln, UK
Hannah Kelly, University of Lincoln, UK
Zoe P. Robinson, Keele University, UK
Sharon M. George, Keele University, UK
David Sands, Independent Researcher, UK

This chapter examines the application of new geography pedagogic practices in relation to remote teaching and learning. The authors report how the co-creation of knowledge between students and lecturers through existing research projects (with a particular focus on disciplinary-based educational research) can still occur through remote learning. Through a case study example of geography teaching practice, they contribute to the pedagogic literature on digital online learning as a form of co-creation between lecturer and students, showing how students were still able to demonstrate their geographical knowledge in a very applied and real-world context despite the move to remote learning. The application of co-creation in these applied contexts demonstrates that the applied perspectives of geographical knowledge can still be delivered through remote learning. They conclude by discussing how applied and critical geography teaching can be delivered and future-proofed in higher education settings, regardless of unforeseen circumstances.

Section 2
Resources in Geography Teaching

Chapter 6
Landscape and Heritage: Didactic Resources for Sustainability Education 117
María Casas Jericó, University of Navarra, Spain
Luis Erneta-Altarriba, University of Navarra, Spain
Ana Villarroya Ballarín, University of Navarra, Spain

The growing technical, political, social, and cultural interconnectedness of our world helps advance societies. However, globalization also entails some negative environmental and cultural effects, which are usually intertwined. For example, the spread of Western consumption models accelerates ecological and social degradation worldwide. A sustainable development model is necessary to face these problems, which needs a conscious and competent society. This chapter explores how heritage and landscape are, at the same time, integrative concepts and didactic resources that promote interdisciplinary learning. Both aspects have great potential for advancing sustainability education. The chapter presents some didactic proposals that use diverse strategies and resources related to heritage and landscape, adopting a local focus whenever possible to make the learning experience more significant. Some of these activities address environmental and social challenges described in the United Nations Sustainable Development Goals.

Chapter 7
Identifying Challenges in Openstreetmap That Potentially Prevent New Users From Contributing: Implications for Addressing OSM in Education................144
Robert Vogler, Paris-Lodron-University of Salzburg, Austria
Sabine Hennig, Paris-Lodron-University of Salzburg, Austria

OpenStreetMap (OSM) is known as the most successful application of VGI. The fact that it is based on user-generated content often leads to the common assumption that OSM data is neutral. These assumptions are legitimate, but the vast majority of OSM contributors are middle-aged, white, male users. Accordingly, OSM data represents the worldview of a delimited social group and therefore lags behind its actual potential regarding the richness and quality of data with regard to diversity. To overcome this, new and diverse user groups have to be recruited. It is therefore necessary to identify the challenges and barriers specific new user groups face. This chapter elucidates these mechanisms based on literature. Subsequently, the authors present a study with teacher training students that identifies challenges of new OSM users. The results indicate that there are three dimensions of potential barriers: the complexity of OSM, the user experience in OSM, and the user interface of OSM.

Chapter 8
A Didactical Approach of Rhumb Line vs. Great Circle in Web Mercator Projection for 6th Grade Pupils...168
Nikolaos Lambrinos, Aristotle University of Thessaloniki, Greece
Ioanna Repanidou, Aristotle University of Thessaloniki, Greece
Georgia Intzidou, Mandoulides Schools, Greece

The chapter presents two studies on the integration of map projections in Greek geography curriculum. Perceptions and misconceptions about world maps of 6th grade elementary school students were investigated and analyzed by qualitative and quantitative methods. Through inquiry-based methods, students learnt about the distortions in distance, shape, and size of areas that are caused by transformation of the spherical Earth into a flat surface. Didactic interventions were carried out where students attained knowledge about 1) world map projections and 2) the overseas flightpaths and the measurements of the distance between two points in the Mercator projection. Through guided research, comparisons, experimentation on digital world maps, and experiential actions, students found that 1) all world maps are wrong and 2) the shortest distance between two points in Mercator projection can be a curved line (great circle) and not a straight line (rhumb line).

Chapter 9
Reading and Spatial Analysis Through Mental Maps in School Geography.....203
Denis Richter, Universidade Federal de Goiás, Brazil

This chapter aims to highlight the mental map as a relevant pedagogical proposal for the work of school geography linked to the process of construction of the students' geographic thinking. To develop this idea, at first, the relationship of geographic science with the issue of spatiality as a way of thinking is highlighted, and, subsequently, the contribution of cartographic language as a means to systematize reading and spatial analysis. In this sense, the mental map is required as a proposal for a cartographic representation that enables the subject who builds it to highlight different contexts, interpretations, and perceptions about a given space or place. For this, some examples of the construction of mental maps from the perspective of school activities will be presented and discussed, highlighting the potential of this material for the development of a more critical reading and analysis of the geographic space.

Chapter 10
River Landscape: A Didactic Proposal for Primary Education 228
Elena María Muñoz Espinosa, Universidad Autónoma de Madrid, Spain
Celia Laguna Mora, University of Castilla-La Mancha, Spain

This educational proposal is built up from a varied set of disciplines with different viewpoints which manage different approaches to rivers: hydrology, ecology, land uses, cultural heritage, or fluvial landscape. Nowadays, traditional approaches are misleading, and they promote conflicts between the required knowledge and what society demands. Thus, didactic innovation on an integrated base is required from early educational ages. Here are presented some didactic strategies that could be implemented from elementary schools, with students from 6 to 12 years old. The aim is to establish an approach and procedures for river teaching and help to develop proposals for teachers, which eventually could be used in class. The development of these guidelines is implemented by the identification of fluvial functions along specific items suggested.

Chapter 11
The Spatial Experience for Teaching Geography in the Anthropocene.............259
Sylvie Joublot Ferré, Haute École Pédagogique de Lausanne, Switzerland

The contemporary period is disrupted by multiple and interdependent crises. The Anthropocene is the concept designating the fact that, for the first time, a geological epoch would be defined by human action. It would be a matter of rethinking the relationship to knowledge and the methods of its transmission, particularly relevant to geography and its teaching in schools, both at primary and secondary levels. What strategies should be developed for teaching geography in a way that is civic-minded, responsible, participatory, and forward-looking, but also embodied, sensitive, and humanistic? The question that now arises is: Why and how can scholars take into account juvenile spatiality when thinking about learning? What avenues could be favoured to bring the relational process and the environmental paradigm into the geography classroom? The idea is to revive a traditional geography practice of field investigation, while allowing participants to have spatial experiences.

Compilation of References 282

About the Contributors 318

Index 324

Preface

In Geography treatises, a chapter is always dedicated to geography education. Whatever the school of thought the treatise's author belongs to or the paradigm he/she advocates, geographic education is always addressed in the treatise. In his Geography treatise, Harvey (1969) showed the full panorama of theoretical approaches and paradigms in the discipline of geography. Among the geography theoretical approaches stand out the cause-and-effect models and the temporal modes to explain the spatial concept in Geography. Through temporal modes, Geography processes can be explained. Harvey (1969) also approaches psychology to understand the acquisition of the notion of time in childhood and youth in order to establish a relationship among the economic, social and cultural processes of land development. Sttodart (1986) offers a review of the original contributions to education by the discipline of Geography to raise concerns about the teaching of Geography towards the end of the 19th century. Unwin (1992) highlights the contrast between American and British geography teaching. In the USA, the discipline of geography is included in the course of studies of the Social Sciences, while in the United Kingdom, Geography occupies a single prominent position among the academic disciplines.

There have been contributions questioning certain geographical paradigms by critical authors who reformulate geographical concepts, such as the social construction of "space" (Massey, 1984), or who overemphasize other, like "landscape" as a term coined by Humboldt (Farinelli, 2003). Still, if the majority of concepts scaffolding the geographical terminology are associated with the human activity, that means that we are facing the priority objective of the discipline, that is, the relationship between humans and the territory. In consequence, the cultural and artistic dimension of Geography (Martínez de Pisón, 2017) evolves from the relationship between the nature and the human action.

Around the 1980s, pedagogical and geographical postulates combined their efforts in an intense educational activity to renew approaches in geography teaching. The educational concern of teachers in the 80s is reflected in the pedagogical trends of the time, adhering back to the principles formulated by authors in the 1930s. Dewey, Decroly, Vygotsky, Freinet, among others, reappeared to set critical educational trends, in an attempt to renew traditional teaching. Graves (1979) proposes substantial changes in the Geography curriculum. Graves' work was devoted to cover all aspects of Geography Education, in terms of disciplinary contents, and teaching strategies and resources (Graves, 1982).

The leading figures of the geographical paradigms avoided being left out of the realm of geography teaching. Like Sttodart, or Unwin, other geographers took a stance on the issue of geographical education from their academic perspectives. The social commitment and political radicalism of Lacoste offers a different view of Geography. Lacoste and Ghirardi (1984) provides a description of the variety of landscapes in the world and of the economic and social inequalities between countries. This handbook includes an innovative and powerful array of graphical resources, like graphs, images, photographs and figures. Such a new handbook stands for a different geography specialized text that is aimed at capturing the student's attention.

In the USA, Gersmehl (2008) proposes a handbook on geographical education in relation to the development of spatial thinking. Gersmehl forged his career in researching into spatial thinking applied to geography teaching (Gersmehl, 1991). In the UK, Best (2011) approaches and reviews Geographical learning models and practical classroom matters. By this hands-on classroom-oriented approach, Best tries to solve key issues of geography teaching. Jones and Lambert (2012) provide a handbook from a theoretical approach to Geography teaching through different topic debates.

The present handbook is designed for geography students, students of teachers' training, and lecturers. The authors participating in this volume have contributed to the teaching and learning of Geography with guidelines of trending strategies and innovative resources in geography teaching for compulsory (primary and lower secondary education), upper secondary and higher education levels. The book is intended to favor a global approach to geography teaching from different cultures and cross-disciplinary perspectives.

This book is divided into two parts, a Section 1, "Teaching Approaches and Strategies," Chapter 1 to Chapter 5, and Section 2, "Resources in Geography Teaching," Chapter 6 to Chapter 11.

Chapter 1, "Using Inquiry-Based Learning Approaches to Teach Undergraduate Geography: Teaching Using IBL," presents a discussion on the use of inquiry-based approaches in teaching geography. Definitions of inquiry approaches and their uses in teaching are introduced. Two case studies are presented involving similar final-year research methods courses, being highly rated by students and with evidence of students attaining high level learning outcomes, in which students work in groups to address real world issues or problems. A discussion follows on the benefits and challenges of teaching using inquiry approaches. Articles are provided that focus on using inquiry-based approaches in other geography contexts. Finally, some recommendations are given teachers willing to incorporate small inquiry activities in their courses, through to the use of inquiry assignments or the design of an inquiry course.

Chapter 2, "Application of Inquiry-Based Science Education, Anticipatory Learning Strategy, and Project-Based Learning Strategies," addresses some global challenges primarily associated with climate change. It is discussed how geographical education should provide knowledge about the changes taking place, shape practical skills, and foster the acquisition of appropriate attitudes towards the geographical environment. The chapter presents three educational strategies, pertinent to the problems of the modern world: teaching through scientific inquiry (IBSE), anticipatory learning strategy (ALS), and teaching through Project Based Learning (PBL). They are characterized, and a proposal for their application to geography education classes is suggested with content contexts for each strategy: global warming for IBSE, ice sheet changes for ALS, and water resources and retention for PBL.

Chapter 3, "The GeoScenario: A Methodological Resource From Problem-Based Learning (PBL) in Landscape Teaching," introduces the geographical scenario as a methodological resource in Geography teaching applied to problem-based learning. The geoscenario is characterized as the minimal significant landscape framework unit in landscape teaching that embodies a synthetic representation of a real-world acute issue involving a particular geographical problem. The geoscenario is addressed in relation to landscape analysis and interpretation, integrated and interdisciplinary curriculum in Geography, constructivist learning environment, and geographical abilities and skills for lifelong learning. PBL is conceptually and methodological framed from the constructivist approach. PBL implementation steps and progression levels for problem-solving are presented. The geoscenario for landscape teaching in PBL is dealt with, including types of geoscenarios, reasoning strategies, teamwork techniques and small-group dynamics.

Chapter 4, "Importance and Educational Possibilities of the Place-Based Education Through Virtual Field Trips," deals with place-based education possibilities through virtual field trips that are based on the environmental educational value of the experience in a certain place, the field-trip geography, and the merits of an environmental education method. The implementation strategy of video content production is presented, where place selection, story composition and content narration can substitute virtual field trips using video content for the actual field trip by including the fieldwork skills of the outdoors and by combining various media types. A discussion is opened on the inherent limitations of virtual field trips using video content compared to outdoor field trips, and their use for place-based education by preparing a plan and support system for more effective blended learning.

Chapter 5, "Future-Proofing Geography Teaching and Learning Using Remote Learning and Co-Creation Approaches," examines the application of new geography pedagogic practices in relation to remote teaching and learning. Co-creation of knowledge between students and lecturers through existing research projects through remote learning is reported. A case study on digital online learning is presented as a form of such co-creation, showing students' ability to demonstrate their geographical knowledge in a very applied and real-world context despite the move to remote learning. The chapter concludes with a discussion of how applied and critical geography teaching can be delivered and future-proofed in higher education settings, regardless of unforeseen circumstances.

Chapter 6, "Landscape and Heritage: Didactic Resources for Sustainability Education," opens with a discussion on points in common of landscape and heritage, including heritage assets, protection figures, and social products. This chapter explores how heritage and landscape are integrative concepts and didactic resources that favor interdisciplinary learning. The education potential of heritage and landscape for advancing sustainability education is further detailed. Didactic proposals using strategies and resources related to heritage and landscape are presented that include environment interpretation through texts and pictures, flipped classroom, comparative and retrospective orthophotograph analysis, GIS and online maps, analysis of an interview with an expert, online questionnaires, case studies and service-learning. Some of these activities address environmental and social challenges described in the United Nations' Sustainable Development Goals.

Chapter 7, "Identifying challenges in OpenStreetMap That Potentially Prevent New Users From Contributing: Implications for Addressing OSM in Education," presents OpenStreetMap (OSM), the most successful application of VGI, based on user generated content. The chapter challenges the common assumption that OSM data is neutral since main OSM contributors are mid-aged, white, male users, and OSM data is the worldview of a delimited social group. The richness and quality of data with regard to diversity is thus questioned. Mechanisms based on literature are elucidated on how new and diverse user groups have to be recruited, and on the challenges and barriers facing specific new user groups. A study with teacher training students is presented, which identifies challenges of new OSM users. The results indicate that there are three dimensions of potential barriers: the complexity of OSM, the user experience in OSM and the user interface of OSM.

Chapter 8, "The Spatial Experience for Teaching Geography in the Anthropocene," addresses the concept of the Anthropocene designating the fact that for the first time a geological age would be defined by human action. the chapter is intended to rethink the relationship to knowledge and its transmission methods, pertinent to the discipline of geography and geography teaching in schools, both at primary and secondary level. Both spatial experience and fieldwork promote the acquisition of the concept of the Anthropocene. The spatial experience is determined by the context of place and time, by the immediate or wider social context, the temporal context, and by the individual context and the student's own social beliefs and cultural interpretations. The fieldwork is the main geographical technique for recognizing and interpreting the real world. The idea is to revive a traditional geography practice of field investigation, while allowing participants to have spatial experiences.

Chapter 9, "River Landscape: A Didactic Proposal for Primary Education," builds up an educational proposal from a variety of disciplines having different viewpoints and approaches to river landscape: hydrology, ecology, land uses, cultural heritage, or fluvial landscape. Traditional approaches are found misleading, because of the conflict arising between expected knowledge and the society educational demand. Didactic innovation on an integrated basis is required from early educational ages. Some didactic strategies are presented to be implemented at elementary schools, for students from 6 to 12 years old. The aim is to establish an approach and procedures for river landscape teaching and help to develop proposals for teachers to use in the classroom. The development of these guidelines is implemented by the identification of fluvial functions along with specific items suggested.

Chapter 10, "A Didactical Approach of Rhumb Line vs. Great Circle in Web Mercator Projection for 6th Grade Pupils," presents two researches on map projection integration in Greek Geography Curriculum. The research and analysis conducted by qualitative and quantitative methods is reported on perceptions and misconceptions about world maps of 6th grade Elementary School students. Implementation of inquiry-based methods is addressed, where students learnt about the distortions in distance, shape and size of areas which are caused by transformation of the spherical earth into a flat surface. Didactic interventions are shown where students gained knowledge on world map projections, and overseas flightpaths and distance measurements between two points in Mercator projection. Research results are that all world maps are wrong and the shortest distance between two points in Mercator projection can be a curved line (great circle) and not a straight line (rhumb line).

Chapter 11, "Reading and Spatial Analysis Through Mental Maps in School Geography," addresses the mental map as a relevant pedagogical proposal for school-geography work linked to the construction process of students' geographical thinking. The association of geographic science with spatiality as a way of thinking is examined, and the contribution is highlighted of cartographic language as a means to systematize reading and spatial analysis. The mental map is explored as a proposal for a cartographic representation that enables the subject who builds it to highlight different contexts, interpretations and perceptions about a given space or place. Some examples of the construction of mental maps from the perspective of school activities are presented and discussed. The potential of this material is dealt with for the development of a more critical reading and analysis of the geographic space.

Alfonso Garcia de la Vega
Universidad Autonoma de Madrid, Spain

REFERENCES

Best, B. (2011). *The Geography Teacher's Book*. Continuum International Publishing Group.

Farinelli, F. (2003). *Geografia. Una introduzione ai modelli del mondo*. Giulio Einaudi.

Gersmehl, P. J. (1991). *The Language of Maps*. Indiana University of Pennsylvania.

Gersmehl, P. J. (2008). *Teaching Geography*. The Guilford Press.

Graves, N. J. (1979). *Curriculum Planning in Geography*. Heinemann Educational Books Ltd.

Graves, N. J. (Ed.). (1982). *New UNESCO Source Book for Geography Teaching*. Longman, UNESCO.

Harvey, D. (1969). *Explanation in Geography*. Edward Arnold Ltd.

Jones, M., & Lambert, D. (Eds.). (2012). *Debates in Geography Education*. Routledge.

Lacoste, Y., & Ghirardi, R. (1983). *Géographie Générale, Physique et Humaine*. Fernand Nathan.

Martínez de Pisón, E. (2017). *La montaña y el arte*. Fórcola.

Massey, D. (2004). Geography matters. In D. Massey & J. Allen (Eds.), *Geography matters!* (pp. 1–11). Cambridge University Press.

Sttodart, D. R. (1986). *On Geography*. Basil Blackwell Ltd.

Unwin, T. (1992). *The Place of Geography*. Longman.

Acknowledgment

My acknowledgment to those authors who have dedicated their research and experiences on geography education for publication in this handbook. I also express my deep acknowledgment to the blind-peers in each chapter of the book. Thanks to them the final result of the book has turned out to be more relevant in the field of Geography.

Section 1

Teaching Approaches and Strategies

Chapter 1
Using Inquiry-Based Learning Approaches to Teach Undergraduate Geography

Rachel A. Spronken-Smith
University of Otago, New Zealand

Simon Kingham
https://orcid.org/0000-0002-2869-3756
University of Canterbury, New Zealand

Ralf A. Ohlemüller
University of Otago, New Zealand

ABSTRACT

This chapter discusses the use of inquiry-based approaches in teaching geography. Following an introduction of definitions for inquiry approaches and how they might be used in teaching, two case studies are presented. The cases involve similar final-year research methods courses in which students work in groups to address real-world issues or problems. Both courses are highly rated by students, and there is evidence of students attaining high-level learning outcomes. The benefits and challenges of teaching using inquiry approaches are discussed, and articles for the use of inquiry-based approaches in other geography contexts are provided. Finally, recommendations are provided for teachers wishing to use such approaches, whether they wish to incorporate small inquiry activities in their courses through to the use of inquiry assignments or the design of an inquiry course.

DOI: 10.4018/978-1-7998-9598-5.ch001

INTRODUCTION

This chapter discusses the use of inquiry-based learning in teaching geography to undergraduate students. Inquiry-based learning (IBL) has been described as a 'signature pedagogy' (Shulman, 2005) in geography, due to this approach being used for many years in field teaching (Spronken-Smith, 2013). The IBL approach encompasses a range of active teaching methods where the learning is stimulated by a question or problem, and learning is based on a process of constructing new knowledge and understanding. The teacher is a facilitator of learning, akin to a guide on the side, rather than a sage on the stage, and the aim is to foster self-directed learning in students. Levy and Petrulis (2012) proposed that IBL could be either 'information-oriented', where the students explore existing disciplinary knowledge, or 'discovery-oriented', where students build disciplinary knowledge. In a further classification, Spronken-Smith and Walker (2010, p.727) drew on the work of Staver and Bay (1987) to make distinctions between modes of IBL depending on the amount of scaffolding:

- Structured inquiry: where teachers provide an issue or problem and an outline for addressing it
- Guided inquiry: where teachers provide questions to stimulate inquiry, but students are self-directed in terms of exploring these questions
- Open inquiry: where students formulate the questions themselves as well as going through the full inquiry cycle.

Spronken-Smith and Walker (2010) noted that IBL is an umbrella term, which includes problem-based learning (PBL), and the terms are often used interchangeably. Problem-based learning is particularly prevalent in medical education contexts and is often used as an active learning approach to help develop clinical reasoning skills. The term IBL is often used alongside 'undergraduate research', as students learning by either approach develop inquiry and research skills. Undergraduate research can be similar to open inquiry, with students undertaking original research, providing students have autonomy in decisions about the research process.

The decision over which type and mode of IBL to use will depend on the intended learning outcomes of the course. For example, if teachers want students to learn a particular body of knowledge on say, migration patterns in a country, or causes of air pollution in a local environment, they would likely adopt either structured or guided information-oriented IBL. They would choose questions that provoke the exploration of existing bodies of knowledge. Students might work individually or in groups to explore that body of knowledge to answer the questions. The teacher likely knows the answer, but the knowledge will be new to the students. Such an approach promotes active learning, especially when students work with peers to find and discuss resource and reference material.

If however, the teacher wants the students to develop higher level learning outcomes such as reflecting, creating, evaluating, applying and analysing, and if they are not requiring the students to learn a certain body of knowledge, they should use open, discovery-oriented IBL. This would mean the students have to develop a question to research, and complete an inquiry cycle (Justice et al., 2002, p.19), involving determining what needs to be known, identifying resources and gathering data, assessing data, synthesising, communicating new understandings, and evaluating success. Central to the cycle is the requirement for self-reflection and self-evaluation. Although students can do individual inquiry projects (and indeed graduate theses involve this), it is often beneficial at undergraduate level to get students to work in groups, so they can benefit from peer learning (e.g., Boud & Cohen, 2013).

But are these different types and modes of IBL fostering higher level learning outcomes? Spronken-Smith et al., (2012) evaluated 15 case studies of IBL courses which used different modes and types of IBL. They found that all types of IBL courses were rated favourably by students. Analysing, applying and understanding were rated highly by students, as well as taking responsibility for their own learning. But more significantly, learning outcomes such as creativity, reflection, evaluating, applying and analysing were enhanced for discovery-oriented and open IBL. Open IBL was rated more highly than guided IBL, which in turn was rated more highly than structured IBL for promoting higher level learning outcomes.

This chapter continues by considering two case studies of discovery-oriented IBL in two different geography departments in Aotearoa New Zealand universities. Then we place these case studies in the wider context of the use of IBL in geography teaching, elaborating upon the benefits and challenges – for both students and teachers.

CASE STUDY 1: GEOG380 FIELD RESEARCH STUDIES IN PHYSICAL GEOGRAPHY

Overview

The University of Otago's main campus is in Dunedin, on the south-eastern coast of Aotearoa New Zealand's South Island. GEOG380 Field Research Studies is a 300-level, full-year, open and discovery-oriented IBL course in the School of Geography, University of Otago, Aotearoa New Zealand. It is a capstone and compulsory course in the final year for students in the Bachelor of Science in Geography and the Bachelor of Applied Science in Environmental Management.

Currently, in a typical year the course is taken by c. 80 students and is team-taught by c. 10 staff.

Intended Learning Outcomes

GEOG380 offers students an opportunity to choose a topic and a path of inquiry and to conduct a research project. The course aims at deepening and applying knowledge of previously learned concepts and methodologies, while also providing an opportunity to work in a group in a setting resembling a real-word work environment. The intended learning outcomes for GEOG380 are:

- identify and formulate a relevant research problem in Physical Geography;
- formulate testable research hypotheses and objectives;
- design an appropriate field or experimental methodology;
- communicate the research project process and outcomes verbally and in writing;
- develop skills in working with a team;
- apply organisational and project management skills;
- critically reflect on the group's and your own research process.

Teaching and Learning Methods

The two fundamental principles underlying teaching in GEOG380 are inquiry-driven research and group work. Students work on one research project in groups of six students and are supervised by a member of staff. Although staff propose research topics well aligned with their own interests, the course attempts to allow students to fully own their project. Group formation is deliberately not done through self-selection (to avoid pre-existing cliques to form groups). Instead, students choose a project according to their interests; staff 'pitch' their project topics to the class at the start of the year and students rank their top four preferred projects. Students are then assigned to projects by the course coordinator and typically all students get either their first, second or third choice of project. While assigning groups, considerations are also given to gender and cultural balances in the groups.

In line with IBL principles, the supervisor's role is that of a facilitator rather than an instructor. After choosing a research topic, the students identify a research problem and formulate appropriate research hypotheses or questions, before designing and executing field/laboratory research. The data obtained are then analysed and interpreted and discussed before the final report is written up. Throughout the entire process, the supervisor provides help and guidance without leading the process. A key tenet of the course is that the supervisor will not organise group meetings but instead expects to be invited to join the group's meetings; this really drives home the fundamental point that the students own and drive the project and the supervisors are there to help.

In addition, six lectures/workshops at the start of the year introduce students to the key generic skills and concepts relevant to conducting a research project such as research design, presentation skills, health and safety, and ethics.

Assessment

The course is 100% internally assessed. The key assignment is the final project report worth 75% of the total grade and with a word limit of 5000 words. At the start of the project, a Theoretical Review assignment (15%) allows the students to become familiar with the research topic and to identify a relevant research problem. After submission of the project report towards the end of the year, a short critical reflection of the research process is worth 5% and a peer assessment is also worth 5%. The peer assessment consists of numerical scores that students assign to all members of the group for various aspects of group participation; the total grade of students will then be adjusted by a small amount up or down depending on whether they are above or below the average group score (Conway et al., 1993).

In addition, there are two non-assessed assignments. Firstly, presentations of the project proposal. Each group gives a 10-min presentation to all students and staff on their project proposal with the aim to a) consolidate and verbalise their plans, and b) to get feedback from peers and staff. Secondly, towards the end of the year and after submission of the project report, a conference is held at which each group presents their project outcomes. While this assignment was graded in the past, in recent years there has been a move away from grading these presentations and instead see this as an end-of-year celebration of the completed research projects.

The Learning Experience and Student Feedback

A typical annual dynamic of the student experience throughout the course of GEOG380 is that of:

- Excitement and anticipation (selecting the research topic);
- Enthusiasm (designing and executing field and lab work);
- Curiosity and being overwhelmed (exploration and analyses of often large datasets);
- 'A-ha' moments (results produced and research questions answered);
- Dread and apprehension (making sense of own results and writing up the final report);

- Pride and satisfaction (completion of first full independent research project).

Feedback from students on the two key facets of GEOG380 (undertaking a complete, applied research project and working in a group) is generally very positive. For example, in relation to the IBL approach:

A research skill that I have learned and developed is the importance of research design. In previous courses and practicals, the methods section was given and we just had to experiment. However, I have gained an insight on how to carefully plan to execute a research project by analysing the research question and working around what data collection needs to be obtained.

The course has changed how I thought I learned as well as how I would like to learn in the future.

The deep focus onto the one topic of […] really did assist my learning.

Regarding group work, the feedback reveals a transformative impact:

Throughout this course, my perspective on how to learn has changed to incorporate more group work instead of making it individual and the confidence to ask questions during groups discussions to help improve my learning.

I learnt throughout this research process how important it is that I express my thoughts to the group rather than go with the majority, this has changed my thinking on my importance in group work and is a skill that I am glad I gained from this process.

Challenges

While the IBL principles and group work structure of GEOG380 offers opportunities for deep approaches to learning, there are also challenges relating mainly to the emphasis on group work, and between group differences in project logistics and timing.

Working as and in a group is a key feature of the GEOG380 teaching approach, but at the end of the project the students are asked to submit an individual project report. This means that at some point a change from group to individual work has to occur, and this can be challenging to navigate for both students and staff. While staff encourage discussions and exchange of ideas, literature and expertise on possible analyses, students are then asked to produce their own project report. Students are often unsure as to what counts as 'their own' and the group supervisor has to be very clear during this part of the project. More often than not, especially when students have worked well as a group, the Results sections in the final report are conceptually very similar between students within a group. It is then up to the examiners (supervisor and another academic) to differentiate reports according to the quality of other parts of the reports such as the Introduction and the Discussion.

The second challenge is the between-group differences in project logistics and timing. In a typical year, 10 members of staff supervise projects for GEOG380. While this offers considerable choice for students, it also introduces some variation in timing, length and logistics of field work activities. While every attempt is made to ensure as much equality between groups as possible, it is unavoidable that some groups, for instance, finish their field work earlier than others. This can cause some concern for students as it is perceived as a disadvantage because the later groups will have less time for data analyses and write up. It is up to the supervisor to ensure that proper workflows are in place and that students have the opportunity to realise that while all projects are different, activities can be managed in a way that equity between groups with regard to assessments is ensured.

CASE STUDY 2: GEOG309 RESEARCH FOR RESILIENT ENVIRONMENTS AND COMMUNITIES

Overview

The University of Canterbury is located in Christchurch on the east coast of the South Island in Aotearoa New Zealand. Its Geography programme staff operate a final year (300 level) research methods course that has acted as a capstone course for geography majors (including physical, human and environmental geography) for many years.

Initially, GEOG309 was traditionally taught with a series of lectures on research methods accompanied by laboratories (e.g. on statistical analyses) and tutorials. However, student feedback was not very favourable, and it was clear the course needed revising. Consequently, in the early 2000s, under the leadership of the first author, the course adopted a guided, discovery-oriented IBL approach revolving around group work on real world problems (see Spronken-Smith, 2005). In consultation with local government or an agency, ideas for student projects were developed. Students then selected projects of interest and were placed into groups of 4-6 for the semester. The course included, and still does, a residential weekend workshop at the start of the course where the students learn about a range of research methods, research, transferrable and employability skills.

In 2008, Eric Pawson and Simon Kingham took over coordination of the course. In that year New Zealand's retiring Parliamentary Commissioner for the Environment organised a roadshow of speakers talking about some of the environmental challenges we were facing, including a series of talks at the University of Canterbury. One student asked something along the lines "why can't we learn AND do something useful and make a difference". In response to this, Pawson and Kingham started to investigate whether GEOG309 could become more sustainability focused. The result was a shift to focus on projects on topics related to creating resilient environments and communities. In addition, more responsibility was placed on the community partners to engage in an overtly active way. This includes attending an introductory session where they talk to the class about their community and project. These expectations formed what is now a 'GEOG309 protocol' which outlines the roles of the students, community partner and staff tutor. This modus operandi has continued to this day. In 2019 the course became compulsory for all students taking a new Environmental Science major, but interestingly, it is only due to become compulsory for Geography majors in 2023. Course numbers have varied over the years from 35 up to just over 100.

Intended Learning Outcomes

The primary goal of this course is to develop a students' ability to undertake research that supports resilient environments and communities. It aims to do this whilst also helping students gain an appreciation of how to work in the community on 'real' issues, and develop teamwork skills, while also fostering a critical appreciation of the research of others. The intended learning outcomes of the course are that upon successful completion of the course students should have an understanding of:

- How to conceive, develop and execute a research project using problem-based learning[1]
- How to work successfully with others using service learning
- How research is framed by broader conceptual and analytical frameworks
- Effective research design

In addition, students will also be able to:

- Undertake and deliver a research programme in a specified amount of time
- Work effectively in groups
- Select and use appropriate geographical techniques in research
- Reflect critically on their research and that of others.

Teaching and Learning Methods

The course operates over a 12 week semester; two six week terms with a two week break in the middle. It operates over the middle of winter, which has some implications for field work. The first few weeks of the course are about setting up the course and the project. The key timings are as follows:

- Week 1: Course introduction and logistics. Students are also presented a summary of the community partners and their projects. These have been developed over the preceding days/weeks/months by the course coordinator and the community partners. The students are asked to rank their top five projects. Students are then allocated a project based on the rankings; 65-75% get their first choice, 25-35% their second and less than 10% their third choice project.
- Week 2: Meet the community partners. All the community partners come to the University and talk to the whole class about their community and project. The students then meet with their community partner, and academic supervisor to start the conversation about the project.
- Week 3: At the end of week three, the whole class and available staff, head away for a residential training weekend. This consists of a series of interactive workshops on engaging Māori in research, how to set up a project, preparing a literature review, working in groups, working with community groups, health and safety, ethics, interviews, sampling, mapping, the different types of data (primary and secondary e.g. spatial, questionnaires, social, physical) and analysis. There are also team building exercises, and times where the students can start thinking about and refining their research question.
- From then on, including in the two week break, the students work on their research projects with their academic supervisor, and community partner. The whole class meets on two occasions during this phase to present updates on their progress to the whole class.
- Week 10: A conference is held on the Saturday at the end of the week, where the student groups present their research. They all attend, as do their community partners. In addition, the conference is a public event, so others can attend too. This sometimes includes other interested parties and research participants. One year a local Member of Parliament attended.
- The students then have two weeks to finalise their reports based on feedback received at the conference.
- Week 12: In the last week of term, after the final report is submitted, is a final review and reflection session. After some course housekeeping, students are asked to reflect on, discuss, and feedback on three questions:

- "*What you will take forward into employment and/ or further study (transferable skills and insights)?*"
- "*What you would say to next year's class about how to do a good job in GEOG309?*" The second question is fed back to next year's class (and the previous year's feedback was provided to the current class as part of the course introduction in week 1).
- "*What you would like us to say from you to the community partners at our wrap up meeting with them next month?*" This is fed back to the community partners. The community partners are also asked what they would like us to tell next year's students, and this feedback is also included in the Introduction class in week 1.

Assessment

The course is fully internally assessed and includes five items:

1. Group mana whenua[2] engagement report (5%). In New Zealand it is essential that research engages with the needs of its indigenous Māori people. This assignment is designed to encourage students to think about the relevance of their project to Māori communities, how they might engage with mana whenua. If appropriate, the group work with Māori staff at the University to ensure suitable processes to facilitate this.
2. Individual critical review of existing research on topic (20%)
3. Group conference presentation (15%)
4. Written group report (40%)
5. Individual critique of project research and students' learning (20%).

The assessment is therefore 40% individual and 60% group work. The group component is weighted by a peer moderation process (after Conway et al., 1993), where students grade the contribution of each of their group members (including themselves) according to their contribution in three areas; generating ideas, researching ideas, and communicating ideas.

The Learning Experience and Student Feedback

Research was carried out on this class in the late 2000s O'Steen et. al, 2010. Among many things, it found that students studying these IBL courses were more engaged in their studies, and unsurprisingly, did comparatively better in this course than their other courses. In 2021, the students were surveyed in relation to how well they felt the course's learning outcomes were successful (Figure 1). Over 50% of students felt the course achieved the learning 'extremely' or 'very well', and in most cases 90% if the response 'moderately well' is included. The problem-based and service-learning learning outcomes were perceived to have been achieved particularly well.

In addition, students were asked to indicate how much they agreed/disagreed with a series of statements in relation to their experience and learning in the course (Figure 2). The degree of agreement with the statements is high for all. The majority of students felt the course has helped develop their skills in communication (85% agreed or strongly agreed), leadership (71% agreed or strongly agreed) and working with others (79% 'agreed' or 'strongly agreed'); all important transferrable and employability skills; and 80% 'agreed' or 'strongly agreed' that they feel better prepared for the workplace. Seventy-six percent 'agreed' or 'strongly agreed' that they "*felt more engaged in my learning than in a more traditionally taught course*", which strongly supports the earlier wok of O'Steen et. al (2010). Moreover, 67% of the students 'agreed' or 'strongly agreed' that they had had "*an increased sense of social responsibility and citizenship*", supporting the suggestion of Bandy (2011). It was also interesting to note that the students agreed that working with a community partner enhanced their learning, compared to if they had done a research project without one.

Challenges

There are number of challenges with the course:

- *Finding partners and projects.* Initially finding suitable projects and partners can be challenging. This includes managing community partner expectations. Over time this becomes less of a problem, especially with community partners who have previously worked on the course. In fact, the problem now is sometimes telling partners they are not needed! Increasingly, partners approach the School with projects, not the other way round.
- *Suitable projects.* Community partners are not always research experts and sometimes their ideas need to be developed into academically appropriate projects for the course, and in a time frame that fits the teaching semester. This can be challenging especially with a new community partner who has not previously been involved in the course.
- *Student buy in.* Some students struggle to adjust to the different way of learning.
- *Engaged staff.* The course relies on staff to supervise groups. At its largest, the course needs 22 groups to be supervised. To get willing 'volunteers' you either need a good department/school culture or an effective workload model to acknowledge the work involved. Getting the 'right' people can also be

Figure 1. Students' level of agreement of how well the learning outcomes have been met in GEOG309

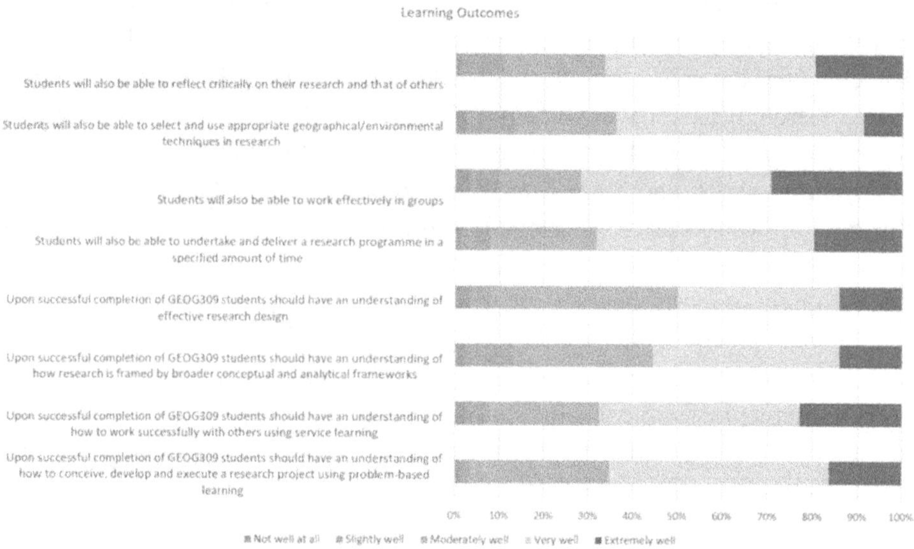

Figure 2. Students' level of agreement with statements about the nature of learning in GEOG309

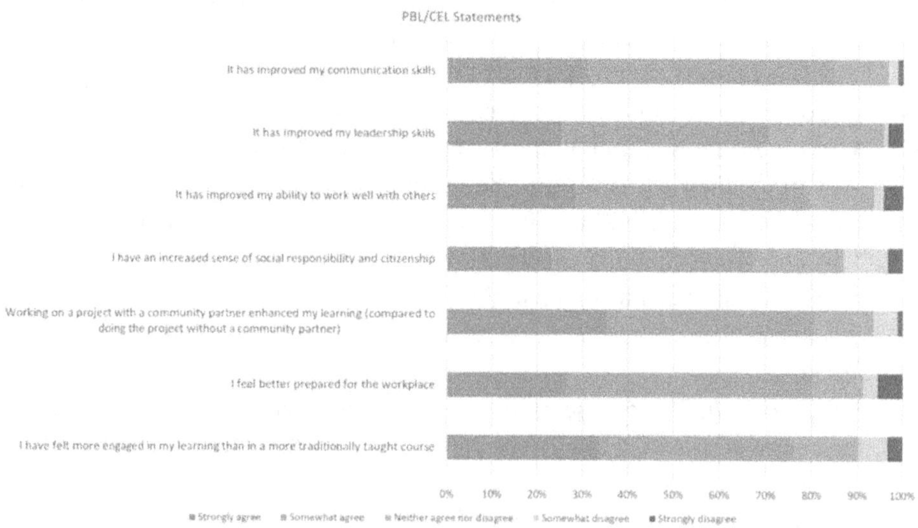

challenging; like the students', some staff struggle adjusting to this type of learning. GEOG309 has increasingly turned to engaged graduate students rather than try and convince unwilling colleagues to help.

- *Accepting occasional project fail.* Every now and then projects do not work out well (estimated 3-4 out of 150+ projects). Occasionally, they are not great projects, but generally the students just do not do a great job. It is hard for a supervisor to not try and take control, and tough for a community partner to not get a good outcome. Managing expectations is really important here.
- *Having a Plan B.* This is important for all research projects, but COVID-19 has really emphasised it. In a field discipline, it can be academically and practically difficult, but increasingly essential to have a back-up desk-based way of doing the research in mind.

DISCUSSION

The two case studies of using IBL in geography are quite similar: both are a final year course and focus on research methods in geography and both involve group work. Moreover, both focus on group meetings with a supervisor as the dominant learning approach, and both have supplementary learning opportunities through lectures and workshops (GEOG380) or a residential training weekend (GEOG309). However, GEOG380 involves students choosing to research topics closely aligned with staff interests, while GEOG309 has students researching topics for community partners.

The rationale underlying the GEOG380 approach is to provide a transformative learning experience for students. Moreover, the approach allows staff to attract students to graduate study. Otago staff commonly report recruiting graduate students from the group project work in GEOG380 (Spronken-Smith and Hilton, 2009). Indeed, Spronken-Smith and Walker (2010) recognised the power of guided and open discovery-oriented IBL to promote strong links between research and teaching, and provide a fertile ground for recruiting graduate students.

In contrast, GEOG309 is more focused on students undertaking research to help create resilient environments and communities. Although this model began operating prior to the devastating earthquakes in Christchurch in 2011, the rationale was no doubt strengthened by this natural disaster, as communities struggled to recover from the catastrophic damage. The use of service learning in geography has been promoted by researchers and educators (e.g. see Mohan, 1995) and was viewed very favourably by students.

Both models deliver powerful learning opportunities for students, and importantly develop high level learning outcomes, as well as a range of research, transferable and employability skills. Metacognitive skills, which as Kriewaldt (2006) argues, are key to an IBL geography curriculum, are developed through the inquiry cycle, the formative experiences of presenting project proposals and/or findings, and the reflective assignments, which form assessment items in both courses. These assignments assist the students to recognise the skills and knowledge they have acquired through the course and help stand them in good stead for pitching themselves to potential employers or graduate school administrators.

The two case studies are positioned at the most powerful end of the spectrum of IBL approaches. As such, although students have the most to gain from this approach, using guided and open discovery-oriented IBL is not without its challenges. The staffing required to resource these courses is high, as each group needs a supervisor. Yet paradoxically this high resourcing may be invisible to students who are more used to high contact hours through lectures and tutorials or laboratories. The first author can distinctly recall a comment in a reflective assignment by a GEOG309 student about 20 years ago, who relayed what he had paid for the course and was wondering if he had got his money's worth! Such views point to the need to carefully pitch the rationale and design of such courses to students, to ensure they know how this teaching differs from more traditional courses, and the benefits to them as learners.

Teaching staff also need to be oriented to this approach, as teaching using IBL for the first time can generate feelings of fear, anxiety and uncertainty (Spronken-Smith and Harland, 2009). An important aspect for staff and students alike is how to deal with the project 'failing'. Staff need to recognise that, fundamentally, the process of conducting all aspects of the research project is the important outcome for the students – not necessarily getting a high-end publishable output of results. In field research settings there is plenty of scope for data collection to not go as planned and for protocols and expectations needing to be adjusted. For students, evidence has shown that there is fairly widespread perception that if no significant relationship between measured factors emerges, the project has failed. It should be a fundamental message for students to take away from IBL activities that an activity that does not result in an expected outcome is not a failed activity. This is because teachers may be more teaching-centred in their approach, focusing on what they want students to know and do (the sage on the stage), whereas in IBL a student-centred approach is required, focusing on what students need to know and do, with the teacher being a guide on the side. Such a shift can in teaching approach can be very uncomfortable for teachers. Thus, as Spronken-Smith and Harland (2009) suggest, the course coordinator should ensure the vision for the course is known amongst the teaching team, and provide opportunities for open discussion about the teaching and learning process.

Designing the assessment regime is another challenge, and this was apparent in the GEOG380 case. Finding a balance between individual and group work is challenging and may be subject to institutional norms, restricting what is possible.

For courses with an intended learning outcome of the ability to work in groups, it is essential that this ability is assessed. The scheme by Conway et al., 1993) was used in both cases, and offers a fair and transparent way for self and peer evaluation of contribution to the group. Courses involving varied group projects can also be challenging for staff, as noted in the first case (also reported by Spronken-Smith and Hilton (2009) in an earlier version of GEOG380), and indeed other authors report increased preparation time for staff and resourcing costs (e.g., Harris and Tweed, 2010; Pawson et al., 2006).

The two cases involve an IBL approach as the design feature for the whole courses, but IBL can also occur as activities within lectures or tutorials or laboratories or fieldwork. Spronken-Smith et al. (2008) provide a range of examples of IBL in geography curricula including in-class activities, components of field courses, assignments as part of a course, and e-learning modules within a course through to full IBL courses and inquiry-based degree programmes. Also, *Planet* (2001, Volume 4, issue 1) provides a series of cases studies of the use of problem-based learning in geography, earth and environmental sciences. Moreover, the higher education literature provides examples of IBL use in a range of subdisciplines of geography (Table 1), and other articles focus on the ability of IBL approaches to provide integrative geography experiences (e.g. Hupy et al., 2008).

RECOMMENDATIONS FOR PRACTICE

Irrespective of the level, type or mode of IBL, there are key aspects which apply in teaching using this approach. Pawson et al. (2006, p.113) provide a helpful checklist, which is adapted here.

Preparation

- Prepare well in advance and, if possible, negotiate teaching release time
- Search for similar courses or assignments, and share ideas with colleagues, including those outside the discipline and academy
- Assemble resources for student use: access to pertinent library and online resources is necessary
- Consider level and training of students. Are they first years or students in a seminar or non-traditional students?
- Consider experience of teachers. Do they need induction and support for this type of teaching?

Table 1. Examples of studies examining the use of inquiry and problem-based approaches in subdisciplines of geography

Subdiscipline/ Topic area	Authors	Comments
Disability and access	Treby et al., (2007)	Reports on a third year course on geographical design and application by group work addressing real world issues– one of which is disability and awareness.
Environment – air pollution	Spronken-Smith & Kingham (2009)	Reports on a third year undergraduate inquiry project exploring factors that influence personal exposure to nitrogen dioxide.
Environment - water	Halvorson & Wescoat (2002)	Reports on a project-based course focusing on world water problems for introductory geography students
Geomorphology	Harris and Tweed (2010)	Reports on a field-centred IBL module on glacial geomorphology for third year students.
GIS	Balram (2019)	Chapter reviewing the literature on pedagogical approaches (including IBL) for teaching GIS
	Jakab et al., (2016)	Chapter on the use of IBL and blended learning to explore the impact of human activity on the environment using GIS techniques in a first year Master's programme
Hydrology	Lyon & Teutschbein (2011)	Discuss a range of non-traditional assessment forms to help facilitate the use of PBL in hydrology courses.
Political, and economic geography and environmental science	Alexson & Kemnitz (2015)	Case study of the use of the World Bank as a scenario for a PBL lesson that can be used in both human geography (politics and economic spatial activities) and in environmental sciences on topics of economic development and sustainability.
Regional geography	Fournier (2002)	Reports on the use of problem-based approaches and group work in a first year course introducing the world's geography to appreciate the complex interactions between the human and physical dimensions of the globe
	Keeling (2008)	Reports on a team-based and problem-based course on Canadian regional geography for third year students
Soil and water science	Al-Maktoumi et al., (2016)	Reports on a final year course using guided and open research projects to explore the hydropedology of arid zones in Oman
Sustainability	Pretorius et al., (2016)	Discusses the benefits and challenges of using IBL approaches for open and distance learning and in educating for sustainability
Tourism	Blazques Salom & Blanco-Romero (2020)	Reports on inquiry-based fieldwork projects to explore the complex spatial consequences of the Anthropocene on the tourist industry.
Urban studies and GIS	Serrano-Estrada et al., (2021)	Reports on a collaborative problem-based activity for third year students using social media datasets and GIS to enhance knowledge of urban phenomena.

Issue/Problem Design

- The issue or problem being addressed should be authentic and of significance. Consider involving external agencies or community partners if appropriate.
- The issue/problem should be compelling, so as to draw students in and generate self-directed learning
- The issue/problem should be complex enough to involve multi-dimensional solutions, requiring students to work as a team
- The issue/problem should be focused enough to be resolvable in the time available

Implementation

- The IBL experience should begin at or near the beginning of a term. Ideally this type of work should be infused into the class culture, not dropped in at random. Students should be informed about this learning approach, including how it differs from traditional teaching and the benefits and challenges involved
- Students should be formed into teacher-assigned, permanent groups of 4–6; Permanent groups allow students to develop team-building skills. In the first group meeting, ensure activities are used to allow students to get to know one another and ascertain the strengths and weaknesses of the group. Discuss roles and responsibilities of group members and teachers.
- Teachers should work with groups to provide clear guidelines for conduct and expectations; they should facilitate by asking probing questions, and help guide student inquiry (depending on student level)
- If the course is team taught, the course coordinator should arrange meetings of the teachers to share experiences about teaching and provide mentoring for teachers new to this approach.
- There should be a clear product or outcome for the issue/problem. Students must know what is expected of them. A report? A poster? An oral presentation? A decision supported by references?
- Ensure reflective activities are built into the course and include an assessment item that demonstrates reflection

Assessment

- Develop authentic assessment mechanisms that mirror problem-solving processes

- Teachers should have clearly established marking criteria that are transparent to students
- Work accountability into the assessment process. Consider both individual and group components to marks
- Students may be asked to do a self-assessment of their learning process, and be assessed by peers both for their research abilities and for contribution to team effort

REFERENCES

Al-Maktoumi, A., Al-Ismaily, S., & Kacimov, A. (2016). Research-based learning for undergraduate students in soil and water sciences: A case study of hydropedology in an arid-zone environment. *Journal of Geography in Higher Education*, *40*(3), 321–339. doi:10.1080/03098265.2016.1140130

Alexson, R., & Kemnitz, C. (2001). The World Bank Scenario - A Problem-Based Learning Activity in Human Geography and Environmental Science. *Planet*, *4*(1), 25–26. doi:10.11120/plan.2001.00040025

Balram, S. (2019). Teaching and learning pedagogies in higher education Geographic Information Science. In S. Balram & J. Boxall (Eds.), GIScience Teaching and Learning Perspectives (pp. 1-8). Springer. doi:10.1007/978-3-030-06058-9_1

Bandy, J. (2011). *What is service learning or community engagement?* Vanderbilt University Center for Teaching. Retrieved 4th Dec 2011 from https://cft.vanderbilt.edu/guides-sub-pages/teaching-through-community-engagement/

Blazques Salom, M., & Blanco-Romero, A. (2020). Fieldwork in ground zero, Mallorca. A methodological fieldwork proposal for a geographical analysis of tourist areas. *Investigaciones Geográficas*, *75*(75), 43–59. doi:10.14198/INGEO2020.BB

Boud, D., & Cohen, R. (2013). *Peer learning in higher education.* Routledge., doi:10.4324/9781315042565

Conway, R., Kember, D., Sivah, A., & Wu, M. (1993). Peer assessment of an individual's contribution to a group project. *Assessment & Evaluation in Higher Education*, *18*(1), 45–56. doi:10.1080/0260293930180104

Fournier, E. J. (2002). World regional geography and problem-based learning: Using collaborative learning groups in an introductory-level world geography course. *The Journal of General Education*, *51*(4), 293–305. doi:10.1353/jge.2003.0011

Halvorson, S. J., & Wescoat, J. L. Jr. (2002). Problem-based inquiry on world water problems in large undergraduate classes. *The Journal of Geography, 101*(3), 91–102. doi:10.1080/00221340208978480

Harris, T., & Tweed, F. (2010). A research-led, inquiry-based learning experiment: Classic landforms of deglaciation, Glen Etive, Scottish Highlands. *Journal of Geography in Higher Education, 34*(4), 511–528. doi:10.1080/03098265.2010.486851

Hupy, J. P., Aldrich, S. P., Schaetzl, R. J., Varnakovida, P., Arima, E. Y., Bookout, J. R., Wiangwang, N., Campos, A. L., & McKnight, K. P. (2005). Mapping soils, vegetation, and landforms: An integrative physical geography field experience. *The Professional Geographer, 57*(3), 438–451. doi:10.1111/j.0033-0124.2005.00489.x

Jakab, I., Grezo, H., & Sevcik, M. (2016). Inquiry based and blended learning using geographical information system. In J. Novotna & A. Jancarik (Eds), *Proceedings of the 15th European Conference on e-Learning ECEL 2016* (pp. 287-295). Academic Press.

Justice, C., Warry, W., Cuneo, C., Inglis, S., Miller, S., Rice, J., & Sammon, S. (2002). *A grammar for inquiry: Linking goals and methods in a collaboratively taught social sciences inquiry course. In The Alan Blizzard Award Paper: The Award Winning Papers*. Society for Teaching and Learning in Higher Education and McGraw-Hill Ryerson.

Keeling, A. (2008). "We are scholars": Using teamwork and problem-based learning in a Canadian regional geography course. *MountainRise, 4*(3), 1–13.

Kriewaldt, J. (2006). The key role of metacognition in an inquiry-based geography curriculum. *Geographical Education, 19*, 24–30.

Levy, P., & Petrulis, R. (2012). How do first-year university students experience inquiry and research, and what are the implications for the practice of inquiry-based learning? *Studies in Higher Education, 37*(1), 85–101. doi:10.1080/03075079.2010.499166

Lyon, S. W., & Teutschbein, C. (2011). Problem-based learning and assessment in hydrology courses: Can non-traditional assessment better reflect intended learning outcomes? *Journal of Natural Resources and Life Sciences Education, 40*(1), 199–205. doi:10.4195/jnrlse.2011.0016g

Mohan, J. (1995). Thinking local: Service-learning, education for citizenship and geography. *Journal of Geography in Higher Education, 19*(2), 129–142. doi:10.1080/03098269508709297

O'Steen, B., Perry, L., Cammock, P., Kingham, S., Pawson, E., Stowell, R., & Perry, D. (2010). Engaging teachers and learners through service-learning. In K.Weir & P. Coolbear (Eds.), *Good Practice Publication Grants*. Ako Aotearoa. https://ako.ac.nz/assets/Knowledge-centre/Engaging-teachers-and-learners-through-service-learning/Engaging-teachers-and-learners-through-service-learning.pdf

Pawson, E., Fournier, E., Haigh, M., Muniz, O., Trafford, J., & Vajoczki, S. (2006). Problem-based learning in geography: Towards a critical assessment of its purposes, benefits and risks. *Journal of Geography in Higher Education*, *30*(1), 103–116. doi:10.1080/03098260500499709

Pretorius, R., Lonmbard, A., & Khotoo, A. (2016). Adding value to education for sustainability in Africa with inquiry-based approaches in open and distance learning. *International Journal of Sustainability in Higher Education*, *17*(2), 167–187. doi:10.1108/IJSHE-07-2014-0110

Serrano-Estrada, L., Martin, T. J., & Marti, P. (2021). Understanding city dynamics: Using geolocated social media in a problem-based activity as an investigative tool to enhance student learning. *Journal of Geography in Higher Education*, 1–22. Advance online publication. doi:10.1080/03098265.2021.2004390

Shulman, L. (2005). Signature pedagogies in the professions. *Daedalus*, *134*(3), 52–59. doi:10.1162/0011526054622015

Spronken-Smith, R. (2013). Towards securing a future for geography graduates. *Journal of Higher Education in Geography*, *37*(3), 315–326. doi:10.1080/03098265.2013.794334

Spronken-Smith, R., & Kingham, S. (2009). Strengthening teaching and research links: The case of a pollution exposure inquiry project. *Journal of Geography in Higher Education*, *33*(2), 241–253. doi:10.1080/03098260802276813

Spronken-Smith, R., & Walker, R. (2010). Can inquiry-based learning strengthen the links between teaching and disciplinary research? *Studies in Higher Education*, *35*(6), 723–740. doi:10.1080/03075070903315502

Spronken-Smith, R. A., Bullard, J. E., Ray, W., Roberts, C., & Keiffer, A. (2008). Where might sand dunes be on Mars? Engaging students through inquiry-based learning in Geography. *Journal of Geography in Higher Education*, *32*(1), 71–86. doi:10.1080/03098260701731520

Spronken-Smith, R. A., Walker, R., Batchelor, J., O'Steen, B., & Angelo, T. (2012). Evaluating student perceptions of learning processes and intended learning outcomes under inquiry approaches. *Assessment & Evaluation in Higher Education, 37*(1-2), 57–72. doi:10.1080/02602938.2010.496531

Staver, J. R., & Bay, M. (1987). Analysis of the project synthesis goal cluster orientation and inquiry emphasis of elementary science textbooks. *Journal of Research in Science Teaching, 24*(7), 629–643. doi:10.1002/tea.3660240704

Treby, E., Hewitt, I., & Shah, A. (2006). Embedding 'disability and access' into the geography curriculum. *Teaching in Higher Education, 11*(4), 413–425. doi:10.1080/13562510600874169

ADDITIONAL READING

Hensel, N., & Blessinger, P. (Eds.). (2020). *International perspectives on undergraduate research: Policy and practice*. Palgrave Macmillan. doi:10.1007/978-3-030-53559-9

Lee, V. S., Greene, D. B., Odom, J., Schechter, E., & Slatta, R. W. (2004). *What is inquiry-guided learning? In Teaching and learning through inquiry: A guidebook for institutions and instructors. Sterling*. Stylus.

Walkington, H., Hill, J., & Dyer, S. (Eds.). (2019). *Handbook for learning and teaching in geography*. Edward Elgar. doi:10.4337/9781788116497

KEY TERMS AND DEFINITIONS

Community of Practice (CoP): A community of practice is a group of people with a shared concern or goal, or interests, who interact regularly to develop a shared practice. For an introduction to CoPs, see https://wenger-trayner.com/introduction-to-communities-of-practice/.

Course: A defined period of learning, often referred to as a paper or module.

Inquiry-Based Learning (IBL): An umbrella team that encompasses pedagogical approaches in which active learning is stimulated by questions, problems, or issues, and in which the teacher is a facilitator of learning. IBL can involve students understanding existing knowledge, or discovering new knowledge, and can range from heavily scaffolded approaches where the teacher provides a question or issue and an outline for how to address it, through to student-driven approaches whereby students identify a question and determine a way to explore it.

Problem-Based Learning (PBL): A pedagogical approach akin to IBL in which students usually work in small groups to respond to trigger scenarios such as problems, scenarios or issues. Like IBL, the teacher is a facilitator. The approach was initially developed for medical curricula, but is now widely used across a range of contexts. PBL is seen as a subset of IBL, since it is usually (but not exclusively) focuses on students understanding existing disciplinary knowledge, rather than building new knowledge.

ENDNOTES

1 The course is pitched to students as being problem-based, but we see it as more in keeping with an IBL approach (often problem-based approaches are used for information-oriented inquiry, but here the mode is discovery-oriented, open IBL).

2 Mana whenua are the local Māori who have historic and territorial rights over the local area.

Chapter 2
Application of Inquiry-Based Science Education, Anticipatory Learning Strategy, and Project-Based Learning Strategies in the Context of Environmental Problems of the Contemporary World

Iwona Piotrowska
https://orcid.org/ 0000-0003-3249-8870
Adam Mickiewicz University in Poznań, Poland

Jakub Sypniewski
https://orcid.org/ 0000-0001-9312-5109
Adam Mickiewicz University in Poznań, Poland

Małgorzata Cichoń
https://orcid.org/ 0000-0002-2666-1873
Adam Mickiewicz University in Poznań, Poland

Dawid Abramowicz
https://orcid.org/0000-0003-0523-8379
Adam Mickiewicz University in Poznań, Poland

ABSTRACT

The global challenges of melting ice caps, rising global ocean levels, increased methane release from permafrost, and restrictions on access to water resources are primarily associated with climate change. Geographical education is of particular importance in this regard. At all levels of education, it should therefore provide knowledge about the changes taking place, shape practical skills, and foster the acquisition of appropriate attitudes in caring for the state of individual components of the geographical environment. Thus, the mentioned problems of the modern world

DOI: 10.4018/978-1-7998-9598-5.ch002

require the selection of appropriate educational strategies. The chapter presents three educational strategies: teaching through scientific inquiry (IBSE), anticipatory learning strategy (ALS), and teaching through project-based learning (PBL). A general characterization was made, and then a proposal for their application to geography education classes was prepared. Content contexts were proposed for the respective strategies: IBSE (global warming), ALS (ice sheet changes), and PBL (water resources and retention).

INTRODUCTION

The global challenges of melting ice sheets in the northern hemisphere, rising global ocean levels, increased methane released from permafrost, limitations in access to water resources, are primarily associated with climate change, influenced by humans and their various activities. The intensity and dynamics of these phenomena require appropriate actions in education, in particular, according to G. Butt and D. Lambert (2014) the creation of a unified strategy for geographical education. This is due to the need to properly understand the environmental challenges that affect almost every corner of the Earth. A common vision of geographical education aims to prepare people (not only the youngest generations) to respond to the problems of the modern world (Bednarz et al., 2013). To date, a significant commitment has been made by an international scientific group, The Commission on Geographical Education of the International Geographical Union, aiming to create a common perspective on geographical education (Van der Schee, 2014). The adopted International Charter on Geography Education (2016, pp. 5) emphasizes the importance of "deepening understanding of many contemporary challenges such as climate change, water management, food security, energy choices, overexploitation of natural resources and urbanization". These issues are consistent with the planetary boundaries identified by J. Rockström et al. (2009) as indicators of Earth system resilience. These include climate change, biodiversity loss, the nitrogen cycle, phosphorus, ocean acidification, land use, freshwater, ozone depletion, atmospheric aerosols, and chemical pollutants (Rockström et al. 2009).

Therefore, at all levels of education it is necessary to provide knowledge of ongoing changes, shape practical skills and foster the acquisition of appropriate attitudes to care for the state of individual components of the geographical environment. Therefore, education in the field of the presented issues should be held in an appropriate (multifaceted, creative) way, using modern information and communication (ITC) technologies, adapted to the needs of learners (pupils/students) taking into account the curriculum requirements. These requirements are related to the expectations placed on learners (Cichoń, Piotrowska 2018), and their implementation involves the need to choose appropriate strategies and methods of education (Piotrowska et al., 2019; Markuszewska et al., 2018; Butt, 2010).

The term *method* is derived from the ancient Greek *methodos* and means study, way of study, way to truth, and search for truth. According to W. Okoń (1998), the method of education in the context of the path to truth is a tried and systematically applied system of activities of teachers and students, carried out consciously in order to bring about the assumed changes in the personality of students. It can also be defined as "a systematic way of working between teachers and students that makes it possible to achieve educational goals" (Okoń 2004, p. 243). In turn, learning-teaching strategies are organized approaches to task completion that include plans, methods, tools, or techniques to facilitate learning (Okoń 1987). The basic strategies in education are associative knowledge (A), problem-based knowledge (P), emotional experience (E) and operational knowledge (O) (op.cit). Thus, educational method is a logically structured set of principles and procedures used by the teacher to achieve goals in a theoretical or practical way (Morandi 1997). In geographical education it is very important not only to methodically reproduce certain tasks, to practice according to instructions, but first of all to accompany learners in the process of creating new knowledge. The creation of new knowledge, according to the authors of this article, requires the use of not only single educational methods, such as working with a map, discussion or practical exercises, but strategies involving a variety of methods. The strategies with the greatest educational benefit according to the authors include: Inquiry Based Science Education (IBSE), Anticipatory Learning Strategy (ALS), and Project Based Learning (PBL). These strategies, while diverse, are action-oriented for students and aim to develop and improve important 21st century competencies such as critical thinking (Abrami et al. 2015), collaboration (Care et al. 2016), communication (Wood and Hartshorne 2017), and self-direction (Butcher and Sumner 2011). Several characteristics distinguish them: 1) they are based on authentic problems, global challenges; 2) they are academically demanding; 3) learners can develop skills such as teamwork, communication, creativity, and data analysis in addition to knowledge; 4) learners actively and independently engage in information gathering and processing; 5) they encourage the use of technology; and 6) they allow for formal and informal modes of evaluation.

Today, competency requirements have changed due to the increasing number of jobs undergoing automation, the increasingly important role of technology in all areas of work and life, and the growing importance of social, civic, and entrepreneurial competencies to acquire skills to adapt to change. Therefore, with regard to the competencies to be developed and the ability to observe and explain changes in the world, strategies such as Inquiry Based Science Education, Anticipatory Learning Strategy and Project Based Learning can play a very important role in the process of understanding global environmental challenges, as well as increasing public awareness of the importance of the phenomena and processes taking place. The chapter on each of these strategies offers a substantive context for today's world issues: Inquiry Based Science Education addresses global warming; the Anticipatory Learning Strategy emphasizes changes in ice sheets, while Project Based Learning addresses the very important issue of water management and retention.

Inquiry Based Science Education

Inquiry Based Science Education (IBSE) is the intentional process of diagnosing problems, critiquing experiments and distinguishing alternatives, planning research, investigating conjectures, searching information, constructing models, debating with peers, and developing coherent arguments (Bell, Davis & Linn 2004). It is an investigative approach to teaching and learning where students are provided with opportunities to investigate a problem, search for possible solutions, make observations, ask questions, test out ideas, think creatively, and use their intuition (Gillies 2020). Teaching according to the ISBE strategy is based on the so-called teaching-learning cycle (Sypniewski 2019), which grew out of the historical models of Herbart and Dewey and further evolved. In the 1960s, it functioned as a 3-element model proposed by Atkin and Karplus, and a dozen years later it was popularized as the so-called 5E model: engage - explore - explain - extend - evaluate (Rodger 2006). In the contemporary model (Fig. 1), there are two additional elements: *introduction* (i.e. an engaging introduction to the topic, encouraging the student to make a research effort) and *deepening/broadening* (extending the knowledge gained in a classroom setting, especially applying the newly acquired knowledge to everyday life situations). Although the focal point of scientific inquiry is to conduct research, it is important emphasize that it is not an end in itself. Experimentation becomes a means, one of the tools to achieve the goal of answering the research question.

Figure 1. Phases of inquiry based science education
Source: compiled from Van Uum et. al. 2017

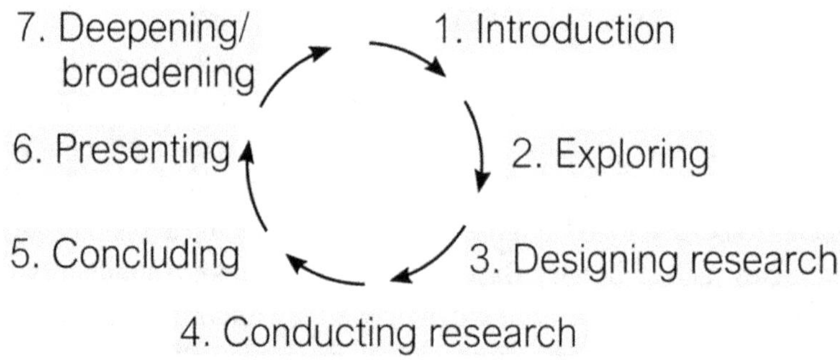

The application of scientific inquiry itself can be introduced during a lesson in one of four ways, called levels of *inquiry*. Each is characterized by different student and teacher participation in the scientific inquiry process (Table 1). An "X" put in successive cells of the table indicates the information that students receive from the teacher. As the level increases, the students' participation increases and the teacher's participation decreases (in terms of posing the research question, proposing methods and giving the solution to the problem). The teacher can introduce the strategy of scientific inquiry in his/her lessons, starting from the simplest level (confirmation) and moving on to higher stages where the teacher's participation decreases. However, this is not a prerequisite. Such a procedure is only aimed at familiarizing students with working according to the IBSE strategy and teaching them how to plan and execute its successive stages. Such a process is challenging, requiring teachers to play an active role in helping students learn the steps in the inquiry process (Gillies 2020).

Table 1. Levels of inquiry

Level of inquiry	Information given to students		
	question	**methods**	**solution**
1. confirmation	X	X	X
2. structured inquiry	X	X	
3. guided inquiry	X		
4. open inquiry			

Source: Bell et al. 2005

The transition between the different stages of the learning cycle is seamless. Each stage is characterized by specific student and teacher activities. The following is a summary of the most important information concerning all seven elements of the cycle; specific student and teacher activities are suggested for each (Table 2). The following is a proposal for a structured inquiry lesson where the teacher presents a research question and proposes methods for solving it (see Table 1), and the students themselves carry out the procedure for performing the experiment, analyze the results and draw conclusions.

Table 2. Characteristics of the stages of the teaching-learning cycle

Stage of the teaching-learning cycle	Brief characteristics of a stage in the teaching-learning cycle
Introduction	Engaging students at the beginning of an inquiry activity can be accomplished in a variety of ways. In younger grades, students can be offered to read a fable (story) that precedes the inquiry or is narrated throughout the inquiry. In older grades, students may find an excerpt from a popular science article to be an intriguing introduction to the topic.
Exploring	This is the first stage of information processing, in which students identify their knowledge of a topic, discuss with each other, and present knowledge and opinions. This stage demonstrates the importance of group work in the IBSE strategy. This is the stage of asking questions about what students do not know and choosing the one (or ones) that will become the research question.
Designing research	The design of the inquiry begins. Students analyze what materials, tools, and instruments will be needed to find the answer to the question posed. Each step must be thought out and designed before conducting experiments.
Conducting research	It is the central stage of scientific inquiry. Experiments are conducted according to a pre-designed instruction/procedure and subsequent steps are subject to observation, which is reliably documented by the inquiry participants.
Concluding	In this phase of the study, the results of previously conducted research are used to formulate conclusions. Students analyze the experiments conducted and the results obtained to answer the research question posed.
Presenting	Individual student groups present the results of their work. This is a very important stage when the results are presented to a wider audience (the whole class or school community). If other groups have worked on the same issue, it is worthwhile to confront the results of each group's work. The differences in the results can become an important element of the discussion. At this stage, it is also worth evaluating the work of the individual groups in order to draw conclusions for their work during subsequent scientific inquiries.
Deepening/ broadening	Newly acquired knowledge may relate to phenomena or processes that are not so much unattainable as remote from the students' perspective. It is worth looking, however, for the applied nature of students' inquiries, i.e., one whose effects they can observe in their immediate environment or a situation in which they can apply knowledge and skills used in everyday life. This stage can also be a natural extension of the previous one, when new questions arise, thus triggering new inquiries (opening new teaching-learning cycles).

Source: compiled from Van Uum et al. 2017

AN EXAMPLE OF USING IBSE IN A GEOGRAPHY LESSON AT THE *STRUCTURED INQUIRY* LEVEL ON THE IMPACT OF CLIMATE CHANGE ON BIVALVES LIVING IN THE SEAS

(based on: Kapusta A., Milewska E., Sypniewski J., Śwital P., CO_2 się stanie w oceanie? [CO_2 will happen in the ocean?], accessed at http://e-chemia.nazwa.pl/amgen/co2/)

Introduction

Introductory text for students:

In 2019, the world's first official retreat of a mountain glacier in Iceland took place. According to contemporary research, 2015 was the first year in which the increase in global temperature due to human activity since the so-called pre-industrial period exceeded 1°C. The increase in the concentration of CO_2 (one of the greenhouse gases) has directly increased the absorption of this gas by the oceans. Carbon dioxide forms a weak carbonic acid with water, which lowers the pH of ocean waters. Ocean acidification affects the building of new shells by bivalves. And what effect does the increased pH caused by the increased supply of carbon dioxide to the atmosphere have on these bivalves living in the oceans and their shells?

Exploring

Students discuss the introductory text. The teacher presents a research question: *How does acid reaction affect the shells of different species of sea bivalves?*

Designing Research

Students are given a card with instructions explaining how they should conduct their experiment and prepare the necessary materials. If the scenario is at a higher level of scientific inquiry (guided inquiry) students design the experiments themselves. The teacher reminds them of the purpose for which limewater is used.

Materials needed:

- Research material: shells of different species of bivalves (preferably characteristic of the latitude; in Poland, typical species of bivalves inhabiting the waters of the Baltic Sea, whose shells can be easily found on the beach, include: *Cerastoderma glaucum, Limecola balthica, Mya arenaria*)
- Instruments (the number of materials depends on the number of shell species): 8 test tubes, 16 clothespins, 4 corks with a hole, 4 straws, 4 pipettes, vinegar, limewater.

Conducting Research

Students conduct the experiment as instructed, recording observations after 5, 15, and 30 minutes.

Figure 2. Control sample and test sample
Source: the authors

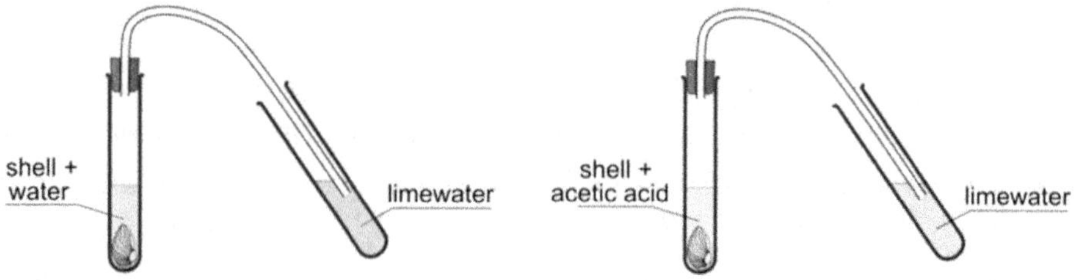

Concluding

Students draw conclusions about what the effect of increased ocean acidification is on the shells of bivalves of different species.

Students should notice a loss of shell mass, washout of some fragments, or their complete disappearance (although the rate may vary between species). At the same time, the limewater into which the outlet of the test samples (i.e., shells immersed in vinegar, at low pH) was directed became cloudy, indicating the release of additional CO_2 during this process. Thus, a so-called positive coupling is observed; an increase in carbon dioxide dissolved in the ocean can accelerate the dissolution of bivalve shells, and additional CO_2 is released during this process, which may further strengthen the process.

*The article uses a simplified inference adapted for elementary school students. In reality, under current ocean conditions, nearly 90% of carbon-containing molecules are bicarbonate ions, 10% are carbonate ions, and less than 1% are carbon dioxide (Popkiewicz et al. 2019).

Presenting

Students present results, discuss and debate observed effects and conclusions. They evaluate the activities and group work.

Example of evaluation - the method of unfinished sentences:

1. I was most surprised by...
2. I remembered that ...
3. I feel more confident in ...
4. I proved that ...
5. My procedure for improving today's experience is: ...

Deepening/Broadening

Kettle scale, like sea bivalve shells, is composed of calcium carbonate. The observed regularity (the action of acetic acid) can be used to clean the kettle of scale deposits. Students can collect pieces of scale and follow the procedure, putting them into a glass and pouring vinegar over them.

Although the topic of climate change has been around for some time, no coherent set of techniques, tools, and strategies for teaching about climate has yet taken shape in pedagogical science (Fazlagić 2021). One proposal in clime education is the inquiry approach, in which students learn about global issues by asking questions and seeking answers. This type of education reflects the actual process of conducting research by scientists. Moreover, conducting experiments is one of the methods of studying geography as a science. This makes it all the more justified to introduce IBSE from an early age in the process of geographical education. Teachers who work with IBSE strategy note a number of benefits, including learning outcomes in terms of students' knowledge and skills, achieving results, combining learning with fun, and finally transferring responsibility for the task to the student (Sypniewski 2015).

Anticipatory Learning Strategy

Anticipatory Learning Strategy (ALS) emphasizes learner autonomy and activity in learning the curriculum outside the classroom. It provides students with opportunities to prepare creatively for class by skillfully searching for information from a variety of knowledge sources, completing assignments, and learning before class, including using Internet resources, computer applications, and reading texts. This approach develops critical and analytical thinking and research skills. The issue of learners' independence in building their knowledge and limiting the dominant role of the teacher by shifting the focus to the role of advisor and organizer of the learning process has been the subject of many studies, including those by F. Herbart (who emphasized the activation of apperceptive masses), L. Vygotsky, J. Piaget, D. Ausubel and K. Kruszewski (Dylak 2013a). D. Ausubel (1960) in reference to J. Piaget, developed a theory which he called *advance organizer*, which means referring during the learning process to what the learner already knows about a given topic. At the beginning of the 21st century in the USA, J. Bergmann and A. Sams (2012) proposed the implementation of teaching in a different way, as *reverse classroom, flipped classroom, flip teaching*, through classroom instruction, but with significant use of computer support in the form of educational videos. This allowed students to learn outside the school building, anywhere, anytime, at their own individual pace. Subsequent studies have deepened the initial strategy, as exemplified by R. Talbert's

(2017) study, in which he proposed an approach to course preparation that allows students to learn about new topics before class.

In Poland, during the implementation of the three-year project "Kolegium Śniadeckich - innowacyjny program nauczania przedmiotów matematyczno-przyrodniczych" (The Śniadecki College - Innovative Teaching Programme in Mathematics and Natural Sciences) in partnership between Adam Mickiewicz University in Poznań and the Polish Nationwide Foundation of Computer Education (Poznań Branch), co-financed by the European Union under the European Social Fund, the idea of developing the Anticipatory Learning Strategy was born (Table 3). It is a concept of education, which through the activity of students also in the online environment puts them in the role of active participants in designing and experiencing their own educational situations (Dylak, Duda 2010; Dylak 2013b). As a result of the project, the methodology of teaching science and mathematics was modified, and the dynamic development of information and communication technologies caused changes in science teaching. In the school year 2010/2011, the project produced 15 thematic modules with scenarios for teaching geography in upper secondary school (Piotrowska et al. 2011).

ALS draws on the concepts of D. Ausubel (1960) and the constructivist theory of knowledge theory and knowledge inquiry. The essence of ALS is that learners actively organize and assimilate knowledge prior to class, in the process of collecting information on their own and by seeking references in their own existing knowledge in relation to the topic to be eventually discussed in class (Dylak 2013b). There is a paradigm shift in teaching in ALS. The role of the teacher changes, who ceases to be a manager and becomes a facilitator, not teaching but supporting learning, not presenting the topic but explaining and clarifying it. The role of the learner also changes, who from a passive recipient becomes an active creator, constructor of his or her own knowledge (Dylak 2000, 2005, 2013a, Kaipa 2000, Klus-Stańska 2000, Sadoń-Osowiecka 2009). Consequently, the classes themselves also change – they cease to be a unit where learners are introduced to new material and during which they are taught. To a large extent, the environment also changes, from classroom to online, much richer and more intellectually challenging. In addition, an equally important feature of ALS is the involvement of students and teacher in a digital environment using information and communication technology.

It is assumed that as the topic of the class is addressed prior to the core activities, questions related to the material being mastered will arise in the mind of the learner. In practice, learners independently collect information and show activity in preparing for classes outside school in advance, organize them into news, build personal subject knowledge by creating news, performing tasks, with the help of the teacher systematize the acquired knowledge. On the other hand, the purpose of the classes becomes consolidation of knowledge, correction of understanding, improvement of certain skills and solution of possible problems.

Table 3. Characteristics of the stages of anticipatory learning strategy

Anticipatory Learning Strategy phase	Phase characteristics
ACTIVATION Tacit knowledge, dialogue	In this phase, students activate their collective knowledge of objects, phenomena, and problems. Students reflect on the topic of the lesson, both independently and with the help of the teacher. The teacher plays a supporting role, mobilizes students to think creatively about a given problem, inspires and mobilizes individual search for information on a given topic from various sources: scientific sources, popular science, colloquial opinions, own thoughts and judgments.
PROCESSING Higher intellectual processes and the construction of knowledge	Students complete specific tasks related to the topic using a variety of instructional materials. They work independently, in groups, and with the teacher, trying to organize and filter the collected material. On the basis of the materials provided by the teacher, students solve instructions, assignments, answer questions and prepare (create) materials (e.g. portfolios, multimedia presentations, films or animations). This phase activates cognitive processes such as analysis, synthesis, comparison, generalization, definition and comprehension processes: justifying, verifying, proving, explaining, inferring. The *Processing* phase can last up to several days, which depends on the learning objectives and content area. According to the assumptions, this is the most important phase for ALS because it is when the idea of the whole strategy is implemented, i.e., anticipatory, prior preparation for the class, before the class itself, in an out-of-class setting.
SYSTEMATIZATION Building new links in the knowledge structure	This stage is carried out during direct teaching, and its idea is to systematize the knowledge acquired by students while searching for answers to questions and solving assignments. Students become aware of their own cognitive constructs and place them in a system of scientific categories, according to the categories of educational objectives chosen by the teacher, which also relate to pragmatic thinking. This allows them to construct their own knowledge and use it to solve problems. The teacher completes, systematizes the news. This stage ends with the students dealing with the given issue.
EVALUATION AND GRADING Reflecting on the structure of knowledge	Students deal with the new knowledge externally. The teacher feedback, peer assessment and self-evaluation are supposed to make the learner aware of the level of new knowledge and acquired skills. At the same time, this stage is supposed to shape in learners the belief that answers to the questions posed generate new questions and that knowledge can be deepened all the time.

Source: compiled from S. Dylak (2013b)

A PROPOSAL FOR THE USE OF ANTICIPATORY LEARNING STRATEGY IN GEOGRAPHICAL EDUCATION TO THE TOPICS OF THE INFLUENCE OF CLIMATE CHANGE ON THE EXTENT OF THE ICE SHEET

Activation Phase

The introduction to ALS implementation is stationary and lasts 45 minutes in a teacher-student relationship. Students/learners activate their prior knowledge and associations about the conditions that determine the formation of ice sheets and the distribution of glaciers on the globe. The activation of prior knowledge is crucial for the learning process, it allows the verification of existing concepts or associations and is a prelude to their proper formation, and above all, there is an interaction of perception, thinking and language (Arends 1995, Galloway 1988, Mietzel 2002).

Processing Phase

This is the most important phase of the entire strategy, involving learning and preparation before the lesson. In this phase, the teacher formulates questions, commands and tasks related to the search in various sources of information, news and data about glaciers on the globe.

Sample questions and commands organizing student work during the search for answers to the problem *Global climate change affects the occurrence and extent of ice sheet:*

1. Explain the mechanism of ice sheet formation on the globe, considering the necessary natural conditions;
2. Demonstrate the effect of increasing air temperature on changes in ice sheet in selected regions;
3. Will research conducted at stations in Spitsbergen, Alaska, and Antarctica help to better understand the effects of climate change on glacier disappearance?
4. What is the influence of ENSO circulation on the mechanism of ocean warming?
5. What is the human impact on climate change since the industrial revolution?
6. Give an example scenario of ice sheet change with increasing air temperatures.

In order for the student to answer the instructions, the teacher prepares teaching materials, which can be scientific texts, popular science texts, diagrams, tables, illustrations, maps, statistical data, graphs, videos or worksheets. These materials should present and explain the mechanism of formation of glaciers depending on natural conditions, the course of changes in their range, the importance of water resources in the world and human economic activity in selected regions. The teacher sends these materials to students directly, or places them on the learning platform, also providing sources of information that are important and necessary to solve the problem and find answers to the questions posed. Sources of information (analogue or digital) should relate to the questions and instructions and be appropriate in terms of content.

Sample selected sources of information for the problem *Global climate change affects the occurrence and extent of ice sheet*:

- Conditions for the formation, movement, and study of glaciers
  Antarctic Ice Movement https://www.pbslearningmedia.org/asset/ess05_vid_icesheets/
- Earth Temperature Changes https://data.giss.nasa.gov/gistemp/
- Global Temperature Anomalies from 1880 to 2021 https://svs.gsfc.nasa.gov/4964
- Global Monthly Mean CO2 https://gml.noaa.gov/ccgg/trends/global.html
- Archived climate data and reanalyzes:
  Climate reanalyzer
  https://climatereanalyzer.org/
  Explore Climate Change https://climatereanalyzer.org/clim/explore/
- Arctic sea ice status http://nsidc.org/arcticseaicenews/
- Changes in Alaska ice extent (using Google Earth Timelapse) https://earthengine.google.com/timelapse/?location=alaska
- Greenland surface melting http://nsidc.org/greenland-today/
- The disappearance of glaciers in Europe in the 20th century https://www.gletscherarchiv.de/fotovergleiche/
- Polar Station of Adam Mickiewicz University in Poznań "Petuniabuka", Spitsbergen, Svalbard archipelago https://polar.amu.edu.pl/index.php
- The Intergovernmental Panel on Climate Change (IPCC) https://www.ipcc.ch/
- Global Warming of 1.5 °C https://www.ipcc.ch/sr15/
- Environmental Change Model https://climatereanalyzer.org/clim/ecm/

While working with scientific and learning materials and searching for answers, the teacher also instructs students to create their own portfolio or e-portfolio.

The work of the student in this phase can be done independently or in collaboration with other students. The important thing is the results of this activity, which are verified by a test that checks the knowledge of students: knowledge, understanding and inference relating to the relationship of natural conditions to the formation of glaciers and the impact of human activity and climate change on changes in the ice sheet. In this phase, the learner reads the materials posted by the teacher on the learning platform and sources of information, formulates questions and discusses the solution of the problem in front of the teacher and in front of other learners, solves tasks and answers questions prepared by the teacher; presents a sample entry in the portfolio or e-portfolio, solves a test posted on the learning platform, checking the knowledge acquired from the indicated sources of information and skills. This substantive preparation of the student is the basis for further activities.

Systematization Phase

At this phase, the teacher discusses the tests taken by learners before the class, proposes the performance of tasks systematizing and generalizing the acquired knowledge, verification by imposing general categories such as names, concepts, definitions on the knowledge mastered by learners. The correct process of forming geographical concepts is also extremely important.

Depending on the teacher's inventiveness, even in this phase he/she can introduce the final discussion on the influence of climate change on the decrease of the ice sheet on the Earth, which will be preceded by reading the article by K. Myers et al. from 2021 entitled *Consensus revisited: quantifying scientific agreement on climate change and climate expertise among Earth scientists 10 years later*; https://iopscience.iop.org/article/10.1088/1748-9326/ac2774. The teacher directs students to ask: Do you agree that climate change exists? What is your position? On the basis of the article, formulate regularities concerning the discussed changes. In addition, he/she instructs students to prepare proposals for possible actions that should be taken by man to reduce the observed processes that threaten to reduce the ice sheet on Earth as a result of climate change.

In turn, the student solves the test of knowledge, but already in direct conditions, checking the actual knowledge of natural conditions, climatic changes affecting the disappearance of the ice sheet, performs tasks ordered by the teacher, makes a concept map of the problem presented in the lesson, completes the portfolio. In addition, they read the proposed article and answer the questions posed while providing their interpretation of the observed phenomena and processes related to climate change, and present their position. They create a proposal for a synthetic infographic outlining actions that are possible, necessary, and feasible for humans to take to minimize environmental climate change in the context of impacts on ice sheet disappearance.

Evaluation and Grading Phase

The teacher and learners summarize the activities undertaken during the ALS. The teacher selects and identifies the information resources that were most frequently used by the learners, evaluates the tasks completed by the learners, correlates the evaluation of the learners' work with the suggestions of others (feedback, peer assessment). The learner, on the other hand, lists what they learned (learner success criteria) and what information was most valuable and educational for them, lists the information resources that were most useful in finding answers to questions and tasks, completes the portfolio, and evaluates their knowledge according to the criteria provided by the teacher (self-evaluation).

The Anticipatory Learning Strategy provides ample opportunities for gifted and hardworking learners/students. When learners (pupils/students) independently search for answers to questions or reflect on the problem posed by the teacher, they develop creative thinking, engage in learning, understand and remember more. The teacher's task (anticipatory organizer) is to connect what the student already knows with what they are supposed to know.

Project Based Learning

Project Based Learning (PBL) is derived from the project method. The project concept was introduced to pedagogy by Ch. R. Richards (ca. 1900), who, however, limited its application to practical technical education (Knoll 1991). The project method was initially used in higher vocational schools, and over time it became widespread in all types of schools (Ausz, Ausz 2007). The creator of the project method is considered to be W.H. Kilpatric (1918). The basis for independent activities undertaken by students was their interests, according to the claim that no one should be forced to undertake unwanted activities. In Poland, the project method was popularized in the interwar period, under the influence of the work of J. A. Stevenson (1930) *The Project Method of Teaching*. In the 1960s there was a return to the practical use of this method, which was abandoned in the 1970s and 1980s due to the then educational policy of the Polish People's Republic. The project method returned to schools thanks to the political transformation and the reform of the educational system in 1999 (Cichoń 2015). According to the Regulation of the Minister of National Education of 20 August 2010, the obligation to organize work using the educational project method was imposed. It was established that the project is a team-based, planned activity of students, aimed at solving a specific problem in four stages: preparation, implementation, presentation and evaluation. The project method became mandatory only for students aged 13 to 16 and was implemented in 2010-2019. Around the same time, the project method began to be used at universities. Currently, the implementation of projects in Poland takes place in almost every sphere of human activity and is an increasingly common way of organizing work.

At the time of Poland's political transformation and educational reform in the 1990s, the scientific community around the world began to treat the project method as a PBL strategy. The use of the term PBL, given the variety of approaches evident in both K-12 and higher education, requires explanation. Project-based learning (PBL) is an action-oriented strategy through complex and critical thinking to solve a problem. Because of the need to solve a problem, the boundary between project-based learning (PBL) and problem-based learning (PrBL) is often blurred; however, J.S. Savery (2006) emphasizes that problem-based learning (PrBL) can be part of project-based learning (PBL), but not vice versa. J.S. Krajcik and P.C. Blumenfeld (2005) define four main tenets of PBL: active construction, situated learning, social interaction, and cognitive tools. It is assumed that during project-based learning, ready knowledge is not provided by the teacher or materials. It is done through experiencing the environment. Besides, just like scientists want to solve a problem, knowledge is created by combining new and old ideas through discussion and action. Integrating and the using technology to create data visualizations, build and test models plays an important role in PBL. J.S. Krajcik and P.C. Blumenfeld (2005) identified five key features of PBL: driving questions, situated inquiry, collaborations, using technology tools to support learning, creation of artifacts. On the other hand, M.S. Szymański (2000) mentioned: progressivist role of the teacher and learner subjectivity, holistic approach and departure from traditional evaluation.

Research on the effectiveness of PBL conducted around the world indicates that it can develop various competencies. According to M. Cichoń (2015), project-based learning develops skills such as collaboration, communication, critical thinking, and creating high-quality, authentic products and presentations. Because of its many benefits, research has emerged on incorporating PBL into teacher education programs. R.A. Ralph (2015) noted that one of the most important benefits of project-based learning is cooperation and negotiation. On the other hand, J.S. Krajcik and N. Shin (2014) emphasizes knowledge construction. However, T. Henry's (2012) team found that both PBL and PrBL are challenging for many students and they do not always like this way of delivering curriculum content.

WATER RESOURCE MANAGEMENT AS AN EXAMPLE OF AN ISSUE THAT CAN BE ADDRESSED IN PBL

Climate change poses a major planning challenge for water resource managers. According to R.J. Lempert and D.G. Groves (2010), changing patterns of precipitation and temperature will disrupt supply and extensive distribution systems in the coming decades, complicating decisions about needed investments. Adequate infrastructure, driven by the need to prepare for water management, has been discussed for several decades in different parts of the world. Adaptation strategies (Andrew, Sauquet 2017) or indicators such as Water Sensitive Cities (Özerol et al. 2020), among others, are being tested. However, the most commonly proposed adaptation solution in many regions is retention (Stańczuk-Gałwiaczek et al. 2018). Research by the team of Z. Bing (2007) indicates that rainwater should neither be simply discharged as wastewater nor completely harvested as a type of resource. Three basic principles should be adopted in rainwater management, i.e., separation of rainfall from wastewater, limited consumption, and use of small facilities. Four methods can be used to manage rainwater: rainwater harvesting, rainwater infiltration, rainwater storage and rainwater pipes. Water management problem, due to its interdisciplinary nature, can be implemented through PBL.

The First Stage of PBL - Preparation

The preparatory stage includes primarily organizational tasks, although substantive tasks, including the selection of a research problem, are also important. Teachers present the principles and features of PBL and the possibility of using the experience of various external institutions. Most often the school's partners are scientific and cultural institutions, museums or local associations. By inviting them to cooperate, we gain experts in various fields. Research of J.S. Lee et al. (2014) indicates that the biggest challenge is to find and include community partners who will devote enough time and be available. However, it is worthwhile to involve experts because the problems raised during PBL become more authentic and the teaching achievements higher. According to K. Wawrzyniak and K. Banaszak (2014), the undertaking of project-related activities by the teacher and students is an exemplification of the thesis that school conditions should be as close as possible to those in everyday life. It is most effective to teach in the context of real, authentic topics that are global in nature but observed locally. Because of the varied nature of tasks in the preparation stage, a variety of learning/teaching methods can be used (Table 4).

The Second Stage of PBL - Implementation

The second stage is primarily to provide adequate time and conditions for activities related to analysis of collected data, inference and problem solving. Field observations with photographic documentation can be one way to search for information on water management. Among the solutions proposed by the students/graduates, there will probably be e.g. collecting water in reservoirs, but there may also be new, innovative

Table 4. Proposal of tasks and methods at the preparation stage of PBL for the topic of water management

Tasks in the project preparation phase	Learning/Teaching method
Prepare students to work with the project method	Demonstration. Case study.
Select groups for project implementation	Talk.
Choose the topic to be implemented using the project method The exhibition of paintings at the Prado Museum, which, together with the WWF, has prepared a special action entitled "+1.5ºC changes everything" at the end of 2019, is a very interesting idea.	Working with image. https://www.bbc.com/news/world-europe-50646625
Introduce the issue that students want to pursue. Presentation of water retention methods in cities.	Brainstorm "What is the most effective way for rational water management?" Lecture with photo presentation.
Discuss the steps of scientific inquiry	Description.
Identify the research problem "How can rainwater from a school building be managed?"	Work with a scientific text. Bogacz A. et al. 2013, Methods of enhancing water retention in urban areas, Scientific Review Engineering and Environmental Science 22 (1), 27-35. https://www.researchgate.net/publication/287280501 ZeleĖáková M. et al., 2017, Urban Water Retention Measures, Procedia Engineering 190, 419–426. https://www.researchgate.net/publication/317325880
Set a design goal A mock-up design for rainwater management around the school building	Discussion.
Identify with students research methods that can be used in the project	Surveys and interviews. Laboratory experiments. Field observations and measurements. Technical exercises.
Decide with students: time of project completion, how to document the work, the evaluation criteria for the stage: implementation of the project and presentation of the final product	Talk. Portfolio.
Prepare a contract	Description. Talk.

Source: the authors

projects e.g. green roofs. At the stage of data collection there are also questions concerning e.g. the possibility of applying specific solutions. It is worth asking them during meetings with experts: hydrologists, engineers, landscape architects, or officials. At the analysis stage the teacher as a facilitator can propose to conduct the JIGSAW method, so that the participants analyze the opinions of the four experts mentioned above in order to determine the technical, environmental and aesthetic conditions of the designed installation.

Consultations are the basis for projects. Consultations should be conducted mainly by a project supervisor and, if possible, experts. The project should shape the culture of cooperation, therefore listening to opinions, sharing experiences is important. From the student's point of view, project implementation is the most difficult stage, because it requires knowledge of research procedures, appropriate methodology, and hypothesis formation. Due to the involvement and effort put in, it is necessary to systematically observe the participants' progress (Cichoń, 2015). For this purpose, it is worth discussing with students the criteria for evaluating individual tasks and creating an appropriate questionnaire or table.

Due to the team nature of PBL, conflicts may arise in the groups. The teacher, through appropriate educational methods and work forms, should show how to be assertive or how to strengthen self-esteem. P.A. Ertmer and K.D. Simons (2006) noted that learners may show considerable frustration if the teacher does not provide activities to achieve the set goal. They also find it difficult to understand that teachers are not providers of knowledge, but are supposed to guide activities and strive for deeper knowledge processing and this is true in both PBL and PrBL.

The Third Stage of PBL - Presentation

A typical presentation should take place in the classroom or in front of a larger audience. In some schools, a "project day" is organized. Public presentation of the results of one's work is an opportunity to present not only an interesting problem, but also one's organizational, technical or methodical skills. The primary goal of the presentation is to arouse the interest of the audience, especially if the results are of an applied nature. This is achieved, among others, through a multimedia presentation. It can be combined with e.g. experience, drama, self-made videos or models. The presentation can be supplemented with drawings, literary works, and recommendations. However, the greatest interest is aroused by happenings, exhibitions, and trips. For PBL participants, the presentation is a great challenge. On the one hand, it can be associated with demonstrating courage in expressing one's own opinions, computer proficiency, and eloquence; on the other hand, it can be paralyzing. One of the solutions is to precisely define the criteria for evaluating the presented final product (Table 5).

Table 5. Criteria for evaluating a rainwater management mock-up project near your school

Criteria	Very good grade	Good grade	Sufficient grade	Unsatisfactory grade
Degree of water use in landscaping	The design allows for the use of minimum 75% rainwater.	The design allows for the use of minimum 50% rainwater.	The design allows for the use of 25-50% rainwater.	The design allows for the use of less than 25% rainwater.
Diversity of landscaping	The design includes four different forms of landscaping including green roofs.	The design includes three forms of landscaping, including larger forms such as ponds.	The design includes two types of landscaping, e.g. flower beds for herbaceous plants.	The design includes one type of landscaping such as a water tank.
Architecture and design	The design incorporates 90% of the school architecture and blends in with the surrounding space.	The design is very much related to the school architecture and blends in with the surrounding space	The design to a small extent incorporates school architecture, but fits into the surrounding space.	The design does not relate to the school architecture and is not integrated into the surrounding space.
Functionality	The landscaping elements do not require additional funds or involve additional energy costs and will not prevent the operation of the school.	The landscaping elements require additional funds or involve additional energy costs, but will not prevent the operation of the school.	The landscaping elements require additional funds or involve additional energy costs, and it will hinder the operation of the school.	The landscaping elements require high costs at the stage of construction and functioning and will hinder the operation of the school, e.g. noise.
Safety	The landscaping elements do not pose any threat. They will function for years to come.	There may be single difficulties with water circulation during the year. Replacement of structural elements may be necessary.	There may be difficulties with water drainage in case of heavy rains. Additional reinforcements, fasteners, etc. will have to be installed.	The landscaping elements create hazards and the constructions used are prone to failure.
Novelty	The design is modern and includes new structural and architectural solutions.	The design is not completely new, but is completed with new architectural or structural concepts.	The design was based almost entirely on another design, 20% modified.	The design was created on the basis of another design, the author's name was not placed (plagiarism).
Aesthetics	The design is refined and executed very aesthetically.	The design is made aesthetically with minor deficiencies.	The design is made mostly aesthetically, there are minor mistakes in colors or finishing materials.	The design is not made aesthetically.

Source: the authors

The Fourth Stage of PBL - Evaluation

It is difficult to assess and evaluate the project in the traditional manner. In PBL we have first of all "process orientation", which from the point of view of the participants and the teacher should be the most important because it determines the formation of key competences. This evaluation concerns the skills developed during the implementation phase of PBL, in which the teacher practically does not participate. This is because the participants collect information and analyze it outside the school. Therefore, it is worth thinking about meetings during which we observe students' work, ask questions such as: What have you managed to do already? So the teacher has no problem evaluating the work of the student team in his or her presence, the greatest difficulty lies rather in evaluating individual and team work done outside the school. However, it is important that at various stages of PBL the participant receives ongoing feedback on the progress made. Project work is sometimes spread over several weeks, so students may lose interest in the tasks they are undertaking. Therefore, it is important to appreciate students at every stage of work. Additional meetings with experts or contact with the local media can increase the motivation of the students. In "process orientation", anonymous peer evaluation can be very important, as M. Frank and A. Barzilai (2004) wrote about almost 20 years ago. The basic qualities that peers evaluate include: systematicity, commitment, motivating others, creativity, communication, and group cooperation. It might be interesting to have a questionnaire administered to each participant in which they indicate their level of competence prior to PBL. A similar survey can be conducted after PBL to determine the participants' progress. It can be particularly useful in assessing soft competencies, which are extremely challenging for instructors to assess. R.W. Marx et al. (1997) emphasize that there is always a danger of a superficial evaluation of work rather than a deeper understanding of the problem, but content knowledge cannot be the only criterion for evaluating PBL. In PBL, we also have an "outcome (final product) orientation" which is more important to the audience, the audience who are interested in the results obtained, recommendations, or creative solutions. Although the grade obtained for the final product shows only a fragment of the participants' skills, it becomes the most important for many teachers. Whether assessing learning or outcomes, the criteria for these evaluations should be discussed with the participants along with the assumptions in the contracts.

CONCLUSION

The changes observed in the geographical environment take place locally, regionally and globally with varying intensity. Knowledge of contemporary natural phenomena and processes and awareness of global challenges is particularly important. These are, first of all, climate change and its warming, which manifests itself in the melting of the ice caps, the increase in the level of the world ocean, the increased release of methane from permafrost or the restriction of access to water resources. The intensity and dynamics of these phenomena require appropriate measures to be taken in various fields and, above all, in education, because the future of the world depends on the education of societies. Education and information about climate change remains a challenge because the concepts of climate change are often not understood by societies, people are not convinced that climate change will affect their own lives (Pruneau et al. 2010). Therefore, climate change education should be a necessity and a priority. Educational activities whether formal or informal should focus on designing environmental messages that convince people that they can still reduce the scale of the phenomenon by linking actions to their aspirations (Pruneau et al. 2010). In the climate positions taken, the most important discussions relate to actions involving adaptation and mitigation. Mitigation is action focusing primarily on improving energy efficiency, increasing the share of renewable energy. Adaptation is the adjustment of current socio-economic systems to a changed climate, which however will take time and may be costly (COP26, Adaptation Gap Report 2021). It is therefore crucial to increase public awareness of the impact of climate change on human development and ecosystems. This can be facilitated by educational activities visible in appropriately selected educational strategies and methods. Educational strategies and methods play an important role in providing not only knowledge, shaping skills but also attitudes among young people. They can in a real way sensitize to the needs and global, natural problems of the contemporary world.

The Inquiry Based Science Education (IBSE), Anticipatory Learning Strategy (ALS) and Project Based Learning (PBL) proposed in the article can be applied to different approaches of climate education. In relation to climate change mitigation education, the authors proposed the ALS strategy, which can, through reflection, constructive and empirical approaches, critical and analytical thinking skills, and research skills to find answers to the questions posed, correct learners' misconceptions and move them to action. In adaptation education, for which IBSE has been proposed, learners could collectively select a specific problem that may be worsened with climate change, analyse it, propose and implement adaptation actions. An important element of inquiry is conducting experiments, especially laboratory experiments. This style of education relates to the nature of research work that scientists do when looking for answers to their questions. According to the authors, another approach to climate education can be separated - identity education, which can be based on PBL. Learners strengthen skills in problem solving, developing recommendations and taking action for the community. Thus, the proposed strategies emphasize an experiential learning process that develops complex competencies. It is also worth recalling the results of the analysis conducted by M.C. Monroe et. al. (2019) reporting that climate change implementation should include engagement in discussions, interactions with scientists, problem solving and implementation of school or community projects. And it is these elements that are particularly emphasised by the IBSE, ALS and PBL strategy. The authors take the position that geographical education is not limited to the strategies described in this chapter, nor do they claim that they will solve all the problems associated with education about contemporary world issues, much less climate change.

REFERENCES

Abrami, P. C., Bernard, R. M., Borokhovski, E., Waddington, D. I., Wade, C. A., & Persson, T. (2015). Strategies for teaching students to think critically: A meta-analysis. *Review of Educational Research, 85*(2), 275–314. doi:10.3102/0034654314551063

Andrew, J. T., & Sauquet, E. (2017). Climate Change Impacts and Water Management Adaptation in Two Mediterranean-Climate Watersheds: Learning from the Durance and Sacramento Rivers. *Water (Basel), 9*(2), 126. doi:10.3390/w9020126

Arends, R.I. (1995). *Uczymy się nauczać*. WSiP.

Ausubel, D. P. (1960). The use of advance organizers in the learning and retention of meaningful verbal material. *Journal of Educational Psychology, 51*(5), 267–272. doi:10.1037/h0046669

Ausz, M., & Ausz, A. (2007). Lublin moje miasto – propozycja projektu interdyscyplinarnego. *Wiadomości Historyczne, 5*, 55–60.

Bednarz, S., Heffron, S., & Huynh, N. (2013). *A Road Map for 21st Century Geography Education. Recommendations and Guidelines for Research in Geography Education*. Association of American Geographers, National Geographic Society.

Bell, R. L., Smetana, L., & Binns, I. C. (2005). Simplifying inquiry instruction. *Science Teacher (Normal, Ill.), 72*(7), 30–33.

Bergmann, J., & Sams, A. (2012). *Flip your classroom: reach every student in every class every day*. ISET.

Bing, Z., Hai-qiao, T., & Li-juan, W. (2007). A New Approach to Urban Rainwater Management. *J China Univ Mining & Technol, 17*(1), 82–84. doi:10.1016/S1006-1266(07)60018-2

Bogacz, A., Woźniczka, P., Burszta-Adamiak, E., & Kolasińska, K. (2013). Methods of enhancing water retention in urban areas. *Scientific Review Engineering and Environmental Science, 22*(1), 27–35.

Butcher, K. R., & Sumner, T. (2011). Self-directed learning and the sensemaking paradox. *Human-Computer Interaction, 26*(1–2), 123–159. doi:10.1080/07370024.2011.556552

Butt, G. (2010). Which methods are best suited to the production of high-quality research in geography education? *International Research in Geographical and Environmental Education, 19*(2), 103–107. doi:10.1080/10382046.2010.482189

Butt, G., & Lambert, D. (2014). International perspectives on the future of geography education: An analysis of national curricula and standards. *International Research in Geographical and Environmental Education, 23*(1), 1–12. doi:10.1080/10382046.2013.858402

Bybee, Taylor, Gardner, Van Scotter, Powell, Westbrook, & Landes. (2006). The BSCS 5E instructional Model: Origins and Effectiveness. Academic Press.

Care, E., Scoular, C., & Griffin, P. (2016). Assessment of collaborative problem solving in education environments. *Applied Measurement in Education, 29*(4), 250–264. doi:10.1080/08957347.2016.1209204

Cichoń, M. (2015). Metoda projektu edukacyjnego. In M. Beczkiewicz & M. Szyszkiewicz-Golis (Eds.), Ekologia Krajobrazu cz. II. Przewodnik metodyczny (pp. 38-47). IŚRiL PAN.

Cichoń, M., & Piotrowska, I. (2018). Level of academic and didactic competencies among students as a measure to evaluate geographical education and preparation of students for the demands of the modern labour market. *Quaestiones Geographicae, 37*(1), 73–86. doi:10.2478/quageo-2018-0006

Dylak, S. (2000). Konstruktywizm jako obiecująca perspektywa w kształceniu nauczycieli. In H. Kwiatkowska, T. Lewowicki, & S. Dylak (Eds.), Współczesność a kształcenie nauczycieli. WSP ZNP.

Dylak, S. (2013a). *Architektura wiedzy w szkole*. Difin.

Dylak, S. (Ed.). (2013b). Strategia kształcenia wyprzedzającego. Ogólnopolska Fundacja Edukacji Komputerowej, UAM.

Dylak, S., & Duda, L. (2010). *The implementation strategy of Collegium Sniadeckich*. Ogólnopolska Fundacja Edukacji Komputerowej, UAM.

Ertmer, P. A., & Simons, K. D. (2006). Jumping the PBL implementation hurdle: Supporting the efforts of K–12 teachers. *The Interdisciplinary Journal of Problem-Based Learning, 1*(1), 40–54. doi:10.7771/1541-5015.1005

Fazlagić, J. (2021). Edukacja klimatyczna. *Dyrektor Szkoły, 4*, 53–56.

Frank, M., & Barzilai, A. (2004). Integrating alternative assessment in a project-based learning course for preservice science and technology teachers. *Assessment & Evaluation in Higher Education, 29*(1), 41–61. doi:10.1080/0260293042000160401

Galloway, Ch. (1988). *Psychologia uczenia się i nauczania*. PWN.

Gillies Robyn, M. (2020). *Inquiry-based Science Education*. CRC Press Taylor & Francis Group. doi:10.1201/9780429299179

Henry, H., Tawfik, A., Jonassen, D., Winholtz, R., & Khanna, S. (2012). "I know this is supposed to be more like the real world, but…": Student perceptions of a PBL implementation in an undergraduate materials science course. *The Interdisciplinary Journal of Problem-Based Learning, 6*(1), 43–81. doi:10.7771/1541-5015.1312

Kaipa, P. (2000). Knowledge architecture for the twenty-first century. *Behaviour & Information Technology, 19*(3), 153–161. doi:10.1080/014492900406146

Kapusta, A., Milewska, E., Sypniewski, J., & Śwital, P. (2020). *CO_2 się stanie w oceanie?* http://e-chemia.nazwa.pl/amgen/co2/

Kilpatrick, W. H. (1918). The project method. *Teachers College Record, 19*, 319–335.

Klus-Stańska, D. (2000). *Konstruowanie wiedzy w szkole*. Wydawnictwo Uniwersytetu Warmińsko-Mazurskiego.

Knoll, M. (1991). Europa-nicht America: Zumursprung der projekt methode in der pädagogik, 1702-1875. *Pädagogische Rundschau, 45*, 41–58.

Krajcik, J. S., & Blumenfeld, P. C. (2005). Project-Based Learning. In R. K. Sawyer (Ed.), *The Cambridge handbook of the learning sciences* (pp. 317–333). Cambridge University Press. doi:10.1017/CBO9780511816833.020

Krajcik, J. S., & Shin, N. (2014). Project-based learning. In R. K. Sawyer (Ed.), *The Cambridge handbook of the learning sciences* (2nd ed., pp. 275–297). Cambridge University Press. doi:10.1017/CBO9781139519526.018

Lee, J. S., Blackwell, S., Drake, J., & Moran, K. A. (2014). Taking a Leap of Faith: Redefining Teaching and Learning in Higher Education Through Project-Based Learning. *The Interdisciplinary Journal of Problem-Based Learning, 8*(2). Advance online publication. doi:10.7771/1541-5015.1426

Lempert, R. J., & Groves, D. G. (2010). Identifying and evaluating robust adaptive policy responses to climate change for water management agencies in the American west. *Technological Forecasting and Social Change*, *77*(6), 960–974. doi:10.1016/j.techfore.2010.04.007

Linn, M. C., Davis, E. A., & Bell, P. (Eds.). (2004). *Internet environments for science education*. Lawrence Erlbaum Associates Publishers.

Markuszewska, I., Tanskanen, M., & Vila Subirós, J. (2018). New Ways to Learn Geography - Challenges of the 21st Century. *Quaestiones Geographicae*, *37*(1), 37–45. doi:10.2478/quageo-2018-0004

Marx, R. W., Blumenfeld, P. C., Krajcik, J. S., & Soloway, E. (1997). Enacting project-based science: Challenges for practice and policy. *The Elementary School Journal*, *97*(4), 341–358. doi:10.1086/461870

Mietzel, G. (2002). *Psychologia kształcenia*. Gdańskie Wydawnictwo Psychologiczne.

Monroe, M. C., Plate, R. R., Oxarart, A., Bowers, A., & Chaves, W. A. (2019). Identifying effective climate change education strategies: A systematic review of the research. *Environmental Education Research*, *25*(6), 791–812. doi:10.1080/13504622.2017.1360842

Morandi, F. (1997). *Modèles et méthodes en pédagogie*. Nathan.

Myers, K., Doran, P., Cook, J., Kotcher, J., & Myers, T. (2021). Consensus revisited: Quantifying scientific agreement on climate change and climate expertise among Earth scientists 10 years later. *Environmental Research Letters*, *16*(10), 104030. doi:10.1088/1748-9326/ac2774

Okoń, W. (1987). *Wprowadzenie do dydaktyki ogólnej*. PWN.

Okoń, W. (1998). *Wprowadzenie do dydaktyki ogólnej*. PWN.

Okoń, W. (2004). *Nowy słownik pedagogiczny*. PWN.

Özerol, G., Dolman, N., Bormann, H., Bressers, H., Lulofs, K., & Böge, M. (2020). Urban water management and climate change adaptation: A self-assessment study by seven midsize cities in the North Sea Region. *Sustainable Cities and Society*, *5*, 55. doi:10.1016/j.scs.2020.102066

Piotrowska, I., Cichoń, M., Abramowicz, D., & Sypniewski, J. (2019). Challenges in geography education - a review of research problems. *Quaestiones Geographicae*, *38*(1), 71–84. doi:10.2478/quageo-2019-0009

Piotrowska, I., Cichoń, M., Mrula, A., & Nowak-Pierszalska, K. (2011). *Scenariusze zajęć z geografii dostosowane do strategii kształcenia wyprzedzającego – Kolegium Śniadeckich.* Pracownia Dydaktyki Geografii i Edukacji Ekologicznej, Wydział Nauk Geograficznych i Geologicznych, Uniwersytet im. Adama Mickiewicza w Poznaniu.

Popkiewicz, M., Kardaś, A., & Malinowski, Sz. (2019). *Nauka o klimacie. Wyd. Sonia Draga Sp. z o.o. oraz Wyd.* Nieoczywiste.

Pruneau, D., Khattabi, A., & Demers, M. (2010). Challenges and possibilities in climate change education. *US-China Education Review*, *7*(9), 15-24.

Ralph, R. A. (2015). Post secondary project-based learning in science, technology, engineering and mathematics. *Journal of Technology and Science Education*, *6*(1), 26–35.

Rockström, J., Steffen, W., Noone, K., Persson, Å., Chapin, F. S., Lambin, E. E., Lenton, T.M., Scheffer, M., Folke, C., Schellnhuber, H., Nykvist, B., De Wit, C.A., Hughes, T., van der Leeuw, S., Rodhe, H., Sörlin, S., Snyder, P.K., Costanza, R., Svedin, U., Falkenmark, M., Karlberg, L., Corell, R.W., Fabry, V.J., Hansen, J., Walker, B., Liverman, D., Richardson, K., Crutzen, P., & Folley, J. (2009). Planetary boundaries: Exploring the safe operating space for humanity. *Ecology and Society*, *14*(2), 32. https://www.ecologyandsociety.org/vol14/iss2/art32/

Sadoń-Osowiecka, T. (2009). *Konstruowanie wiedzy geograficznej w klasach gimnazjalnych. Możliwości i zaniedbania*. Oficyna Wydawnicza Impuls.

Savery, J. S. (2006). Overview of PBL: Definitions and distinctions. *The Interdisciplinary Journal of Problem-Based Learning*, *1*(1), 9–20. doi:10.7771/1541-5015.1002

Stańczuk-Gałwiaczek, M., Sobolewska-Mikulska, K., Ritzema, H., & Van Loon-Steensma, M. J. (2018). Integration of water management and land consolidation in rural areas to adapt to climate change: Experiences from Poland and the Netherlands. *Land Use Policy*, *77*, 498–511. doi:10.1016/j.landusepol.2018.06.005

Stevenson, J. A. (1930). *Metoda projektu w nauczaniu.* Książnica - Atlas, Lwów - Warszawa.

Sypniewski, J. (2015). *Influence of the strategy of the Inquiry Based Science Education on the development of natural interests of students and acquiring natural-geographical com-petences in the course of extracurricular activities of natural science activity group* [MA thesis]. Adam Mickiewicz University.

Sypniewski, J. (2019). Where the geographical expanse ends – Space education in primary school. Implementation of inquiry based science education (IBSE) in geography lessons in Polish school. *Miscellanea Geographica, 23*(4), 256–266. doi:10.2478/mgrsd-2019-0024

Szymański, M. S. (2000). *O metodzie projektów*. Wydaw. Żak.

Talbert, R. (2017). *Flipped learning: a guide for higher education faculty. Sterling*. Stylus Publishing, LLC.

The Comission on Geographical Education. (2016). *International Charter on Geography Eduction*. International Geographical Union.

Van der Schee, J. (2014). Looking for an international strategy for geography education. *Journal of Research and Didactics in Geography, 1*(3), 9–13.

Van Uum, M., Verhoeff, R., & Peeters, M. (2017). Inquiry-based science education: Scaffolding pupils' self-directed learning inopen inquiry. *International Journal of Science Education, 39*(18), 1–21. doi:10.1080/09500693.2017.1388940

Wawrzyniak, K., & Banaszak, K. (2014). Metoda projektów – dydaktyczną syntezą współczesnej wiedzy i doświadczeń w edukacji. In S. Dylak, G. Barabasz, D. Hejwosz-Gromkowska (Eds.), Metoda projektów w edukacji ponadpodstawowej (na przykładzie przedmiotów przyrodniczych) (pp. 100-119). UAM, OFEK, Samorząd Woj. Wielkopolskiego.

Wood, L., & Hartshorne, M. (2017, March 8). Literacy: The role of communication skills. *The Voice for Secondary Education*. https://www.sec-ed.co.uk/best-practice/literacy-the-role-of-communication-skills

Zeleňáková, M., Diaconu, D. C., & Haarstad, K. (2017). Urban Water Retention Measures. *Procedia Engineering, 190*, 419–426. doi:10.1016/j.proeng.2017.05.358

Chapter 3
The GeoScenario:
A Methodological Resource From Problem-Based Learning (PBL) in Landscape Teaching

Alfonso García de la Vega
https://orcid.org/0000-0001-5767-3984
Universidad Autónoma de Madrid, Spain

ABSTRACT

The geographical scenario (geoscenario) is considered a methodological resource in geography teaching. The geoscenario may be defined as the minimal significant landscape framework unit in landscape teaching that embodies a synthetic representation of a real-world acute issue involving a particular geographical problem. The application of geographical abilities and skills and the selection of real global issues in the geographical scenarios promotes lifelong learning. A broadly-based planning is needed for an integrated and interdisciplinary curriculum in geography. PBL method is aligned with the fundamentals of the constructivist approach and promotes collaborative and cooperative learning in small group work. BPL implementation involves six steps and three progression levels for problem-solving in relation to cognition processes, classroom dynamics, and the identification degree of features and events in the geoscenario. Experiential knowledge and analogical thinking are applied to problem formulation and solving in the geoscenario.

DOI: 10.4018/978-1-7998-9598-5.ch003

INTRODUCTION

This paper presents the geographical scenario (geoscenario) as a methodological resource in Geography teaching and Problem-based learning (PBL) as the methodological approach where a most profitable use of this resource can be made. In the first section the geoscenario is characterized, by providing a definition and identification of integral elements, as a real landscape setting for geographical analysis and interpretation. Landscape is addressed as a curricular content showing major cross-disciplinary features that poses a pedagogical challenge. Then selection criteria are proposed for teacher's suitable selection of geographical elements and events in the scenario, and so is a broadly-based planning for an integrated and interdisciplinary curriculum in Geography teaching. The educational factors are identified in PBL implementation within the constructivist learning environment. Geographical abilities and skills that promote lifelong learning are underlined in the pursuit of creative solutions to real problems as committed citizens. The second section deals with the PBL method, its conceptual background, and methodological framework in line with the fundamentals of the constructivist approach and the constructivist learning environment, in terms of discovery learning, deep learning, meaningful learning and social interaction among peers and between students and the teacher. Also, Geography experiences of PBL are discussed linked to fieldwork and fieldtrip. In the third section 6 steps and 3 levels are suggested for BPL implementation that show strategies for problem-solving, in relation to cognition processes and classroom dynamics. Finally, the fourth section addresses the geoscenario for landscape teaching in PBL, addressing the close relationship between a geoscenario and the construction of a scientific concept, types of geoscenarios, and experiential knowledge and analogical thinking applied to problem formulation and solving.

CHARACTERIZING THE GEOGRAPHICAL SCENARIO

The geographical scenario (geoscenario) is here presented as a teaching resource in landscape teaching that may be defined as the minimal significant landscape framework unit that embodies a synthetic representation of a real-world acute issue involving a particular geographical problem (García de la Vega, 2015). This unit must contain geographical elements in a relevant number to be significantly expressive features to delimit a landscape framework. These features need to be identified in order to understand the links inherent in the landscape. Analyzing these links may lead to a recognition of the geographical events. A geographical event is considered the resulting effect of a particular natural process or/and cultural action on the landscape. The geoscenario provides a real landscape setting for geographical analysis and interpretation.

To illustrate this definition of geoscenario, a real landscape setting is presented below for geographical analysis and interpretation. This scenario is based on a satellite photograph taken from the NASA spatial station in the space. This view is one of the few images having earth features that can be recognized from satellite space. This scenario shows the acute issue of an extensive horticultural exploitation in plastic greenhouses along a vast coastal territory in Andalusia, Southern Spain (Figure 1). This striking photograph of the vast extension of white hoods that are distributed along the Gulf of Almería shows how Campo de Dalías has become a peculiar phenomenon on the Almeria coast, being covered in plastic patches. Based on a study on horticulture in greenhouses (Pulido Bosch et al., 2000), Almería is estimated to have more than 30,000 hectares devoted to fruit and vegetable production under greenhouses permanently, both with plastic and with glass. Around 70,000 hectares are distributed throughout Spain, the second country in the world for "overcast" extension, only behind China, with 82,000 hectares. This geoscenario brings the problem of the outstanding increasing proportion of plastic-greenhouse hectares in a relatively small and delimited area into sharp focus, resulting in lack of nutrients and overexploitation of water resources in an area of very poor annual rainfall that demands intensive water pumping.

Figure 1. Geoscenario of Dalías plastic sea (Campo de Dalías, Almería, Spain). Source: NASA, 2014

Landscape as a curricular content shows major cross-disciplinary features that allow for implementing innovative didactic strategies. In the curriculum, landscape comes mainly into focus in Geography teaching, though it has been approached from other disciplines. There exists a long curricular tradition of Landscape Ecology (Schreiber, 1990 and Forman, 1990) and Geological Landscape (Allen, 2008) that comes from the naturalists in the late 19th Century. Thus, landscape is a content in the curriculum of a variety of disciplines that, far from being a problem, presents a pedagogical challenge. The challenge facing teachers lies in creating methodological strategies to pull all contents on landscape together and provide a global concept of landscape in an integrated curriculum. Since the European Landscape Convention (Council of Europe, 2000) was ratified, landscape is gaining growing appreciation through actions of cultural promotion and education campaigns. These actions have been performed in the context of European projects and recommendations that, in their turn, have given place to directives at a national level of European landscapes (García de la Vega, 2019b). Such European boost to landscape has led to educational projects targeting school activities on landscape in different countries (Castiglioni, 2010 and 2012). Recommendations, such as CM/REC 2008/3 and CM/REC2014/8, aim to lay down and disseminate learning modes on landscape (Council of Europe, 2008, 2009 and 2014). There are numerous attempts to promote knowledge, observation and interpretation of landscape in order to raise awareness and appreciation of natural and cultural elements in landscape (García de la Vega, 2018).

In the panoramic photograph of this countryside landscape, natural and cultural features are clearly recognizable, namely, eroded plateau, creek of Río Dulce and periglacial valley, settlement and ancient medieval castle, and land of cereal crops (Figure 2). Students are able to identify geographical events (e.g. rich farming and low population density) and the links inherent in the landscape (e.g. the relationship between the crops, the valley and the tiny settlement) and to correlate them (e.g. the settlement grew up around the castle, the abundant crops in this cultivated land point to a high level of water table on the valley). Yet, in principle, there are no striking patent elements in this landscape that may draw the students' attention to an acute issue nor enable them to address a geographical problem, and therefore this image does not stand for *a priori* geoscenario. This is obvious unless the students are already acquainted with this natural environment and have learnt on some features from previous direct observation in fieldwork that allow them to find evidences in this landscape. Then, a further analysis of the underlying features may lead them to draw some conclusions, namely, this geoscenario shows an extensive area of rich arable land comparing to the poor population settlement on the hill, and the number of inhabitants that is registered in the population census is between 13 and 15 inhabitants in the last 20 years (INE, 2022), which underlines a problem of rural depopulation.

Figure 2. Countryside landscape of dominant valley landform cutting a slightly folded plateau (Pelegrina, Guadalajara, Spain). Source: A. Garcia de la Vega, 2013

The geoscenario may result a key methodological resource in Problem-Based Learning (PBL), since the scenario embodies a synthetic representation of a real-world acute issue that involves a particular geographical problem. The geoscenario must contain sufficient significant elements and geographical events to show an acute issue. The teacher's suitable selection of geographical elements and events in the scenario is the main determinant of the extent to which the students will be able to identify and accurately formulate the problem. In fact, a successful identification and formulation of the problem depends on how precise the selection of the scenario may be, and so does the further search for an optimal solution to the problem. Wilson (2006) established 11 selection criteria for effective study cases or scenarios for instructors to know how to create problems in Engineering. These are divided into criteria about the goals and about the scenario created. Regarding the goals, the author's proposal is based on three variables: 1) the chosen learning goals are achievable; for single courses (for example in hybrid or conventional programs) about 3 to 5 hours of study for an individual student; about 6 to 10 objectives for a group of 6 students so that each will research/teach the others; 2) the learning outcomes are consistent with the stage of development and builds on and activates prior knowledge; and 3) goals integrate knowledge, skills and attitudes across subjects and disciplines. In relation to the scenario created, the scenario can be a single scenario, or a sequence of scenarios, but each would expect the same 3 to 5 hours of student study. The variables for the scenario created are as follows: 4) the scenario contains "cues" that will trigger the desired search for learning objectives; the learning outcomes expected by the teacher are identified correctly by the students; 5) the scenario includes an appropriate level of complexity; 6) the scenario allows an openness; 7) the scenario is motivational and relevant; 8) the scenario is similar to one we might encounter in professional practice (for example, in Engineering this might include rating, debottlenecking, design, trouble shooting, labor relationships, team work, public, monitoring compliance with legislative regulations); 9) promotes student activity; 10) any data given should be raw data (like we encounter in practice); 11) the scenario identifies the context, gives a concrete scenario and clearly identifies the expected task without spelling out specifics. Figure 1 above shows a geoscenario having significant geographical elements and events to represent such a synthetic representation of a real-world acute issue that involves a particular geographical problem and meets Wilson's selection criteria.

For the geoscenario to be implemented, a progressive guideline is suggested at three levels of difficulty—basic, intermediate and advanced— in the gradual understanding and acquisition of the concepts that may be drawn from the scenario. These levels of difficulty have been established on the basis of first-hand experience coming from the application of the geoscenario to PBL in the professional teaching practice. This personal and hands-on experience has been gained in teaching higher-education student groups in Geography, Tourism and Teachers Training courses of study in European, South-American and South-Asian Universities.

However, should the geoscenario be included in classroom planning, it must be given a new place and didactic meaning. A broadly-based planning is needed for an integrated and interdisciplinary curriculum in Geography. The planning may be centered on key geographical problems that may draw together related curricular contents from different disciplines (e.g. geology, ecology, biology, forestry, economics, environmental sciences, land planning and management, etc.). These contents need to be approached as interdisciplinary problems. This is in line with the proposal of curricular integration by Beane (1995). Thus, geographical scenarios representing a real-world acute issue are chosen to show a pertinent geographical problem that students need to solve by drawing on curricular contents from other disciplines. Geographical scenarios also need to be given a framework of competences and assessment criteria, as well as of skills and abilities to achieve holistic learning. Geographical scenarios also promote meaningful learning. In this sense, the problem posed should provide interest and meaning for the student. Thus, the students attribute meaning to learning to the extent that it belongs to their area of interest (Schmidt, Rotgans and Yew, 2011). Real-world problems encourage geographical skill and competence acquisition. Partoune (1999) provided some valuable contributions of competences in Geography to a cross-disciplinary approach. These geographical competences are, namely, being in contact and understanding a real situation; understanding the spatial structure of the environment; identifying relationships and spatial dynamics; analyzing a reality in reference to a certain number of spatial models; and finally, building diagrams, cartograms, sketches and spatial models (García de la Vega, 2022).

The implementation of problem-based learning approach demands a set of educational factors in order to establish a constructivist learning environment. Such an environment needs to cover all the teaching factors that play a part in the teaching-learning process. These factors are derived from the psycho-pedagogical principles in the constructivist theory but applied to classroom planning. The constructivist learning environment fits well in both discovery learning and significant learning as a platform for implementing problem-based learning. Besides, other principles related to the student's autonomy and self-regulated work must be considered, as well as the teacher's intervention as a learning facilitator. The teacher's role is secondary to the student's learning acquisition, being just limited to fostering learning among equal classmates or peers. Thus, the teacher as a learning facilitator plays a prominent role in encouraging teamwork techniques in order to enhance students' individual and social abilities. Students gradually acquire autonomy through both self-regulated learning and the engagement to sharing tasks, working in small groups and being respectful of peer contributions and of their learning advancement. In PBL, the student's autonomy, interest and creativity strengthen cooperative and collaborative learning, and foster reasoning and critical thinking following the socio-constructivist approach (Savery and Duffy, 1995).

The application of geographical competences and skills, and the selection of real global issues in the geographical scenarios promotes lifelong learning. In lifelong learning, the citizen acquires knowledge to manage geospatial technological tools in searching for and downloading documentation, and by connecting onto and interacting with a variety of networks. The NCR report (2006, ix) explicitly states that "spatial thinking—one form of thinking—is based on a constructive amalgam of three elements: concepts of space, tools of representation, and processes of reasoning". Nevertheless, each of these three elements needs a life's span for development and acquisition. Thus, it would be convenient that the competences linked to acquiring geospatial abilities are incorporated into the core curriculum with a view to enable citizens to cope with everyday spatial situations and to seek effective and efficient solutions in their professional careers (García de la Vega, 2019a). The development of spatial thinking and abilities has been promoted by the European Union through key competence acquisition for lifelong learning (Europpean Commission., 2018). In this respect, the European Union has made some recommendations in their guidelines for lifelong learning (UE, 2006). In particular, the digital competence requires developing further capacities, such as the ability to search, collect and process information, and to use it in a systematic and sound manner, as well as the ability to access, search and use internet-based services. The European Union suggests that using digital tools for producing, presenting and understanding complex information is an essential skill of the digital competence. The European Union sets forth a critical use of Information Society Technology (IST) to support critical thinking, creativity and innovation (UE, 2006 and Carretero et al., 2017).

The material being used in the geoscenario encourages and reinforces geographical abilities and skills for lifelong learning. The use of mobile devices involves being familiar with geospatial concepts such scale, orientation, location and estimation of distances (García de la Vega, 2019a). In Geography teaching, landscapes that provide significant and complete geographical scenarios, even complex ones, are shown by means of satellite and aerial imaging, panoramic photographs. Other geographical scenarios may be drawn from cartography at varying scales, documentary resources, economic and population graphs, short reports and videos. This means that the student should acquire certain geographical skills that are associated with managing such resources. According to Woods et al. (1997), problem solving skill includes the following 12 attributes:

- being aware of the processes used;
- using pattern matching to quickly decide whether a situation is a problem or an exercise;
- applying a variety of tactics and heuristics;
- placing an emphasis on accuracy (as opposed to speed.);
- being active by writing down ideas, creating charts and figures;
- monitoring and reflecting on the process used;
- being organized and systematic;
- yet being flexible (keeping options open, seeing the situation from many different perspectives and points-of-view);
- drawing on the pertinent subject knowledge and objectively and critically assessing the quality, accuracy and pertinence of that knowledge and data;
- being willing to risk and cope with ambiguity, welcoming change and managing distress;
- being willing to spend time reading, gathering information and defining the problem (as opposed to equating problem solving with "doing something" despite its pertinence); and
- having an overall approach that uses fundamentals rather than trying to combine various memorized sample solutions.

Finally, lifelong learning encourages critical thinking. DeVito and Tremblay (Cfr. Boisvert, 2004) provide a skill-set for students to develop critical thinking, namely, to accurately and precisely define the problem; to order, convey and coherently express ideas; and to draw and analyze conclusions.

Problem-Based Learning: An Educational Change

Problem-based Learning arose in the context of the complex diagnosis in medicine, where the solution was reached from a cross-disciplinary approach to a medical problem. PBL is widely used in medicine and the technology disciplines, as a hands-on approach to learning core subjects in Higher Education (Savin-Baden, 2000 and Savin-Baden and Wilkie, 2004). PBL has also gained considerable strength in the teaching of scientific disciplines (Smith and Hoersch, 1995 Maitland, 1998, O'Neill, 2000 and Schmidt, Vermeulen and van der Molen, 2006). PBL is nowadays a widespread teaching methodology in different professional fields and disciplines, namely medicine, business, education, engineering, mathematics and science (Wilkerson and Gijselaers, 1996).

Barrows (1986 and 1996) defines the PBL methodological framework, specially, the teacher-student relationship and the steps to be taken in classroom dynamics. As Barrows claims (1986:6):

"New information is acquired through self-directed learning. As a corollary to the characteristics already described (the student-centered curriculum and the teacher as facilitator of learning), the students are expected to learn from the worlds knowledge and accumulated expertise by virtue of their own study and research, just as real practitioners do. During this self-directed learning, students work together, discussing, comparing, reviewing, and debating what they have learned."

Teacher and students are the key actors in PBL. The teacher is attributed a role as a tutor and a learning mediator, and the student is attributed a role of commitment in his/her own learning acquisition. However, the students need to gradually acquire a certain degree of autonomy in the learning process and in developing certain abilities and skills. Till then, the teacher is more present throughout the entire teaching-learning process. This didactic intervention brings about an educational change in the teacher-student relationship, in the cross-disciplinary approach to learning and in the classroom dynamics. Even PBL may be adapted and partially implemented in the traditional lesson-plan, where the teacher's knowledge is predominant and the students' role is more passive (Stepien, Gallagher and Workman, 1993).

PBL promotes the acquisition and integration of new learning, by providing new challenging contents and by stimulating the student's interest in these contents and their applicability to everyday life. These aspects are associated with deep learning (Egan, 2008). Wilson (2012) points to the fact that students experiencing PBL show a preference for deep learning, instead of surface learning that is mostly based on lecture experience. Deep learners search for meaning in what they are learning. These strategic learners will adapt their style of learning to the expectations of the course.

In this sense, Hmelo-Silver and Barrows (2006) pointed out the aims and strategies of PBL that are linked to deep learning. The contents organized around a theme may lead to the structure of an integrated curriculum, a term coined by Beane (1995). Barell (2006) regarded problem-based learning as an inquiry process that aims to resolve questions and to awaken a students' curiosity. Thus, he considers that, in this research approach to problem solving, students achieve a greater degree of autonomy in the learning process (Watson, 2008). In this sense, Wood, Bruner and Ross (1976) highlighted the value of the teacher as a mediator in the construction of learning and the skills that students should achieve. Weiss (2003) underlines that PBL promotes higher order thinking. Savery (2006) states that the integration of knowledge and skills provides greater clarity to the solution of the problem posed.

Savery and Duffy (1995) showed the close connection between the constructivist approach and problem-based learning. In fact, psycho-pedagogical principles, such as Bruner's discovery learning and Ausubel's meaningful learning, form an essential methodology basis for PBL. Bruner (1978), when studying the mental processes of learning, emphasized the identifying features that generated a cognitive categorization of the disjunctive or conjunctive character of certain elements. This assignment of characteristics may constitute a strategy to discover and distinguish the geographical features in the scenario. Novak (2002) considers that significant learning involves an incorporation of knowledge to the cognitive structure. This author considers three conditions to be necessary for this cognitive incorporation: the deliberate effort to relate new knowledge to another of a higher order located in such a structure; learning is related to facts or objects of experience; and, finally, affective commitment by linking new knowledge to previous learning.

In Geography experiences of PBL, there have been some proposals of PBL linked to fieldwork and fieldtrip. PBL is promoted as a classroom-based preparatory basis for fieldwork (Bradbeer, 1996). Howorka (2009) reinforces this practice while highlighting that "it does so by reconceptualizing 'the field' based on the idea of 'everyday life' as a meaningful entry point within a classroom context, and as a space of learning in which students construct knowledge for themselves. An empirical investigation of student learning experiences explores the possibility of re-creating the benefits of residential field course offerings in a classroom-based field course." In the classroom, PBL may be encouraged by working out on geographical topics and issues by means of photographs (Sidaway, 2002). According to Fuller (2012), fieldwork provides opportunities for students to acquire hands-on subject-knowledge in the real world. PBL can be implemented as a methodological course of action in fieldtrip (García e la Vega, 2012, 2021 and 2022). GIS tools can be applied to PBL

(Drennon, 2005) in a classroom-based field course and also during the fieldtrip for localizing benchmarks, designing routes, correlating waypoints, estimating distances and route times by using digital cartography and apps on mobile devises. In this respect, managing online cartography, cartographical apps on mobile devises and GPS tools can become essential to enhance geospatial skills in fieldtrip, for instance, when developing direct observation of real features and compiling a systematic register of such features on the field notebook. However, Spronken-Smith and Harland (2009) found that the most difficult task in PBL is the teacher's role as a learning mediator. As a consequence, should PBL be integrated into the classroom planning and dynamics, either as preparation of students for field classes, or to be implemented during the fieldwork or fieldtrip, and should the geoscenario be used as the methodological resource in fieldtrip PBL, the distribution of the teacher's and students' roles in this methodological approach is reinforced since their roles must be well defined and assigned.

Barkley et al. (2007) argue that collaboration in teaching raises significant advances in general intellectual skills and in the psychosocial dimension. This is due to the aforementioned interaction among peer students and between students and the teachers. Furthermore, Barkley et al. (2007) propose work techniques related to PBL, namely, dialogue, reciprocal teaching and graphic organizers of information. Dialogue forms the basis of reciprocal teaching (Vygotsky, 1978) and the graphic and conceptual organizers are related to the concept maps of meaningful learning (Novak, 2002). Spronken-Smith (2005) highlights the commitment and enthusiasm shown by the student as a positive value in the implementation of PBL. Although, she considers that overload in academic tasks and group dynamics have turned out to be the greatest difficulties for teachers in the teaching experience following this method. These small-group teaching techniques are problem solving in pairs aloud and structured problem solving. In all cases, students should be introduced to the acquisition of these skills to improve PBL achievements (Exley and Denninck, 2005).

Problem-Based Learning. Steps and Levels of Implementation

Barrows (1986 y 1996) and Hmelo-Silver and Barrows (2006) underline five steps for implementing the PBL method. These steps show how to address the problem and the strategies for problem-solving, in relation to cognition processes and classroom dynamics. In this chapter, some steps are taken from Barrows' proposal that are reviewed and modified, and finally, six steps are proposed. They are essentially to: 1) Identify the problem, 2) Formulate the problem, 3) Build hypothesis and break the problem into minor questions 4) Search and select rigorous information on each minor question 5) Group the solutions to the minor questions and, 6) Find a solution to the main problem on partial solutions (Figure 3).

Figure 3. Steps in the problem-based learning cycle. 1. Identify the problem; 2. Formulate the problem; 3. Build hypothesis and break down the problem into minor questions; 4. Search for and select rigorous information on each minor question; 5. Group the solutions to minor questions, and 6. Find a solution to the main problem based on partial solutions. Source: A. Garcia de la Vega, adapted from Barrows

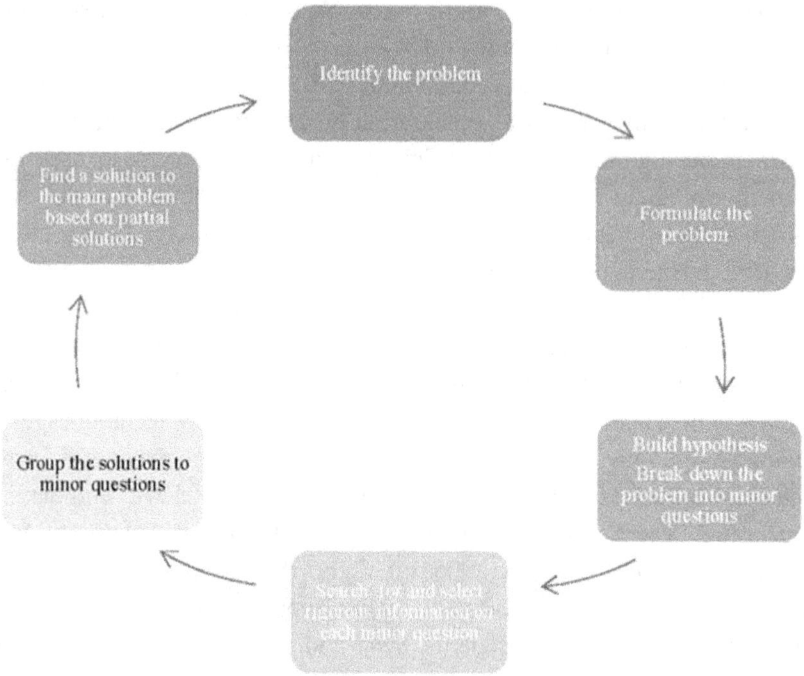

The first step is the most important one for developing this method, since the classroom dynamics is based on previous experiences. The teacher must master his/her role as a learning mediator, and the student must participate with peer students for problem-solving in small groups. When the small groups of students are organized, the teacher presents the geoscenario. The teacher has chosen a clear scenario with significant outstanding features. The geoscenario may be taken from a landscape photograph, a satellite photograph, a map, or other audio-visual and digital material (video, website, etc.) or from documentary resources. The teacher hands out the essential documentation for the students to be able to further formulate the problem in order to pursue an inquiry, and the students should identify the potential problem. The teacher must clearly discriminate between identifying and formulating, being the first two steps in this model that have been taken from Barrows' first step.

Within each small group, the students carry out individual work to observe and identify the geographical features and events. The teacher, as a mediator, ensures the participation of all students, while respecting the autonomy of the groups. Hmelo-Silver and Barrows (2006) argue that the teacher may suggest ideas aloud and help students reflect upon metacognitive approaches. In short, this stage is targeted to approaching a tentative formulation of the question for the geoscenario.

To illustrate a geoscenario, a photograph is chosen of a sky resort, named Baqueira-Beret, located at the Pyrenees. This geoscenario shows an outstanding problem, namely to what extent the impact of tourism on the mountains may be sustainable in the future. This problem is posed because this natural environment is on the threshold of sustainability, and action should be considered for nature preservation (glacier landforms) and a balanced use and/or exploitation of natural resources (water, wood, and grassland) (Figure 4).

Figure 4. Baqueira-Beret sky resort at Vielha, Val d'Arán (The Pyrenees, Spain). Source: A. Garcia de la Vega, 2013

The second step consists of recognizing terms and concepts of known and unknown meaning in the geoscenario. It is about recognizing all significant elements on the geographical setting and establishing possible links between them. This step must show the scope of the learning to be acquired by the students. This step represents the first level in PBL, and may be called "Advance recognition level". The Baqueira-Beret scenario reveals different features, both patent and underlying, cultural and geographical features. For example, land use for skiing activities is a patent cultural feature, like sky tracks, while glacier landforms are underlying geographical features (Table 1). Students should be able to recognize and understand the links connecting patent and underlying cultural and geographical features in order to formulate the problem.

Table 1. The second step, or problem formulation by recognizing geoscenario elements into main and secondary features and by understanding the links connecting patent and underlying cultural and geographical features, in application to the geoscenario of Baqueira-Beret sky resort at the Pyrenees (Figure 1). Source: A. Garcia de la Vega

Problem	Major Features	Secondary Features
Sustainable mountain landscape	Glacier Landforms	Horn, Arête, Cirque
		Hanging valley, moraine, ablation zone
	Land use	Sky tracks, sky lift

The third step is taken to build the problem hypothesis, and consequently, to break down the posed problem into minor questions. From the given list of major and secondary features, the problem is broken down into minor questions. These minor questions are distributed among the members of each group. Likewise, a schedule is planned with the assigned tasks and a feasible strategy is proposed for the resolution of each minor question. This task assignment enables the teacher to carry out specific techniques for learning in small working groups. Essentially, these groups gain collaborative and cooperative learning (Hmelo-Silver and Barrows, 2008). In this step, as claimed by Hmelo-Silver and Barrows (2006), the potentially most creative situations of PBL take place. The elaboration of hypothesis upon the elements and concepts found in the geoscenario furthers exceptionally creative situations from the dialogue among students in peer learning. It is in this most creative and autonomous step where the student's individual progress is appreciated. This corresponds to the most expressive step of the so-called Zone of Proximal Development being coined by Vygotsky (1978).

The fourth step moves onto searching for and selecting relevant and rigorous information for each minor question. The tasks are based on the search for, selection and retrieval of information from both analogical and digital resources and by using information and communication technologies (books, journals, internet, etc.). At this stage, compilation and curation of information on minor questions is performed. The small groups are in pursuit of bringing forth solutions to the minor questions. To do so, they need to assign meaning to the concepts that represent the geographical features and events in the geoscenario. Also, they must work out all those factors that are needed to reach an overall solution to the formulated problem.

Both third and four steps gather momentum and educational effectiveness by students' working in small groups. Barkley, Cross, and Major (2005) promote collaborative learning through pedagogical techniques. Also, Exley and Dennick (2004) underline learning strategies in small groups. Small groups in collaborative and cooperative learning are ideal for the different steps of PBL since, in small groups of students, an open and free dialogue is encouraged. This is one of Vygotsky's fundamentals of constructivism (Vygotsky, 1978). Several authors emphasize the relevance of working in small groups when developing PBL (Myers Kelson and Distlehorst, 2000). The learning experiences in Higher Education taking place in the universities of Maastricht (Netherlands) and McMaster (Ontario, Canada) back up with solid empirical evidence the theory of cognitive processes being stimulated by discussion in small groups and their beneficial effects on learning achievement (Schmidt and Moust, 2000 and Wilson, 2000). Myers Kelson and Distlehorst (2000) advocate the cognitive processes elicited by small group discussion and their effects on achievement by laying great emphasis on problem discussion as a means of activating prior knowledge.

In the fifth step, students have already come to identify and understand all the geographical features and events appearing in the geoscenario, and resolved doubts accordingly. Students are expected to come up with solutions to minor questions, share them with peer students and pull those solutions together to give a group response to every minor question. This is the step in which, after retrieving the data obtained in the documentary searches, students are able to associate all the ideas and concepts to start proposing partial solutions to each minor question. The partial solutions to the different minor questions could be organized by means of a concept map. Novak (1990, 1991 and 2002) promotes the creation of these concept maps for students and teachers. In the example of the sky resort proposed above (Figure 2), once the major and secondary geographical features and events have been recognized through direct observation on the geoscenario, meaning is given to their representing concepts. Then, associations among these concepts are built up from conceptual mapping that enable students to put forward partial solutions to the minor questions. Minor questions are in this case related to a number of factors as follows: the orientation of the sky resort on a shady hillslope with an abundance of snowfall, the average annual quantity of snowfall, the number of sky-track kilometers; which glacial landforms have been left largely intact and naturally preserved by the resort; and the resort's demand on water consumption from the water table level.

Reaching partial solutions to minor questions leads to the final step in this learning cycle. The sixth step consists of finding an overall solution to the formulated problem. To do so, students draw on the set of partial solutions that were agreed among peers on the minor questions. Students can elaborate on and produced a satisfactory and feasible solution by intertwining the variety of elements being identified and worked out in the partial solutions reached (features and events, places, and concepts). In the geoscenario of the sky resort (Figure 2), in relation to the formulated problem of whether the sky resort will be a sustainable landscape in the future, a viable solution may be found to in reducing the skiing season and decreasing the number of sky tracks. These measures may allow for keeping an efficient exploitation of the sky resort, while maintaining a sustainable use of the natural resources (specially, water resources) and preserving the natural environment.

Regarding PBL implementation levels, three levels are herein proposed based on previous learning experiences (García de la Vega, 2018). They are namely, basic, intermediate and advanced levels. Level progression is defined by two criteria: the classroom dynamics and the identification degree of features and events in the geoscenario. At basic level, classroom dynamics involve small-group flexibility, students' commitment to self-learning, and the teacher's role as a learning mediator. In relation to the identification degree of features and events in the geoscenario, students are required to recognize the most evident features on the landscape, among which stand mountains, ramps, valleys, rivers, cities and towns.

At the intermediate level, the class dynamics required in PBL have already been achieved. This means that, when working in small groups or in the large group, the students know what to do in each step. The students are able to acknowledge their active role in their own learning process and their intervention on the learning of other peer students. Also, they can appreciate how the learning of other peer students intervenes in their own learning. These PBL keys of the constructivist learning environment have been fully acquired at this stage, and both the teacher's and the students' roles are clearly defined and assumed.

Regarding the identification degree of features and events in the geoscenario, students should be able to identify and analyze both geographical features and events and their interaction on the landscape. For instance, students may associate the location of towns in risk areas owing to their landform instability (e.g. debris cone) or relief process (e.g. mudflows, floods in the wadis), because of excessive occupation, or on the contrary, due to the abandonment of farmlands. At this stage, if the problem is appropriate to student level, students are able to deduce the facts from the features and events that are outlined on the geoscenario. Students can set relationships among these features and events, which will head towards the identification of the problem to solve.

At the advanced level, students are prepared to be assigned shared tasks and roles that are needed at this stage to develop their own group dynamics. Students also attain a degree of commitment to the process of learning itself that leads them to act on their own initiative in the large group and in the small group. Students are willing to share with peer students the pictures or pieces of information that they have obtained in their search for relevant documentary evidence. The degree of cooperative and collaborative learning alternates along the different steps of the learning cycle, and the performance of the teacher's role as a mediator is clearly delimited and acknowledged within the groups. All patent and underlying geographical features and events are identified, and the problem is formulated with accuracy and rigor. Relationships among the elements appearing in the geoscenario are also precisely recognized and deeply understood. At this level, students are able to propose creative and alternative solutions to complex situations presented on the scenario. For instance, if the scenario presents territorial and social inequality, natural risks, unsustainable use of energy and natural resources, students are able to readily work out potential solutions that are pertinent to such acute issues.

Geoscenario for Landscape Teaching in PBL

Humboldt and Ritter point to the empirical analysis of reality as the basis of the inductive approach to Geography (Unwin, 1992). Interdisciplinarity is a framework to build up a new cross-disciplinary scaffold. Thus, Geography may partially lose its identity to adopt a disciplinary holistic approach. PBL leads empirical learning towards applying real-world everyday situations to problem-solving of acute issues. In fact, empirical observations, jointly with the body of conclusive evidence collected from cartographic, photographic, audio-visual, digital, technological and documentary sources, provide the basis for the analysis and interpretation of geographical scenarios. The geoscenario shows relevant features and their significant intercorrelation for the landscape to be analyzed and interpreted. However, Pawson et al. (2006) have expressed reservations concerning the difficulty involved in acquiring geographic concepts through PBL. Yet, everyday experiences bear meaning to scientific concepts. Cavalcanti (1998) highlights that the experiential dimension is essential in the construction of concepts. Vygotsky (1978) considered that the processes of development of spontaneous and non-spontaneous concepts constantly interrelate. In brief, there is a close relationship between a geoscenario and the construction of a scientific concept.

Figure 5. Volcanic landforms and silent volcano in La Palma Island (Canary Islands, Spain). Source: NASA, 2008

The acquisition of experiential knowledge enables students to apply scientific concepts and skills to real situations. Real-world geographical issues may be addressed in relation to natural risk events, anticipation of future risk events, and training of citizens facing natural risks; sustainable landscapes; heritage conservation and intervention; and land policy-making and administrative decision-making. This sort of contents may provide fundamental knowledge for students to analyze and interpret a landscape, to give an assessment on the landscape's sustainability, and to raise awareness on acute issues. In addition, training in student decision-making abilities is encouraged for students to reach optimal informed solutions, to become committed to landscape sustainability and to envisage a balanced use of natural and cultural resources in their future professional careers.

There follows a real-world geoscenario based on two satellite photographs that presents a recently-occurred natural risk event, the eruption of *Cumbre Vieja* volcano in La Palma Island (Canary Islands, Spain). The geoscenario shows a satellite view of the island with silent volcano and volcanic landforms before the eruption takes place in 2008 (Figure 5) and another view of the island during the eruption in 2021 (Figure 6). Some geographical features may be recognized, namely island landforms, volcano cone, lava flows, settlement, and banana trees covered by ash and lava, and a National Park. Also, there appear a volcanic ash cloud and corresponding atmospheric dynamics. Remarkable relationships can be set among the geographical features and events that define lava flows and settlements, volcanic ash and land uses. For instance, the fact that lava erupted from the volcano resulting in an increase in volcanic land involves new land properties and uses. So, the volcano eruption poses a new problem on land use and exploitation, environmental protection, and land planning and management for the island's inhabitants, farmland owners and policymakers (Santa Cecilia, García de la Vega y Martín, 2021). After solving out the minor questions, this new problem of the increase in volcanic land may allow for a role-play for students' decision-making on the new perimeter of the National Park in relation to its environmental protection and tourist exploitation. This learning technique helps students acquire social abilities and inter/intrapersonal competences because, essentially, students are placed in the complex situation of the social actors involved. In this respect, group competence acquisition including social competences becomes as relevant as individual competence acquisition in responding to some exciting new challenges for problem-solving (OECD, 2002). This type of geoscenario allows for reviewing different geography core-contents and, in turn, for developing some cross-disciplinary competences. Analysis of the Canary Islands' volcano scenario provides deep learning on topics related to volcanism,

Figure 6. The eruption at Cumbre Vieja volcano, La Palma Island (Canary Islands, Spain). Source: NASA, 2021

such as volcanism in the Canary Islands, volcanic rock types, changes in the nature, and land uses and exploitation.

Besides, there are other kind of concepts that point to changes in time, as the volcano was emitting lava, gases and ashes for two months. This sort of geological processes requires a revision of the concept of time in relation to geographical risk events. Similar risk events are hurricanes, floods and earthquakes that cause dramatic changes on the landscape in a few hours. From one day to the next, the landscape is completely changed, and as it may be recalled, this is the case of the 2004 tsunami in Thailand and the 2011 tsunami in Japan. On March 11, 2011, a magnitude 9.1 earthquake jolted the seafloor about 70 kilometers (45 miles) offshore of Japan's Tohoku region. It was the largest quake recorded in Japan and the fourth largest one in the world since seismic recording began around 1900 (Figures 7 and 8).

In similar contexts of changes in time or/and in latitude, a strategy by analogy may be applied to intellectual challenges facing students in problem formulation and solving. Analogical thinking is developed when searching for similarities in an attempt to overcoming a problem. If someone's constraint is a limitation beyond which he/she cannot advance, the student seeks for a solution to a difficult but surmountable problem. One way to approach this problem is by drawing an analogy (Holyoak, 1995). In the present case, if students neither recognize features and events in the volcanic eruption scenario nor understand the relationships among them, students may draw on lessons learnt on vulcanism from previous imaging. Also, students can search for similar images that they do understand to remove this constraint. Likewise, when students have analyzed and understood the main characteristics of a Mediterranean landscape in a geoscenario based on an image of a southern European country, they are able to draw an analogy to deal with a Mediterranean landscape in other latitudes (e.g. Chile, South Africa, Australia).

Holyoak and Thagard (1997) examined certain uses of analogy from the perspective of a multiconstraint theory, which postulates 3 basic types of constraints: similarity, structure, and purpose. They affirm that the operation of these constraints is apparent in naturalistic settings. In this sense, the kind of geoscenario juxtaposing two or three images from the same place promotes analogical thinking. The geoscenario below consists of a series of three satellite images showing the landscape of Japan's Tohoku region, prior to and right after the earthquake and subsequent tsunami, and ten years later. It is by analogy between the images' elements that students will appreciate the dramatic changes and the direct physical impact of the tsunami on the landscape of Tohoku region. In particular, the 2011 image displays the devastating effects of the tsunami on Japan's Tohoku region and the 2007 image shows the region's state prior to the earthquake (Figure 7). By comparing the features appearing in both pictures of 2011 and 2007, students will appreciate that the bay was tinged purple by the flood debris on the 2011 image. This is because the offshore Tohoku was flooded by the tsunami on March 14, 2011. The analogic strategy requires looking for similarities, such as geographic reference points and tsunami coastline, but also, spotting the differences, such as beaches, roads, buildings that disappeared in the tsunami. In the last picture of 2022, ten years later, the students may pinpoint the reconstructions and barriers that were built up to minimize the impact of tsunamis on this area in the future (figure 8).

Figure 7. The upper picture shows the devastating effects of the tsunami on Japan's Tohoku region (14 march 2011) and the lower picture presents the region's state prior to the earthquake (1 march 2007). Source: NASA, 2022

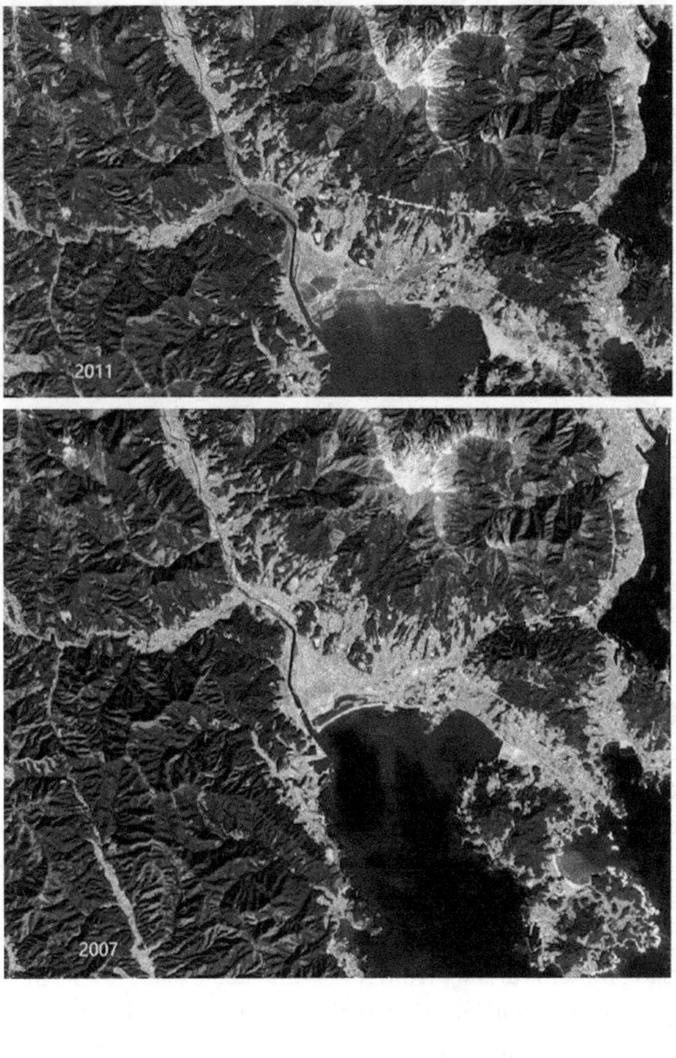

Figure 8. The picture shows the landscape changes in the Japan's Tohoku region (28 February 2021) ten years after the tsunami of 2011. Source: NASA, 2022

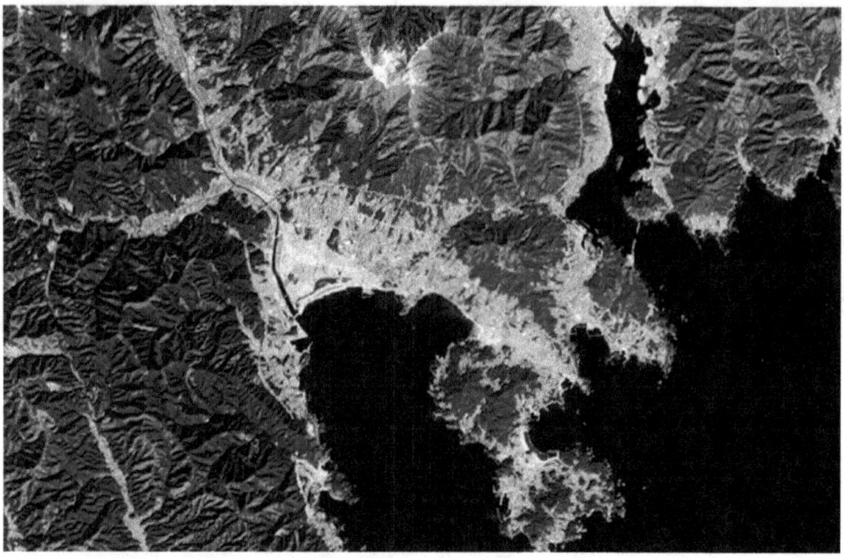

The geoscenario requires to approach a problem on a real-world acute issue, based on the main features and events in the landscape. This is a key factor in the relationship between the geoscenario and PBL. In the geoscenario, experiential knowledge is a requirement for learner understanding, analysis and interpretation of landscape. The geoscenario is a methodological resource that promotes the study of landscape from different reasoning strategies like analogy, by teamwork techniques like role-playing, and in small-group dynamics like collaborative learning. The constructivist learning environment enhances learner analysis and interpretation of the geoscenario.

CONCLUSION

The geoscenario is conceived as the minimal significant landscape framework unit in landscape teaching, having patent and underlying cultural and geographical features and events. The geoscenario embodies a synthetic representation of a real-world acute issue that involves a particular geographical problem that students need to solve by drawing on curricular contents from other disciplines. PBL promotes the search for solutions to problems arising from real-world acute issues. This methodological approach is associated with a constructivist learning environment that involves self-regulated and autonomous learning in small groups. The teacher as a learning facilitator plays a prominent role in encouraging teamwork techniques in order to enhance students' individual and social abilities. PBL fosters analogical thinking, reasoning abilities and spatial thinking by the use of technological resources, and help develop social abilities and inter/intrapersonal competences. PBL may be implemented at three levels (basic, intermediate and advanced levels) of progression that is defined by the classroom dynamics and the identification degree of features and events in the geoscenario. The geoscenario becomes a key methodological resource in geography-applied PBL that enhances concept acquisition and geographical-competence development. The geoscenario promotes the study of landscape from different reasoning strategies like analogy, by teamwork techniques like role-playing, and in small-group dynamics like collaborative learning.

REFERENCES

Allen, P. (2008). From landscapes into geological history. *Nature*, *451*. doi:10.1038/nature06586

Barell, J. (2006). *Problem-Based Learning: An Inquiry Approach*. Corwin Press.

Barkley, E. F., Cross, K. P., & Major, C. H. (2005). *Collaborative learning techniques: A handbook for college faculty*. Jossey Bass.

Barrows, H. S. (1986). A Taxonomy of problem-based learning methods. *Medical Education*, *20*(6), 481–486. doi:10.1111/j.1365-2923.1986.tb01386.x PMID:3796328

Barrows, H. S. (1996). Problem-Based Learning in medicine and beyond: A brief overview. In L. Wilkerson & W. H. Gijselaers (Eds.), *Bringing Problem-Based Learning to Higher Education: Theory and Practice* (pp. 3–12). Jossey-Bass Pub. doi:10.1002/tl.37219966804

Beane, J. A. (1995). Curriculum Integration and the Disciplines of Knowledge. *Phi Delta Kappan*, *76*(8), 616–622.

Boisvert, J. (2004). *La formación del pensamiento crítico. Teoría y práctica*. Fondo de Cultura Económica.

Bradbeer, J. (1996). Problem-based learning and fieldwork: A better method of preparation. *Journal of Geography in Higher Education*, *20*(1), 11–18. doi:10.1080/03098269608709340

Bruner, J. (1960). *The process of education*. Harvard University Press.

Council of Europe. (2000). *European Landscape Convention.* https://rm.coe.int/1680080621

Council of Europe. (2009). *5th Council of Europe Conference on the European Landscape Convention. Education on Landscape for Children.* https://rm.coe.int/16806b081b

Council of Europe. (2014). Recommendation CM/Rec (2014)8 of the Committee of Ministers to member States on promoting landscape awareness through education. Brussels.

Council of Europe. (2015). Recommendation CM/Rec (2015) of the Committee of Ministers to member States on pedagogical material for landscape education in primary school. Strasbourg.

de Souza Cavalcanti, L. (1998). Geografia, escola e construção de conhecimentos (14th ed.). São Paolo: Papirus.

Egan, K. (2008). Learning in depth. *Educational Leadership*, *66*(3), 58–63.

European Commission. (2018). *Proposal for a Council Recommendation on Key Competences for LifeLong Learning*. Author.

Exley, K., & Dennick, R. (2004). *Small group teaching: Tutorials, seminars and beyond.* Routledge Falmer.

Forman, R. T. T. (1990). The Beginnings of Landscape Ecology in America. In I. S. Zonneveld & R. T. T. Forman (Eds.), *Changing Landscapes: An Ecological Perspective* (pp. 35–44). Springer-Verlag. doi:10.1007/978-1-4612-3304-6_3

García de la Vega, A. (2012). El aprendizaje basado en problemas en los itinerarios vinculados al patrimonio. *Educación y Futuro*, *27*, 155–175.

García de la Vega. (2015). The Geographical Scenario. An Innovative Methodological Resource in Landscape Teaching. In *Book of Abstracts 100th Annual Meeting of the National Council for Geographic Education (NCGE)*. Academic Press.

García de la Vega, A. (2018). El escenario geográfico en el estudio del paisaje. Propuestas metodológicas en la enseñanza de la geografía. In A. García de la Vega (Ed.), *Reflexiones sobre educación geográfica. Revisión disciplinar e innovación didáctica* (pp. 129–152). Ediciones UAM y Edições Lagim.

García de la Vega, A. (2919a). Spatial thinking acquisition through geospatial technologies for lifelong learning. In R. de Miguel, K. Donert, & K. Koutsopoulos (Eds.), *Geospatial Technologies in Geography Education* (pp. 21–40). Springer.

García de la Vega, A. (2019b). Perspectivas de futuro en el aprendizaje del paisaje. *Didáctica Geográfica*, (20), 55–77. doi:10.21138/DG.441

García de la Vega, A. (2021). Design-Based Implementation Research for Exploring the Ocean: A Geographical Perspective. In K. C. Koutsopoulos & J. H. Stel (Eds.), *Ocean Literacy: Understanding the Ocean* (pp. 115–147). Springer. doi:10.1007/978-3-030-70155-0_6

García de la Vega, A. (2022). A Proposal for Geography Competence Assessment in Geography Fieldtrips for Sustainable Education. *Sustainability*, *2022*(14), 1429. doi:10.3390u14031429

Hmelo-Silver, C. E., & Barrows, H. S. (2006). Goals and Strategies of a Problem-based Learning Facilitator. *The Interdisciplinary Journal of Problem-Based Learning*, *1*(1), 21–39. doi:10.7771/1541-5015.1004

Hmelo-Silver, C. E., & Barrows, H. S. (2008). Facilitating Collaborative Knowledge Building. *Cognition and Instruction*, *26*(1), 48–94. doi:10.1080/07370000701798495

Holmes, J. H. (2002). Geography's Emerging Cross-Disciplinary Links: Process, Causes, Outcomes and Challenges. *Australian Geographical Studies*, *40*(1), 2–20. doi:10.1111/1467-8470.00157

Holyoak, K. J. (1995). Problem solving. In E. E. Smith & D. N. Osherson (Eds.), *An invitation on cognitive science* (2nd ed., Vol. 3, pp. 267–296). MIT Press.

Holyoak, K. J., & Thagard, P. (1997). The analogical mind. *The American Psychologist*, *52*(1), 35–44. doi:10.1037/0003-066X.52.1.35 PMID:9017931

Maitland, B. (1998). Problem-based learning for an architecture degree. In D. Boud & G. Feletti (Eds.), *The Challenge of Problem-Based Learning* (2nd ed., pp. 211–217). Kogan Page.

Myers Kelson, A. C., & Distlehorst, L. H. (2000). Groups in Problem-Based Learning (PBL): Essential Elements in Theory and Practice. In D. H. Evensen & C. E. Hmelo (Eds.), *Problem-based learning: A research perspective on learning interactions* (pp. 166–183). Lawrence Erlbaum.

Novak, J. D. (1990). Concept maps and vee diagrams: Two metacognitive tools for science and mathematics education. *Instructional Science*, *19*, 29–52. doi:10.1007/BF00377984

Novak, J. D. (1991). Clarify with concept maps: A tool for students and teachers alike. *Science Teacher (Normal, Ill.)*, *58*, 45–49.

Novak, J. D. (2002). Meaningful learning: The essential factor for conceptual change in limited or appropriate propositional hierarchies (liphs) leading to empowerment of learners. *Science Education*, *86*(4), 548–571. doi:10.1002ce.10032

O'Neill, P. A. (2000). The role of basic sciences in a problem-based learning clinical curriculum. *Medical Education*, *34*(8), 608–613. doi:10.1046/j.1365-2923.2000.00629.x PMID:10964207

OECD. (2002). *Definition and selection of competences (DeSeCo): Theoretical and conceptual foundations. Strategy paper.* http://deseco.ch/bfs/deseco/en/index/02.parsys.34116.downloadList.87902

Partoune, C. (1999). *L'apport du géographe dans un projet interdiscipliniare.* http://www.lmg.ulg.ac.be/competences/chantier/contenus/cont_interdis1.html#ancre555514

Pawson, E., Fournier, E., Haigh, M., Muniz, O., Trafford, J., & Vajoczki, S. (2006). Problem-based Learning in Geography: Towards a Critical Assessment of its Purposes, Benefits and Risks. *Journal of Geography in Higher Education*, *30*(1), 103–116. doi:10.1080/03098260500499709

Perrenet, J. C., Bouhuijs, P. A. J., & Smits, J. G. M. M. (2000). The suitability of problem-based learning for engineering education: Theory and practice. *Teaching in Higher Education*, *5*(3), 345–358. doi:10.1080/713699144

Pulido-Bosch, A., Pulido-Leboeuf, P., Molina-Sánchez, L., Vallejos, A., & Martin-Rosales, W. (2000). Intensive agriculture, wetlands, quarries and water management. A case study (Campo de Dalias, SE Spain). *Environmental Geology*, *40*(1-2), 163–168. doi:10.1007002540000118

Santa-Cecilia, F. J., García de la Vega, A., & Martín-Moreno, R. (2021). La erupción del volcán Cumbre Vieja en la Isla de La Palma (2021). El enfoque educativo de un volcán urbano. *Didácticas Específicas*, (25), 7–31. doi:10.15366/didacticas2021.25.001

Savery, J. R. (2006). Overview of Problem-based Learning: Definitions and Distinctions. *The Interdisciplinary Journal of Problem-Based Learning*, *1*(1), 9–21. doi:10.7771/1541-5015.1002

Savery, J. R., & Duffy, T. M. (1995). Problem Based Learning: An instructional model and its constructivist framework. *Educational Technology*, *35*, 31–38.

Savin-Baden, M. (2000). *Problem-Based Learning in Higher Education: Untold Stories.* Society for Research in Higher Education and Open University Press.

Savin-Baden, M. (2004). Understanding the impact of assessment on students in problem-based learning. *Innovations in Education and Teaching International, 41*(2), 223–233. doi:10.1080/1470329042000208729

Savin-Baden, M., & Wilkie, K. (2004). *Challenging Research in Problem-Based Learning*. Open University Press.

Schmidt, H. G., & Moust, J. H. C. (2000). Processes that shape small-group tutorial learning: A Review of Research. In D. H. Evensen & C. E. Hmelo-Silver (Eds.), *Problem-based learning: A research perspective on learning interactions* (pp. 19–52). Lawrence Erlbaum.

Schmidt, H. G., Vermeulen, L., & van der Molen, H. T. (2006). Long-term effects of problem-based learning: A comparison of competencies acquired by graduates of a problem-based and a conventional medical school. *Medical Education, 40*(6), 562–567. doi:10.1111/j.1365-2929.2006.02483.x PMID:16700772

Smith, D., Hoersch, A. L., & Gordon, P. R. (1995). Problem-based learning in the undergraduate geology classroom. *Journal of Geological Education, 43*(4), 149–152. doi:10.5408/0022-1368-43.4.385

Smith, R. M. (1993). The triple jump examination as an assessment tool in the problem-based medical curriculum at the University of Hawaii. *Academic Medicine, 68*(5), 366–371. doi:10.1097/00001888-199305000-00020 PMID:8484850

Solem, M. (2001). Using geographic Information Systems and the Internet to Support Problem-based Learning. *Planet, 2*(1), 22–24. doi:10.11120/plan.2001.00040022

Spronken-Smith, R. (2005). Implementing a Problem-Based Learning Approach to Teaching Research Methods in Geography. *Journal of Geography in Higher Education, 29*(2), 203–221. doi:10.1080/03098260500130403

Spronken-Smith, R., & Harland, T. (2009). Learning to teach with problem-based learning. *Active Learning in Higher Education, 10*(2), 138–153. doi:10.1177/1469787409104787

Stepien, W. J., Gallagher, S. A., & Workman, D. (1993). Problem-based learning for traditional and interdisciplinary classrooms. *Journal for the Education of the Gifted, 16*(4), 338–357. doi:10.1177/016235329301600402

Unwyn, T. (1992). *The place of Geography*. Longman Group Limited.

Vygotsky, L. S. (1978). *Mind in Society. The Development of Higher Psychological Processes* (2nd ed.). Harvard University Press.

Weiss, R. E. (2003). Designing problems to promote higher order thinking. In Problem-Based Learning in the Information Age. San Francisco, CA: Jossey-Bass. doi:10.1002/tl.109

Wilkerson, L., & Gijselaers, H. (Eds.). (1996). Bringing Problem-Based Learning to Higher Education: Theory and Practice, New Directions for Teaching and Learning, No. 68. San Francisco, CA: Jossey-Bass.

Wood, D., Bruner, J. S., & Ross, J. (1976). The role of tutoring in Problem Solving. *Journal Child Psychologist Psychiatrics*, *17*(2), 89–100. doi:10.1111/j.1469-7610.1976.tb00381.x PMID:932126

Woods, D. R. (2006). *Preparing for PBL* (3rd ed.). McMaster University.

Woods, D. R., Hrymak, A. N., Marshall, R. R., Wood, P. E., Crowe, C. M., Hoffman, T. W., Wright, J. D., Taylor, P. A., Woodhouse, K. A., & Bouchard, C. G. K. (1997). Developing Problem Solving Skills: The McMaster Problem Solving Program. *Journal of Engineering Education*, *86*(2), 75–91. doi:10.1002/j.2168-9830.1997.tb00270.x

Chapter 4
Importance and Educational Possibilities of the Place-Based Education Through Virtual Fieldtrips

Okkyong Yoon
https://orcid.org/0000-0003-2794-117X
Cheongju National University of Education, South Korea

ABSTRACT

This chapter deals with the possibilities of place-based education through virtual field trips and strategies to realize them. These are based on the environmental educational value of the experience in a certain place, the geography of the field trip, and the merits of an environmental education method. In video content production, the selection of a place and the composition of the story is very important. In addition to the story of the place, the narration of the content can supplement the actual field trip by including the fieldwork skills of the outdoors. Video content has the effect of combining various types of media so it can be effectively used along with field trips. Virtual field trips using video content have inherent limitations compared to outdoor field trips, but they can be used for place-based education by preparing a plan and support system for more effective blended learning.

DOI: 10.4018/978-1-7998-9598-5.ch004

INTRODUCTION

People occupy a certain range of territory and live in a place as a part of nature and society within the territory. Therefore, the natural and social environments of the place are important factors that constitute human life. But people often don't connect directly to the communities and places where they live now. So people need to think about life in a place and be interested. Place-based education accounts for the social and ecological aspects of the community and place where people live and also takes into consideration the diversity of their experiences in the community and place. Because of this, place-based education has provided valuable insights for geography educators and has been widely implemented in geography education.

This chapter explains how to help students develop strong ties to their community and place, increase students' understanding of the importance of the community and place, and get students to be active citizens on social and environmental issues through place-based education.

This chapter introduces my experience and opinions on conducting place-based education that I have gained through this experience. Every semester, I teach first-year college students a liberal arts course called "Exploring the region." This course aims to help students understand the diversity and scale of the region, identify the dynamics of regional change, develop a sense of place and spatial thinking, and take an interest in local issues. Students taking this course are pre-service teachers of elementary schools.

This course starts with students acquiring geographic information about the region where their university is located. Then, students compare this with the geographic information of their hometown. This activity aims to help students understand their hometowns better and enhance their perspective on the geographical environment of the region by comparing two regions. Through this activity, students can get a sense of security in their lives by obtaining useful geographic information about the place they currently live.

In the course, students conduct discussions, investigations, research projects, and two field trips. The themes of the two field trips are local history and local environmental issues. Students conduct field trips at local parks not far from the university. The field trip on the local history theme is conducted at the park in the city center with a long history. The field trip on the local environmental issues is held at the Wonheungi Ecological Park. This chapter will focus on the field trip to the Wonheungi Ecological Park that explores local environmental issues.

Wonheungi Ecological Park was created according to the urban development plan to expand the residential area of the city. "Wonheungi" is derived from a historical place name and is also known as a "Toad Ecological Park." Before being developed as a residential area, the park was a habitat for toads and other amphibians. The amphibian habitat was once threatened with destruction by development plans that did not prioritize the conservation of animal habitats. However, human settlements and toad habitats have been able to coexist due to citizens' relentless efforts to conserve the amphibian habitat. Therefore, the Wonheungi Ecological Park, which contains the history of development and environmental conservation, is suitable for students to explore local environmental issues. This chapter also introduces the knowledge, feelings, and experiences students acquire by visiting this ecological park.

An outdoor field trip is the best education method for students to understand a particular place because they can learn about the place and experience a sense of place during the field trip. Virtual field trips substituted outdoor field trips when outdoor field trips were not possible. Based on the experience of using this virtual field trip, this chapter examines the elements of virtual field trips dealing with the themes related to a place. This chapter also suggests useful methods to use video content for the success of virtual field trips in place-based education. In conclusion, this chapter summarizes the strategies of blended learning, methods, and considerations for environmental education using virtual field trips as a method of place-based environmental education.

BACKGROUND: PLACE-BASED LEARNING

Place and Sense of Place

Place is a term commonly used in our daily lives. It refers to a space occupied by the body of geographical features. Because people experience countless places and each place has a unique personality, they receive different feelings about each place. Therefore, there can be many manifestations of each place (Cresswell, 2004). Place is also often nostalgic. Agnew (1987) defined a place as a space referring to a specific point on the surface and described a place as a locale where people's daily activities and interactions take place and an object through which people can feel a certain sense of place. People often consider a place the foundation of personal identity or a space where individual or collective feelings and attachments are attached. Place is a space where human values are reflected in the physical environment, a complex of nature and culture, and a site where people conduct their daily lives (Kwon & Hwang, 2005). Therefore, people become attached to a certain place in various ways. Cresswell (2004) stated that "Place is a way of seeing, knowing, and understanding the world. Place is how we make the world meaningful and how we experience the world." Place is a space that is meaningful to someone, and sometimes power invests meaning in a place. Therefore, it is possible to understand the people who lived in the place if looking closely at the place.

Emotions about a place arise through experiences or memories obtained from the place. The emotions are diverse, such as intimacy, comfort, and belonging (Tuan, 1977). Sense of Place can be confirmed in emotional aspects and cognitive, affective, and behavioral aspects. In detail, a sense of place can be distinguished by knowledge of the place, attachment to the place, and practice for the place (Kim & Yoon, 2013). A sense of place is formed through the following steps. People have an attachment to a place at first, then go through the emotional stage of the place, and finally develop into a practical stage of participation, sacrifice, and dedication to the place (Shamai, 1991; Lalli, 1992).

Place-Based Education as an Alternative

In modern education, which focuses on excessively delivering standardized and objective knowledge, it is difficult for students to learn the knowledge that cannot be learned from textbooks through various experiences and logical thinking and put it into practice and action. In recent years, the uniqueness of places is gradually disappearing as the exchanges of goods, and people between regions have expanded with the development of transportation and communication. This trend can be seen worldwide, but it can be easily found within a community. Due to economic integration due to globalization, real-time culture spread through the internet, and a standardized lifestyle, it is hard for the younger generation to connect with the local community and strengthen community spirit. Therefore, a place should be used as the central component of teaching at all levels of education to overcome these problems of modern education.

Place is a space with a locality of special significance. In place-based education, a place has physical and human characteristics which make each place unique. Place is diverse and complex. Therefore, in a practical context, place has the potential to be used as content for alternative education. Place-based education aims for diversity or multiculturalism. Life and experiences in holistic environments, including historical, social, and environmental environments, are the core contents of place-based education. Students acquire the experience within a specific space and time (Greunewald & Smith, 2008). Place-based education focuses on the place as a living space where students live. It aims to develop an education curriculum that contains students' lives, neighborhoods, and local communities (Nam, 2005). A community is a geographical place where education takes place for students. Students interact with members of the local community. Outdoor field trips are an important learning method of place-based education. The main learning topics of place-based education are the historical elements of the region and various issues related to land, the local community, and the environment. Through place-based education, students connect the place in their home community with the places in the community they currently reside. Place-based education uses an ecological approach and spans all generations and cultures of the community (Woodhouse & Knapp 2000).

Place is not only an object of education but also an object of inquiry and a space where people can participate. Therefore, place-based education can be an alternative to today's "one standard" educational practices (Cho & Nam, 2015). Smith (2002) identified five education practices that can be implemented in place-based education, including cultural studies, natural studies, real-world problem solving, internships and entrepreneurial opportunities, and induction into community decision-making processes. Place-based education collects and analyzes authentic data found in everyday life (Smith, 2002; Woodhouse & Knapp, 2000). Place-based education promotes students' learning experiences by covering various content across many topics, designing education practices that stimulate students' curiosity, and facilitating students' community engagement using social networks within the community. Students recognize and solve historical, economic, and ecological issues of the place and the community they are involved in while carrying out education practices included in the school curriculum. Place-based curriculum and education pay attention to nature and humans and address authentic content in students and their communities in broader contexts (Woodhouse & Knapp, 2000; Knapp, 2005).

Although place-based education is a relatively new term, its roots are linked to the ideas of progressive educators that have lasted for more than 100 years. Progressive educators want their students to learn from experience in a local environment. Thus, they introduce traditional outdoor education methods and help students relate to specific places (Woodhouse & Knapp, 2000). Therefore, place-based education can be defined as qualitative experience in a local environment (Knapp, 2005).

Unlike textbooks and classroom-based education, place-based education aims to transform students through pedagogical engagement with place. Place-based education uses the local community and environment as a source of learning to educate students about the region's unique culture, history, economy, art, and literature. Students have the opportunity to directly experience the real world through place-based education and conduct research projects related to it. In this process, students and school officials identify problems in the community and tend to solve them. Each school is located in a different environment. Place-based education begins with education here, then gradually expands the educational content to schools, surrounding schools, and local communities. Through this, place-based education seeks diversity in education and ultimately achieves educational reform (Sobel, 2005).

The use of experiential pedagogical methods entails students learning from their own experiences. Place-based education emphasizes environmental education and uses the local natural environment and ecosystem as the core educational material. It takes an integrative pedagogic framework that spans multiple disciplines. Unlike classroom-bound education, place-based education employs out-of-classroom experiences where students interact with the local natural environment and cultural landscape. It also provides opportunities for students to look at their lives from a communal perspective, helps them engage with environmental and social issues in local places, and helps them grow up as citizens actively participating in the community. As such, place-based education aims to evenly develop students' intellectual capability and functional elements as community members. Place-based education conducted in group cooperative learning also allows students to learn how to collaborate. Place-based education also energizes teachers, changes the way they teach, and helps them engage meaningfully with their communities.

Place-Based Education and Geography Education

What is the relationship between place-based education and geography education?

The ultimate goal of place-based education is to create a fairer and more sustainable society by changing how students feel and behave in their daily lives. This purpose is also pursued in geography education (Fu, 2020; Meadows, 2020; Yli-Panula, Jeronen, & Lemmetty, 2020).

Place-based education is an education in which students meet places of nature and social environment. It values the interaction or connection between school education and the surrounding nature, humanities, and community environment. Place-based education goes further from learning places to making places good to live. The unfair situation of the real places becomes challenging learning resources and materials for education. Place-based education is practiced as a new pedagogy and pursues social change through community welfare and social justice. As such, place-based education does not have a background in just one specific area. It involves approaching a place in a new way beyond the classroom and taking care of it. The place-based education approach to containing nature and local communities is a part that should be noted in geography education as it explores various aspects of the place.

Geography education has also recently emphasized responsibility, expanding the scope of interest to better society, international peace, and environmental justice. Geographic education based on geography focuses on the interrelationships between nature and humans and how places work. In addition, outdoor field trips are a long tradition of geography. In terms of learning methods, geography education has something in common with place-based education. (Israel, 2012).

In place-based education, students are reminded to have a sense of social and ecological responsibility as citizens by experiencing, intervening, and participating in a place. In addition to acquiring knowledge, learning for social change is similar to field-based learning conducted in geography education. During field trips, students confirm the geographic knowledge learned in the classroom, improve their geographic skills, and have opportunities to experience new perspectives and improve environmental ethics through collaboration.

Five aspects of place-based education suggested by Smith (2002) are cultural research, natural research, real problem solving (service education), internship and career exploration, and community decision-making focusing on improving the environment and conditions of the place. Therefore, interest and efforts to solve local environmental problems can be a good point of contact between place-based education and geography education.

As a geographic research topic, a place can provide insight to make the world a better place. Geographers explore how places work and how these factors interact to create regional uniqueness. In geography, the study of places not only reveals the diversity, dynamics, and interrelationships of places but also finds how what happens locally can be understood on a wider scale. Places have many aspects of characteristics and are not stagnant or isolated. Places continue to change, and their contexts vary. Geographic education, which deals with the characteristics of such a changing place, can provide a clue to practicing the social change pursued by place-oriented education.

Exploring Places and Storytelling

Outdoor field trips are an important and effective learning tool in geography education. During outdoor field trips, students can learn how to utilize equipment, tools, and techniques in conducting geographic research. The skills and techniques allow students to observe and examine geographical phenomena occurring in real places. Place-based education was also incorporated in the case of operating the GIS curriculum in American middle schools (Perkins, Hazelton, Erickson, & Allan, 2010). Students learn about places they visit (Sun, Chan, & Chen, 2016). They have more opportunities to better understand the place during field trips by observing and experiencing what they have learned in the classroom. A curiosity that arises while conducting outdoor field trips can be a new start to further learning. As such, outdoor field trips can increase students' academic achievement and energize students.

Israel (2012) stated that outdoor field trips in geography education are a cognitive process of knowledge and skill acquisition. Environmental education starts with experiencing the natural environment. Because students are exposed to the environment through field trips, environmental education can also be achieved in geographic education. Outdoor field trips in geography education examine the place itself, its location, and its relationship to its surroundings. Therefore, outdoor field trips allow students to obtain practical research experience in exploring problems in the natural environment and suggesting solutions for the problems. In particular, urban areas with well-preserved natural environments are suitable places for environmental education to improve students' ecological literacy through field trips.

Storytelling is a strategy to broaden students' interest and understanding of places. Science education seeks logic and universal laws. On the other hand, storytelling education uses expressions and contrasts in stories, history, and fairy tales. Most curriculums adopt scientific education methods to deliver knowledge. Most textbooks are often organized in a way that lists concepts and knowledge. Teachers often use definitions and brief explanations rather than contextual stories to convey knowledge. The present educational environment limits the use of storytelling narratives as a learning tool (Cho, 2020). However, it is difficult for students to concentrate on the learning method without storytelling.

People constantly interact with the environment. They have feelings and emotions about the environment while living in the environment. People usually perceive the experiences acquired in the environment from an integrated, continuous, and interactive perspective rather than analyze the experiences from a segmented perspective and confining them into formality. Therefore, it is necessary to introduce an education method using narrative thinking for environmental education. Storytelling, a type of narrative thinking, is an educational method that captures human experiences and memories. A place tells a story. Thus, storytelling narratively organizes experiences and what was learned at the places and the route during outdoor field trips.

So, which place should you choose for place-based education? For storytelling to be successful as a method of learning, it is important to develop a story systematically connected to the learning objectives or topics. A place with many environmental education elements that can be made into a story is a suitable place for outdoor field trips. If place-based education deals with the interaction between humans and the environment and their changes, it is better to choose a place with a story that spans the past and present.

VIRTUAL FIELD TRIP1

This section will talk about the class conducted using video content containing the story of the place. Students watched video content about environmental issues within the community, focusing on the location. Various feedback from students on this video content helped me gauge the possibility of virtual field trips.

The Elements of Video Content

Students gain various knowledge and experience about the field through outdoor field trips and learn more about the lives of residents in the community and various issues in the community. However, conducting outdoor field trips is often not available due to various reasons, such as cost, time, and conditions. In those cases, virtual field trips using video content can be substituted for outdoor field trips.

The video content must meet the following conditions for virtual field trips to successfully substitute an outdoor field trip. First, video content should present the place realistically by capturing the outdoor scenery, sound, and situation. Video content indirectly helps students experience the field through sensory experiences using various sensory elements. Two, video content should include clues or elements for students to reach the learning goal. The content for outdoor field trips consists of overall information about the place that students can experience when visiting the place in person. The layout of video content should focus on allowing students to explore learning topics independently. If outdoor field trips are to be executed, students can use virtual field trips to obtain information about the place and check the information through outdoor field trips. With a lack of background knowledge of the visiting place, it is not easy for students to observe the natural and human environmental factors of the place in detail during outdoor field trips. When virtual field trips systematically train these elements, students can observe and pay attention to them in detail during outdoor field trips.

The amount and types of information acquired from the internet or media are vast. Various technologies can execute this information on virtual field trips. In particular, the street view applications provided by the web map services are a suitable web service tool for virtual field trips. However, when the web map services do not properly cover a place, video content designated to the place must be produced. In other words, appropriate resources suitable for learning purposes will be useful.

The layout of video content selects visual images that properly explain the place and match the fieldwork's theme. The explanation of the places proceeds in a narrative as storytelling. Narration is useful for a self-directed field trip because it indicates sites to pay attention to and themes to think about when students visit the place privately or in groups.

Selection of the Place

For selecting a place to produce place-oriented environmental education content, it is necessary to consider that a place is still used as a place to think about local environmental issues, not as a one-time environmental issue. Therefore, it is important to produce content that will be used as a motif for content users to transfer and expand their interest in local environmental campaigns.

When producing video content for outdoor field trips or place-based environment education, it is important to select a place suitable for educational purposes. The place suitable for outdoor field trips is a location where there are many learning elements that students can experience only in the field, and there are many learning elements where students can acquire knowledge and experiences that can be applied in daily life.

Outdoor field trips are a good learning method to learn from the place. However, it is sometimes difficult to implement outdoor field trips due to time, distance, budget, and safety problems. Therefore, the place suitable for outdoor field trips is located within the living space where the student's life takes place and is easy for students to access. In addition, it is also a particularly good place where students learn about the history and natural elements of the community and societal and environmental issues embedded in the community.

A good place for students to encounter the natural environment and develop environmental literacy is an ecological park that adds human management to a wild environment. An ecological park provides safety, one of the most important elements of outdoor field trips, and provides various learning programs. A place with a well-maintained natural environment is more suitable for outdoor field trips than a place where wildness is preserved. For example, bulletin boards inside the ecological park contain various information about plants and animals that inhabit the park. The information can be used for students to complete individual or group projects during outdoor field trips. Any educational programs offered by an ecological park service become an important content of outdoor field trips. Above all, it is important to select a place where students can encounter various environments and experience various community environmental issues.

The Wonheungi Ecological Park is built within the city for the coexistence of nature and humans. "Wonheungi bangjuk" is a key site of this park. "Wonheungi" is an old place name in this neighborhood. "Bangjuk" means a small reservoir with banks built to collect agricultural water. "Wonheungi bangjuk" is a spawning site for toads. It was protected by citizens when it was once on the verge of disappearing. The video content about Wonheungi Ecological Park narrates the history of the park. In detail, this video content contains the information on the space of this park, the distribution of facilities in the park and residential areas near the park, the process of the park becoming lower in altitude than the surrounding area, and the fact that other ponds were artificially created in addition to the "Wonheungi bangjuk." In addition, the video content also includes ecological facts that other amphibians live in the pond in the park and that the color of toad tadpoles differs from that of other amphibians.

The Wonheungi Ecological Park is a place where environmental issues such as the conservation of amphibian habitats were highlighted during the development of urban residential areas. Therefore, this park is an appropriate place where students can learn the responses to environmental problems emerging in the community and the environmental issues of the past. This park can allow students to take an interest in local environmental campaigns. Field trip to this park provides a motive for transferring and expanding this interest to other regions.

Layout of the Video Content Based on Place and Story

The video content layout consists of explanations of the scenery of points to be experienced while walking along the trail that looks around the site, stories related to those points, and environmental and ecological elements. The video content storyboard was organized in accordance with the time and speed required to explore the trail in the park during an outdoor field trip. The narration emphasized the elements that can be felt and experienced while walking there.

When an environmental education element is located along the route, the narrator stops the exploration at that site and explains the information related to the site. The narration includes the history of the park, animals and plants living in the park, and the management and use of the park. The content of the park bulletin board, which explains the environmental learning elements of the ecological park and contains precautions related to the use of the park, is also included in the narration.

The total length of video content is 44 minutes and 59 seconds, the average time spent touring the park. However, the video content was divided into three sections (Section 1: 15' 28", Section 2: 18' 46", Section 3: 10' 45"), allowing students to maintain their concentration on the video content. Table 1 itemizes the environmental education elements included in the video content (See Appendix).

The screen layout is as follows. A video illustrating the scenery of the park observed along the trail is located in the center of the screen, and a theme related to the landscape element is located above. The current location where it stops to explain the environmental educational elements in this park is presented with an index map so that the location of the point can be confirmed on the map (Fig. 1).

Close-up images of animals and plants observed while filming video content at the park aroused students' interest. Furthermore, by mentioning and explaining the flowers and trees that the narrator can find while walking in the park, students were encouraged to feel like they were walking together with the narrator during the virtual field trip. In Korea, the urbanization rate is very high, so students mainly grow up in urban environments. Therefore, the tadpoles filmed at the ecological park were interesting enough to attract the attention of students who only observed tadpoles in textbooks or brochures.

Figure 1. Screen capture of video content layout used in a virtual field trip

Effects of Video Content and Virtual Field Trip

Students taking The course "Exploring the region" must participate in two outdoor field trips during the semester: a field trip at a historical park and a field trip at an ecological park. Virtual field trips replaced outdoor field trips when face-to-face meeting classes were limited due to COVID-19. As face-to-face classes are allowed, this course conducted an outdoor field trip at a historical park, but a virtual field trip using video content replaced the outdoor field trip at an ecological park. This study investigated whether virtual field trips can produce the same learning effects as outdoor field trips. This study conducted a survey of 18 students to examine the effectiveness of virtual field trips. The questions of the survey were open-ended type. The questionnaire contains questions about the potential of replacement with video content, the learning effect of the video content, the student's preference for the video content of virtual field trips, and the suggestion for virtual field trips using video content. This study also explores the method and possibility of blended learning using video content.

A Comparison between Virtual Field Trips and Outdoor Field Trips

This study investigated the advantages and disadvantages of virtual field trips using video content compared to outdoor field trips for students. The advantage of virtual field trips is their video content that conveys a sense of the reality of the place and the convenience of video content that can be used repeatedly. On the other hand, the disadvantages of virtual field trips are that they do not convey a sense of place vividly as much as outdoor field trips, provide fewer opportunities to see the site, and cannot make memories of visiting the place, so the memory of the site does not remain for a long time. However, students stated that virtual field trips were a good alternative when outdoor field trips were difficult during the pandemic, but viewing video content was not effective enough to replace outdoor field trips. The advantages of the virtual field trip revealed by Student A are as follows.

- I was very sad that I couldn't go to the ecological park. However, I was able to know the ecological park in detail through the video content. Also, I could feel like I was actually experiencing the ecological park. So I thought that the video content had the effect of making me experience the place indirectly. The video content contains live sounds, such as birds and wind, and even the sound of the narrator's footsteps so that I could feel the sense of reality. (Student A)

Students gave comments on the potential of virtual field trips. Students stated that virtual field trips would gradually replace outdoor field trips if access to the place becomes difficult. After going through this virtual field trip, students found that virtual field trips were useful in terms of the quality of learning, cost and constraints, and technology.

Students said that virtual field trips do not convey the sense of place as much as outdoor field trips. However, because virtual field trips describe places more clearly, they are more effective in identifying and remembering places. In particular, in studying unknown places, virtual field trips are more effective than outdoor field trips because they can accurately introduce important information about the place through narration. In particular, they revealed that narration enables efficient and meaningful learning in a short time.

Although it takes a lot of time and cost to create video content for virtual field trips, students pointed out that the advantage of virtual field trips is that it is much less than it takes many students to visit the place in person. The external environment, such as the weather, is an important factor that determines the success or failure of outdoor field trips. On the other hand, high-quality video content has a great advantage in that students can safely and remotely enjoy the effects provided by outdoor field trips regardless of external restrictions.

The video content contains scenes produced when a video producer visits the place. By comparing scenes produced at different seasons or periods, students can identify the seasonal change in the environment of the place. For example, videos produced in spring show blooming flowers and todd tadpoles, while ones produced in fall include autumn foliage and adult toads.

Students pointed out the difference between virtual field trips and outdoor field trips. If outdoor field trips are conducted on a large scale, it is difficult to provide individual students with sufficient time to observe field trips. In addition, various sounds are mixed outdoors, making it difficult for students to understand the instructions or explanations of educators. Thus, while some students do not fully understand the subject matter of outdoor field trips, virtual field trips can provide students with the convenience of learning experiences that are easily accessible to auditory explanations. Students answered that an outdoor field trip is better than a virtual field trip if they do the field trip alone. It is because personal outdoor field trips allow students to explore the field for as long as possible and look at things that cannot be seen in virtual field trips.

Meanwhile, the students specifically pointed out the drawbacks of virtual field trips. Because video content cannot convey the atmosphere, feel, and smell of the field, it is difficult for students to feel the visual, tactile, and olfactory senses that they can acquire when conducting an outdoor field trip. Students answered that virtual field trips were a little boring because they could not see the entire area of the place and had to observe the scenes shown by the video content according to the narrator's movement. As such, students were well aware of the advantages and disadvantages of virtual field trips and outdoor field trips. They answered that both methods have something in common in learning about a place, but there are differences in how to memorize and sense the place.

Learning Effects through Virtual Field Trips

After watching video content, students were asked what field trip skills they learned and whether they were interested in learning topics, such as local environmental problems, to examine the learning effect of virtual field trips.

Remarks on the significance, field research methods, and effect of the fieldwork were included in the narration of the video content. The video content also includes obtaining information, observing animals and plants, and finding facilities installed in the park on various signs. Students learn about Wonheungi Ecological Park through virtual field trips. It introduces the history of park development, environmental conservation activities currently underway in the park, and projects carried out with the parks and villages around the park, such as library operation. Conflicts with residents of neighboring villages opposing the preservation of ecological parks are also mentioned. Students were also able to learn the fieldwork sequence and field through virtual content. They were able to acquire field learning skills using video content as follows.

- I believed that outdoor field trips were the only way to explore places. However, after experiencing virtual field trips, I learned that there are many different ways to explore. This virtual field trip was effective because the necessary information could be obtained without visiting the site in person (Study B).

- Before conducting an outdoor field trip, It is needed to know the background of the field trip, the story related to the place, and the meaning of the place in advance and conduct an outdoor field trip (Student C).

Students answered that they learned about the appearance and spawning ground of toad tadpoles, the growth process of toads and other amphibians, and the types of plants and animals inhabiting the park. Students remembered what they learned from the virtual field trip in relatively detailed and impressive ways. After the virtual field trip, the student's description of the ecological park environment is as follows.

- The street view is not the latest photo. The time the photo was taken did not match the season when the virtual field trip was performed, so it was hard to feel the correct seasonal image of the park. However, during the virtual field trip, I was able to see the inside of the park and experience the lake or road inside the Toad Park indirectly. The map included in the virtual field trip allowed me to understand what facilities were around and how it was built. (Student D)

- Compared to what the brochure introduced about the ecological park, the virtual field trip conveyed a sense of realism of the ecological park better. The virtual field trip broke my prejudice about ecological parks when I realized that the ecological park was managed cleanly enough to be a local infrastructure for recreation. Through a virtual field trip, I found the effectiveness of the ecological park by observing that many toad tadpoles actually live in the park created to protect the toads. (Student E)

When asked whether they became interested in lecture topics such as local environmental issues through watching video content, the students' responses were divided into two. Some students focused only on the scenery as a convenience facility for residents' leisure. It failed to pay attention to the conflict between the community group and the residential area development entity.

However, most students agreed that watching content provides an opportunity to be interested in local environmental issues. They were interested in the ecosystems destroyed during the development of the city. He also wondered if there were any more similar cases in his hometown or across the country. In addition to the toad, they wondered if any other ecological parks were built to conserve animals other than toads. In this way, students' interest in environmental issues has expanded.

On the other hand, in response to the question of whether they became interested in learning topics, such as local environmental issues after watching the video content, there was a student who remembered the specific story related to this place and mentioned the need for a "balanced development through nature conservation in regional development." Given that the students did not know exactly some terms and concepts related to development, it is necessary to include accurate explanations of the terms and concepts in the video content and teach students the terms and concepts not only in field trip-related lectures but also in other lectures.

Evaluation for the Improvement of Virtual Field Trip

To evaluate the content included in the virtual field trip, I investigated whether the virtual field trip using video content can replace the outdoor field trip and whether the virtual field trip motivated students to visit the site.

Of the 18 students, 16 students (88.9%) responded that virtual field trips could replace on-site visits. However, the percentage of students who preferred outdoor field trips (13 students, 72.2%) was much higher than that of students who preferred virtual field trips (5 students, 27.8%). In addition, all students responded that they wanted to visit The Wonheungi Ecological Park after conducting the virtual field trip. The virtual field trip induces students' curiosity about the field trip and motivates them to visit the site. It turns out that the virtual field trip can be used as learning preparation material for the outdoor field trip.

The students' responses suggest a glimpse of the type of fieldwork students prefer. Students want to take a leisurely look at the place and feel the place rather than explore the place as a pre-planned, time-based educational program. Therefore, they selected "Nature Sound" as good content to be added to the video content of the Virtual Field Trip. Students responded that if natural sounds heard in the exploration area are added to the video content, the virtual field trip can be closer to the outdoor field trip. In other words, video content using quality sound and image quality is a necessary condition for a successful virtual field trip.

Students wanted to add a map showing the path of the exploration from beginning to end in the video content, not edit the scene at each point, and provide the entire process of moving from beginning to end. It is because it was difficult for the students to properly imagine the overall appearance of the site by watching the video content. Compared to outdoor field trips, it is a fundamental drawback of video content.

There was a suggestion that "It will be great if video content uses a wide view camera and changes the camera's angle to add more realism to the place." It seems difficult for students to see the park scenery as a whole through the video content. It is a major drawback of virtual field trips compared to outdoor field trips. In addition, the students suggested that it would be good to add interactive functions such as bookmarks to the virtual field trip. It reflects the desire of students to participate in the use of video content directly.

Discussion

The previous section explains the effect of virtual field trips using video content. Then, it compares the pros and cons of virtual field trips and outdoor field trips based on students' responses. This section discusses implementing virtual field trips to blended learning and points to note when implementing it. In addition, this section presents opinions on the necessary components to be included in video content and the production and proper use of video content for effective blended learning.

Virtual field trips are useful for blended learning strategies. In particular, virtual field trips can effectively replace outdoor field trips when access to field trip places is limited. This study confirms the effect of blended learning that integrates virtual field trips and face-to-face field-based education. In particular, video content helps educators achieve the learning objectives pursued through outdoor field trips. Students can familiarize themselves with the field trip places through virtual field trips before conducting outdoor field trips.

This study confirms that the effect obtained through outdoor field trips can be maximized when virtual field trips are used complementary to outdoor field trips rather than as a substitute for outdoor field trips or as an auxiliary means of outdoor field trips. Effective blended learning can be achieved by including a narration explaining a description of the field trip place and what to explore at the field trip place in the video content. The narration helps students focus on the virtual field trips because it explains the social and environmental significance of the field trip places and emphasizes what should be observed with particular care in the field trip places. The students emphasize that the content conveyed by the virtual field trips is not significantly different from that delivered by the field trip. The students also agree that virtual field trips can replace outdoor field trips. However, they point out that virtual field trips cannot properly convey the sense of realism and charm of places obtained through outdoor field trips. As a result, students prefer outdoor field trips, although they acknowledge the advantages of virtual field trips.

Virtual field trips enable students to visit the field trip place in advance and help students learn the objectives of outdoor field trips, content and learning methods, and activities to be performed during the outdoor field trips. Students achieved their engagement with local social justice issues and environmental issues addressed by virtual field trips. Students also had the chance to think about their role in the communities where they live and are involved. Through the virtual field trips, students met local residents who wanted to conserve the Wonheungi Ecological Park, learned about their efforts to conserve nature and nurture the ecological environment, and had an opportunity to think about how to maintain the relationship between humans and nature. This park is located not far from the university where the students attend. Students are likely to visit the park several times during their college years. If students come to know the significance of a place and visit it, they can better understand the place and have an emotional attachment to the place, which can provide students with an emotionally positive effect. The Center for Wonheungi Ecological Park provides various educational programs related to the park's unique environment. In addition, it provides recreation for the public while conserving the natural environment of the park. Interaction with the park's program operators and participants helped me create virtual field trips to the park.

The video content of the virtual field trip used in this study was produced by recording the narration while filming the field trip place. This video content is a one-way media. Students cannot engage in interactive education through virtual field trips by streaming and watching one-way media. Therefore, virtual field trips implementing the blended learning strategy must incorporate interactive educational methods to utilize video content. For example, students can obtain the actual appearance and geographic information at the spot simultaneously through video content that contains an image of a landscape and a map that can identify the location and the route of the narrator. The video content used in this study presented a map in the lower right corner of the video screen. The map indicates where each spot is located in the Wonheungi Ecological Park, a field trip place. However, the video content did not give students the sense of realism that outdoor field trips can provide because the map did not display the narrator's location and route in real-time mode.

For more effective blended learning, virtual field trip needs to contain video content for an extended period, such as season footage, to keep students interested in the place beyond the period a course covers the themes. Therefore, periodic update of video content is necessary. It is also necessary to study how to use video content for blended learning for outdoor field trips.

CONCLUSION

Place-based education can deliver students specific content related to the environment of a place and helps students develop sensitivity to the environment. Therefore, it is necessary to develop content and educational programs with storylines related to the history and unique characteristics of the place and the intervention of people in the place.

In geography, a place is where people directly experience the environment and connect with other places on a multi-layered spatial scale. Therefore, it is necessary to implement the place study methods used in geography to conceptualize the pedagogical ideas in place-based education. Place-based education should allow students to experience places. Therefore, it is better to select a place rich in environmental stories currently in progress. However, the following should also be considered important selection criteria in choosing a place for place-based education: whether a place contains environmental education elements, whether students can acquire environment-related experiences, whether educational activities can take place, and whether the place is easy for students to access. Video content is a good alternative to use when outdoor field trips are limited. It is also a useful educational resource when preparing or supplementing an outdoor field trip. Virtual field trips using video content and outdoor field trips are complementary.

When various types of media, such as visual content and audio content, are embedded with video content, the effect of virtual field trips is great. Therefore, it is useful to implement techniques such as sensory elements and index maps that help students obtain the locational information of the place to maximize the effect of virtual field trips. And a blended learning strategy should be established to enhance the effect of the produced video content and sustain the effect.

ACKNOWLEDGMENT

This research was supported by Cheongju National University of Education (Support for video content producing) [RB2020011].

REFERENCES

Agnew, J. (1987). Place and Politics: the Geographical, Mediation of State and Society, Boston: Allen and Unwin.

Cho, C. (2020). *Place-Based Geography Education*. Kyongbuk University Press.

Cho, S., & Nam, S. (2015). An exploration on the significance and the effect of place-based education (PBE) in Social Studies Education. *The Journal of The Korean Association of Geographic end Environmental Education*, *23*(1), 1-17.

Cresswell, T. (2004). *Place: a short introduction*. John Wiley & Sons, Ltd.

Fu, B. (2020). Promoting geography for sustainability. *Geography and Sustainability, 1*(1), 1–7. doi:10.1016/j.geosus.2020.02.003

Gruenewald, D. A., & Smith, G. A. (Eds.). (2008). *Place-Based Education in the Global Age: Local Diversity.* Routledge.

Israel, A. L. (2012). Putting geography education into place: What geography educators can learn from place-based education, and vice versa. *The Journal of Geography, 111*(2), 76–81. doi:10.1080/00221341.2011.583264

Kim, M., & Yoon, O. (2013). Development and application of a sense of place test instrument: A case study of gender differences of elementary students. *The Journal of The Korean Association of Geographic and Environmental Education, 21*(2), 17–28. doi:10.14697/jkase.2013.33.1.017

Knapp, C. (2005). The "I-thou" relationship, place-based education and Aldo Leopold. *The Journal of Environmental Education, 27*(3), 277–285.

Kwon, Y., & Hwang, M. (2005). The significance of the sense of place in environmental education. [in Korean]. *Environmental Education, 18*(2), 55–65.

Lalli, M. (1992). Urban-related identity: Theory, measurement, and empirical findings. *Journal of Environmental Psychology, 12*(4), 285–303. doi:10.1016/S0272-4944(05)80078-7

Meadows, M. E. (2020). Geography education for sustainable development. *Geography and Sustainability, 1*(1), 88–92. doi:10.1016/j.geosus.2020.02.001

Nam, H. (2005). The nature and approach of place-based learning in geography education. *Social Studies Education, 44*(3), 195-210.

Perkins, N., Hazelton, E., Erickson, J., & Allan, W. (2010). Place-based education and geographic information systems: Enhancing the spatial awareness of middle school students in Maine. *The Journal of Geography, 109*(5), 213–218. doi:10.1080/00221341.2010.501457

Shamai, S. (1991). Sense of place: An empirical measurement. *Geoforum, 22*(3), 347–358. doi:10.1016/0016-7185(91)90017-K

Smith, G. A. (2002). Place-based education: Learning to be where we are. *Phi Delta Kappan, 83*(8), 584–594. doi:10.1177/003172170208300806

Sobel, D. (2005). *Place-based Education: Connecting Classrooms & Communities.* The Orion Society.

Sun, Y., Chan, R. C. K., & Chen, H. (2016). Learning with geographical sensitivity: Place-based education and its praxis. *The Professional Geographer*, *68*(4), 574–583. doi:10.1080/00330124.2015.1121835

Tuan, Y. F. (1977). *Space and Place: The perspective of experience*. University of Minnesota Press.

Woodhouse, J., & Knapp, C. (2000). *Place-based curriculum and outdoor and environmental education approaches*. ERIC Digest.

Yli-Panula, E., Jeronen, E., & Lemmetty, P. (2020). Teaching and learning methods in geography promoting sustainability. *Education Sciences*, *10*(1), 5. doi:10.3390/educsci10010005

ADDITIONAL READING

Choi, S., Choi, J., & You, J. (2020). Development and application of ecological education program based on place-based education (PBE). *Environmental Education*, *33*(3), 262–277.

Clark, D. (2016). *Learning to Make Choices for the Future- Connecting Public Lands, Schools, and Communities through Place-based Learning and Civic Engagement.* The Center for Place-based Learning and Community Engagement. Available online: https://promiseofplace.org/sites/default/files/pdfs/PBE_Manual_2012.pdf

IGU CGE. (2016). *InterInternational Charter on Geographical Education.* Available online: https://www.igu-cge.org/wp-content/uploads/2019/03/IGU_2016_eng_ver25Feb2019.pdf

Reynolds, H. L., Brondizio, E. S., & Robinson, J. M. (Eds.). (2010). *Teaching Environmental Literacy: Across Campus and Across the Curriculum.* Indiana University Press.

Smith, G. A., & Sobel, D. (2010). *Place-and Community-Based Education in Schools*. Routledge.

Wattchow, B., & Brown, M. (2011). *A Pedagogy of Place: Outdoor Education for a Changing World.* Monash University Publishing.

Yoon, O. (2016). Practices of place-based environmental education in a general education course for pre-service elementary teachers. *Journal of The Korean Association of Geographic and Environmental Education*, *24*(1), 131–150.

KEY TERMS AND DEFINITIONS

Blended Learning: Method of teaching that integrates technology and digital media with traditional instructor-led classroom activities.

Field Trip: Visit made by students to study something away from their school or college.

Fieldwork: Process of observing people, cultures, and natural environments and collecting data about the dynamic places, people, and species around them in the wild of our everyday surroundings rather than in the controlled environments of a lab or classroom.

Outdoor Field Trip: Field trip that takes place outdoors, not indoors, used in contrast to a virtual field trip.

Place-Based Education: Education that utilizes the local community and the overall local environment and emphasizes direct learning experiences in the real world.

Sense of Place: Thoughts and emotions about a place.

Video Content: Content created through filming and editing of moving pictures and sound, especially as a digital file, DVD, and others.

Virtual Field Trip: Exploring on screen with video content as an alternative to being outside or exploring the areas online prior to an outdoor field trip.

ENDNOTES

1 This main part is based on the experience of conducting virtual field trip in the course of "exploring the region".

APPENDIX 1

Table 1. Composition of video content for virtual field trips in Wonheungi Ecological Park

	Scenes of video content and narration	Related topics and concepts
Field trip 1 (15¢ 28^{2})	• The location of Wonheungi Ecological Park • Environment around the park • Toad ecological culture center • The ruins of Wonheung-ri (village) • Solar streetlights • Life of a toad (in the bulletin board) • Outdoor concert hall (stage for Toad festival)	• Development of housing district (Sannam district 3) • Legal complex (legal business cluster) • Site for ecology educarion • Community movement (e.g., Environmental campaign) • Excavation of historical sites • Eco-friendly energy • Toad's ecology and ecological passage • Symbiosis with nature • Experience oriented education programs
Field trip 2 (18¢ 46^{2})	• Playground with sound-related facilities • Prohibited activities in the park (bulletin board) • Aquatic plants of the frog pond (bulletin board) • Safety rules when exploring (bulletin board) • Stamp tour (ecological park tour with senses such as sight and hearing) • Plants (flower, tree): iris, reed, weigela, willow tree, acacia tree, oak tree, and others • A toad tadpole	• Food chain • Ecosystem • Habitat • Seasonal plants • Reservoir facility (Bangjuk) • Ecological park
Field trip 3 (10¢ 45^{2})	• Campaign to prevent birds from colliding and dying on soundproof walls • Aamphibians in Korea (bulletin board) • Wonheungi Bangjuk (small reservoir) • Ecological passage • The zelkova tree in Wonheungi village • Bird observatory	• Environmental conservation • Public-Private Cooperation Project • Amphibians • History of the village • Ecological passage

(Data Source: Author)

Chapter 5
Future-Proofing Geography Teaching and Learning Using Remote Learning and Co-Creation Approaches

Theresa G. Mercer
University of Lincoln, UK

Andrew P. Kythreotis
University of Lincoln, UK

Hannah Kelly
University of Lincoln, UK

Zoe P. Robinson
Keele University, UK

Sharon M. George
Keele University, UK

David Sands
Independent Researcher, UK

ABSTRACT

This chapter examines the application of new geography pedagogic practices in relation to remote teaching and learning. The authors report how the co-creation of knowledge between students and lecturers through existing research projects (with a particular focus on disciplinary-based educational research) can still occur through remote learning. Through a case study example of geography teaching practice, they contribute to the pedagogic literature on digital online learning as a form of co-creation between lecturer and students, showing how students were still able to demonstrate their geographical knowledge in a very applied and real-world context despite the move to remote learning. The application of co-creation in these applied contexts demonstrates that the applied perspectives of geographical knowledge can still be delivered through remote learning. They conclude by discussing how applied and critical geography teaching can be delivered and future-proofed in higher education settings, regardless of unforeseen circumstances.

DOI: 10.4018/978-1-7998-9598-5.ch005

INTRODUCTION

The impact of the COVID-19 pandemic and enforced move to remote learning for many HE institutions at various times over the past two years has been a catalyst for the exploration of novel pedagogies around remote learning (Müller et al., 2021). As such, there has been a proliferation of remote teaching and learning best practice materials that have arisen from the COVID-19 pandemic. These have been in the form of guides, blogs, online webinar events, books and journal articles published at individual university levels (e.g. see Solent University, 2020; University of Kent, 2020; University of Lincoln, 2020) and national platforms such as the UK Staff and Educational Development Association (SEDA, 2020). The material ranges from best practice online teaching tools to use for increasing interactivity with students (Khuri & Reed, 2020) to making sure material remains accessible to students (Gibbs & Woods, 2020). Academics are finding that they must re-evaluate their teaching material and delivery modes, to meet the wide and varied needs of students whilst also meeting the learning outcomes and objectives of their modules.

To select the most appropriate teaching approaches and delivery platforms, it is important to consider the different approaches to, and definitions of, remote learning. In the wake of the COVID-19 pandemic and move to online teaching and learning, a whole new language has developed around new modes of delivery (Bryson & Andres, 2020). In a previous study, Moore, Dickson-Deane and Galyen (2011) demonstrated that differing terminology can lead to different expectations and perceptions of the online learning environment. The Quality Assurance Agency (QAA) for Higher Education in the UK has since created a glossary of common terms and approaches as a way of reaching a common terminology for describing how students engage with digital education (QAA, 2020a). For the purposes of this chapter we refer to remote learning as 'activities that happen away from the physical site of a provider with students using digital means to engage with a programme, rather than having any requirement to visit the provider' (QAA, 2020a). 'Remote learning' was the term that was most widely used to describe the initial move to digital delivery during the COVID-19 pandemic.

In response to the long-term outlook for remote learning, the QAA has released several key considerations for maintaining quality in a digital environment moving forward beyond the immediate COVID-19 crisis (QAA, 2020c). These include ensuring that digital pedagogy is adequately considered as opposed to focusing solely on technologies and tools for delivery, and that student-centred learning, teaching and assessment as well as the promotion of co-creation of learning materials between academics and students is at the heart of the development of teaching in a digital environment. Co-creation refers to "students' active participation, learner empowerment, shared decision-making, student agency and negotiation of learning and teaching" and can be manifested in a number of ways, including co-researching, participating in course design committees, designing assessments and co-assessing work (Bovill, 2020; p1024). Co-creation is highlighted as a tool that facilitates the creation of 'an engaging digital experience and places students at the heart of their own learning' (QAA, 2020c). This could be particularly useful for remote learning where the student experience of learning can feel even more remote to the experience of the educator teaching.

With the uncertainty that COVID-19 has brought to the higher education environment, academics are finding that they now need to be more adaptable and prepared for every eventuality and possible level of remote learning. There is now the need to ensure that teaching design and delivery is approached strategically and innovatively to ensure that high quality teaching and standards are preserved while remaining adaptable to teaching across both physical and remote learning environments (Adedoyin & Soykan, 2020; QAA, 2020b). One such approach is through co-creation of learning materials through co-produced research which results in educational assets that can be used for different cohorts and different purposes when unforeseen circumstances occur that disrupts the (once) normalised traditional models of HE pedagogy.

This book chapter discusses the benefits of co-productive research between students and course instructor. We present examples from our teaching practice that demonstrate how co-production of applied knowledge between students and lecturer through existing research projects can still occur through remote learning. In fact, their inclusion in teaching can help 'future-proof' courses in times of uncertainty and change as demonstrated during the UK lockdown period of the COVID-19 pandemic. It can also offer new opportunities for university teachers to be more flexible in curriculum design, and lead to greater independent learning by students (Müller et al., 2021). We present a case study that relates to a third-year environmental management module in which students remotely produced pro-environmental educational tools for teaching secondary stage school pupils. We discuss some of the key lessons learnt and future opportunities for remote teaching in Geography, Earth and Environmental Science (GEES) disciplines. We specifically present our reflections on the approach taken within the wider literature of co-production and remote learning in the hope that it is a useful case study and resource that other academics and practitioners can easily adopt into their future teaching and pedagogic practice when unforeseen circumstances arise.

REMOTE LEARNING AND CO-CREATION APPROACHES

Co-creation in teaching has tended to focus on developing teaching materials and strategies (QAA, 2020c), but research-based teaching strategies in which students are engaged as partners in research are also forms of co-creation. Walkington (2015) distinguishes between four different categories of involvement with research in teaching, ranging from teaching to students about current research (research-led) to students actively participating in research and inquiry (research-based). Co-creation challenges traditional power relationships and didactic knowledge transfer of research-based teaching (Bell & Pahl, 2018) . In the case study presented below, we demonstrate how we have inverted the normal sense of expertise and authority by inviting students to act as game developers and researchers, through a form of research-based teaching.

Discipline-based educational research (DBER) describes research that is linked to learning and teaching within a discipline and augmented with research based on human learning and cognition (Henderson et al., 2017); as such DBER sits across multiple fields of study - including the sciences, social sciences and education (Dolan et al., 2018), and uses disciplinary content to focus research with an emphasis on 'improving what is most important in moving undergraduates towards expertise in the discipline' (Henderson et al., 2017). The benefits of DBER (particularly with an inter-disciplinary focus) are still largely unrealised across the entire HE sector (Henderson et al., 2017). We have used this approach in some of our related GEES research since 2012 (Kythreotis et al., 2019; Kythreotis & Mercer, 2020; Mercer et al., 2017; Mercer & Kythreotis, 2020) and have recently called for academics who haven't tried it out yet to 'step out of their comfort zone' and embrace it in their approach to teaching (Mercer et al., 2019).

DBER was explored through GEES programmes at two different institutions. In 2012, as part of a second year Human Impacts on the Environment module at Keele University, UK, students were introduced to a novel assessment that involved designing, creating, playing and delivering games that encouraged pro-environmental behaviours and sustainability amongst school children. This exercise gave students the opportunity to demonstrate the knowledge gained over the module, their degree programme, as well as their own personal experiences. In summary, student groups selected an environmental topic that they wanted to educate on to promote pro-environmental behaviours. They then designed and developed age-appropriate games related to this topic and demonstrated them to the rest of the class in a final practical session. Students produced a written report based on an activity plan which was assessed. The games were subsequently used on university outreach days (outside of the module) with local schools in the surrounding area. Feedback from the schools and HE students were reported through the discourse lens of Education for Sustainable Development (ESD) (Mercer et al., 2017).

This educational game design approach was included within a new third year Environmental Management module running for the second half of the academic year in the UK (January-May 2020 and January-May 2021) (School of Geography, University of Lincoln). The module was designed to examine a range of interdisciplinary environmental management techniques that are used to address critical environmental, human and planetary health issues. Students gained an understanding of the most appropriate techniques and solutions through exploring a range of global, regional, and local environmental case studies. There was a strong emphasis on practice-based skills with guest lectures from practitioners in the field, practicals and seminar classes. The educational game-design was included as a summative assessment, giving students the opportunity to be more involved in actively influencing sustainable behaviour in the younger generation and giving them a chance to become the 'teacher' (see Table 1 for assessment brief).

Research was conducted both ***on*** the students in terms of how the development of the games and tools affected their knowledge and skills acquisition as well as their own pro-environmental behaviours. Research was also carried out ***with*** the students in terms of researching how local schools responded to playing the games and the extent to which they promoted pro-environmental behaviour. Figure 1 demonstrates the flow of work for the assessment and where HE students were involved in research. In the first (in person) iteration of this assessment, students reported improvements to key skills including teamwork, problem solving, discussion and negotiation. They enjoyed the student-centred assessments and chance to make an impact with the local community (Mercer et al., 2017). They also felt greater ownership of the co-created knowledge in terms of the knowledge they had gained on the course and how they translated these into games. This learn-by-design model of learning (Baytak & Land, 2011) kept students engaged, motivated and invested in their own learning, with evidence of a perceptible move to deeper learning compared to traditional teacher-based approaches based on student feedback (Manduca et al., 2017; Mercer et al., 2017). The wider community also benefitted from this approach through students supporting the capacities and knowledge of teachers within the field and intergenerational learning through school children passing on the messages of pro-environmental behaviour, sustainability and climate change to their families and wider communities (Kythreotis & Mercer, 2020; Mercer et al., 2019; Mercer & Kythreotis, 2020).

The first UK lockdown was introduced in March 2020 just as students were getting together in groups and developing ideas for games. Introductory material on Education for Sustainable Development and an example of a previously created educational game were used to provide context for the assessment. The assessment

Table 1. Educational game assessment brief

Description of Assessment Task and Purpose: You are required to create an educational tool/game in groups of 5-6. This needs to be based on an environmental topic that you want to educate on and change behaviour for to reduce human impacts on the environment. In the final session of the module, you will present your activity to the class and demonstrate how it works. Pairs of groups can run through each other's activities. Each group must also hand in a 2000-word (maximum) report at the session outlining the lesson plan for the activity produced. An overall group mark will be given on the tool/game that you produce and the report.
Learning Outcomes Assessed: LO2: Contrast developing solutions to current environmental problems that cross-cut a wide range of scales from local to global. LO4: Demonstrate educational and outreach practices relating to environmental issues.
Knowledge & Skills Assessed: *Subject Specific Knowledge, Skills and Understanding:* Referencing, literature searching, project planning, subject-specific knowledge. *Professional Graduate Skills:* Independence and personal responsibility, groupwork skills, adaptability, verbal communication, written communication, creativity, critical thinking, IT skills, outreach skills, science communication skills, problem solving, effective time management, working under pressure to meet deadlines, negotiation skills, leadership. *Emotional Intelligence:* Self-awareness, self-management, awareness of others, managing and supporting others, motivation, resilience, self-confidence. *Career-focused Skills:* An understanding of the range of graduate careers, skills and attributes required by employers, a range of strategies to present skills and attributes to employers. Direct engagement with external environmental organisations.
Format for Assessment: The report should include a detailed lesson plan with the following: • Aims and objectives of the activity • Intended learning outcomes • Intended participants • Facilitators instructions and debrief sheet • Worksheets/instructions for participants
Marking Criteria for Assessment: For the **report** portion of the task, your group will be assessed on: • Clarity of the exercise that you produce • Suitability for the intended audience • Ease of running the exercise • Appropriateness of the selected topic • Fulfilment of the original intended learning outcomes • Group performance (related to the presentation of the games)

involved working remotely in groups to design and produce their games with summative assessment in the form of a report that included full activity plans for their games. However, with the complete move to remote learning, groupwork had to take place online and engagement with the local schools was delivered remotely in 2021 (Lincolnshire County Council, 2021).

We consider that basing the summative assessment on the activity itself is central to the success of our approach. It is well documented that methods of active learning

Figure 1. Educational game assessment workflow process and research opportunities. At each stage, activities have been able to be carried out in person or remotely depending on restrictions on teaching delivery at the time.

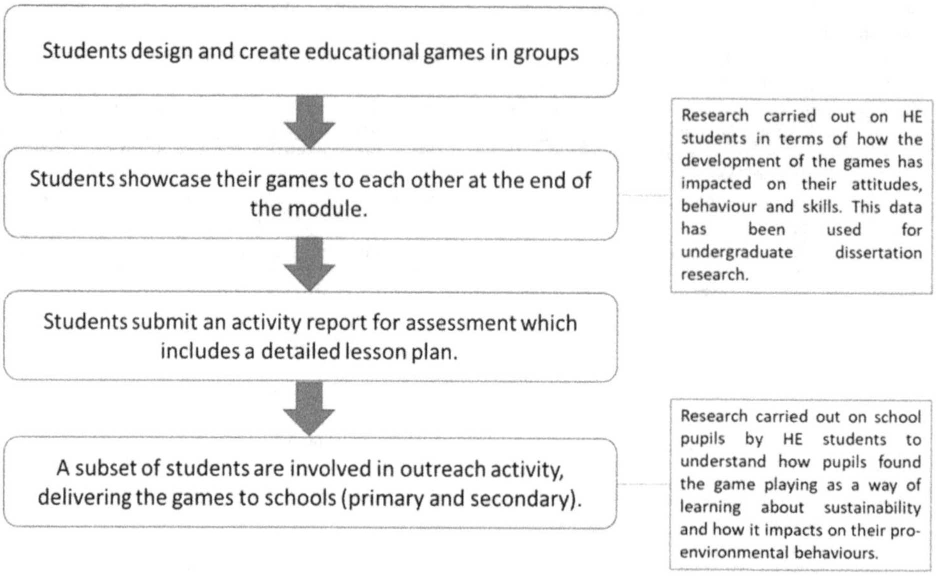

are more effective than lecturing for fostering a deeper understanding of subject material, but that raises the question of how to assess the learning. A conventional end-of-course exam undermines the active learning approach because students are motivated to memorise a lot of information in the expectation of being able to recall it as required with an assessment based largely on knowledge, rather than assessing a wider range of skills developed. Moreover, as students prepare for such assessments they are susceptible to the feeling that active learning has not prepared them well for the exam, even if this is not the case (Deslauriers et al., 2019). Learning gains can be measured using pre-instruction and post-instruction conceptual tests (Sands et al., 2018), but this depends on having a suitable validated test. Another approach is to use the activity as formative assessment (Adkins, 2018), but that doesn't solve the problem of the summative assessment. We solve that problem here by having students produce a report on the game itself. This is an essential, and therefore authentic aspect of the co-production and is in large part responsible for the success of the approach.

Student feedback on the process of designing and delivering a remotely delivered educational game was collected at the end of the module by a questionnaire survey containing a 5-point Likert scale and open-ended questions that was designed to evaluate if they felt the assessment had allowed them to develop new skills and improve

Figure 2. Extent of changes to skill sets because of the game creation assessment

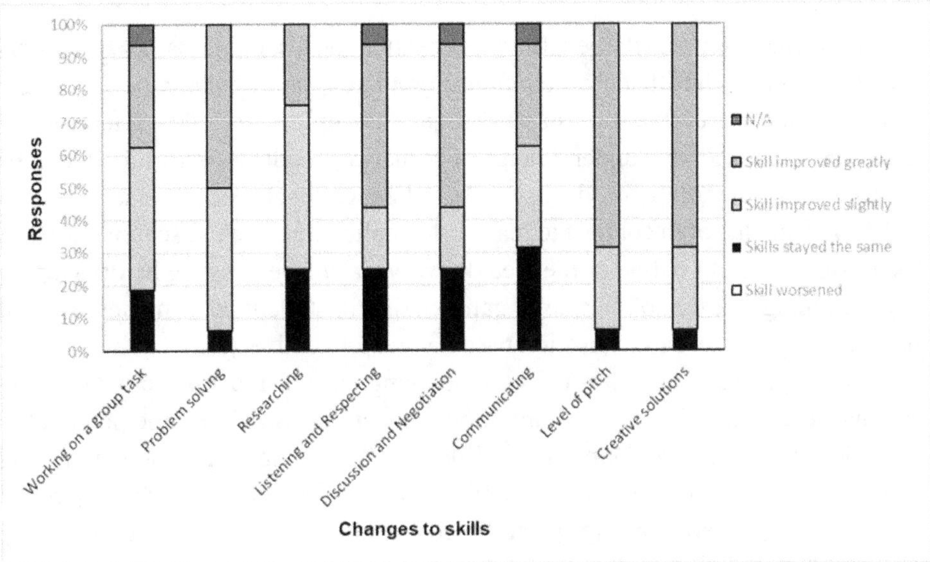

any pre-existing skills despite the move to remote learning. Of the 16 (out of 22) students who completed the questionnaire survey, 63% felt they had developed new skills, whilst 31% were unsure and only 6% felt they had not. In terms of improving pre-existing skills, none of the skills were reported to have 'worsened', whilst the majority were reported to have either 'improved slightly' or 'greatly' (Figure 2). Some students commented that they felt they had greatly improved their skills in areas such as problem solving, creative solutions and discussion skills. Additionally, one student suggested that in the future environmental games could be made for online use as it would be *"useful in these times of Covid-19"*.

Healey et al. (2016) further discuss the benefits of giving students a significant amount of independence, choice and autonomy in their learning and assessments in order to create true partnerships that result in high impact independent learning through active and experiential learning. By being involved in game creation, development and eventually delivery, this approach presents an integrative learning experience for HE students that addresses the full learning cycle and model of Kolb et al. (2014; also see Mercer et al., 2017). This type of innovative and non-traditional assessment type that addresses real world problems (in this case the need to promote pro-environmental behaviours) is increasingly being called for by both students and academics, particularly as the COVID-19 disruption has demonstrated the need for diverse assessment that helps support career skills for graduates at a time when there are less graduate jobs available (Gibbs, 2020).

LESSONS LEARNT

This chapter has discussed the benefits of co-creation and discipline-based educational research as approaches that can help future-proof geography teaching and learning provision during uncertain times and change. The approaches allowed teaching to be adaptable and flexible under a range of scenarios. Students were still able to be involved, albeit remotely, on discipline-based educational research and research-based teaching that both connect to real world applications such as school outreach. The move to entirely online or blended delivery was made possible by drawing on a long-running study into the role of game design in student learning and allowed for a new angle of research question based on remote delivery.

Crucially, these approaches to online learning ensure students' active learning and engagement despite the medium of online learning which has the potential to lead to lower engagement (Dumford & Miller, 2018). By transferring active learning strategies from the face-to-face to online environment, such as enquiry-based learning, game-based learning, place-based learning and project-based learning, students continue to learn by doing through applied and real world contexts that produce meaningful outputs that go beyond the assessment process despite, the online learning environments (Mintz, 2020). For the educational activities to be successfully executed we argue that designing the educational activity through the lens of research-based teaching and DBER is a valuable starting point. This shows that good face-to-face pedagogy can be successfully transferred to generate good online pedagogy, despite some of the initial concerns of staff faced with the need to rapidly convert their teaching to online formats in the face of the pandemic (Dhawan, 2020). It also highlights that such approaches may have benefits to be utilised beyond the temporary online teaching delivery requirements catalysed by the pandemic. This chimes with recommendations of the UK's Quality Assurance Agency, which suggest thinking beyond the immediate COVID-crisis, by seeing this as an opportunity to consider more generally student-centred learning, teaching and assessment, improving teaching delivery in the long term (QAA, 2020c).

What has become clear through the experience of having to change the way in which courses are delivered to students in a time of unforeseen exogenous circumstances (like the COVID-19 pandemic has catalysed), is that academics need to work more collaboratively with professional university staff to ensure that teaching delivery maintains the highest of pedagogic standards whilst also meeting the diverse needs of the student user, particularly the type(s) of learners. The examples of pedagogic adaptation for remote learning presented in this paper were only made possible by working closely with the technical, digital education and student administration teams.

OPPORTUNITIES

There is considerable scope for improvement for future online or blended delivery within the educational game activity approach discussed in this chapter and within the wider GEES disciplines. There are also opportunities posed by considering more blended forms of research engagement beyond pandemic response, with the potential to produce more 'collaborative, ethical and diverse' research (Littlejohn et al., 2021), overcoming issues linked to research travel and access to participants such as through the remote school outreach work carried out by HE students (Bothwell, 2020), as well as the potential to reduce carbon emissions. The COVID-19 pandemic has also realised opportunities and benefits for diversifying teaching delivery. Crawford et al. (2020) discuss the benefits of novel pedagogical approaches resulting from the COVID-crisis but with the caveat that academics need to be conscious of monitoring the quality of their learning design and provision through quality assurance activities. By incorporating co-creation teaching principles and DBER principles into online learning development, evaluation of these activities must meet the rigours required of full academic research, acting as a form of quality assurance monitoring as well as generating new educational research knowledge. This moves far beyond the usual student engagement through feedback and evaluations that come at the end of module delivery. According to the four areas of student engagement in learning and teaching discussed by Healey et al. (2014), student feedback and evaluations tend to be the most widely used but also the least transformative in terms of impact (Gibbs & Wood, 2020).

The ever-growing accessibility, in terms of cost and ease of use, of technology to underpin remote learning, continues to create new possibilities of remote learning delivery. For example, in future years we will be proposing the development of digital games that can be developed and delivered remotely in groups if necessary but also delivered remotely to schools. For example, 'dungeon master' type games can be delivered where the game is facilitated by a game leader remotely or creating board games in Excel and sharing the file for simultaneous playing through the Cloud where game players can move their pieces around in real time. E-games can complement the more hands-on interactive games that students in previous cohorts have developed, thereby contributing to a future-proofed game repository for school outreach activity. With the right kind of mindset, digital support and innovative ideas (as demonstrated during the COVID crisis and rapid move of HE to remote learning), it is entirely possible to achieve this. There are also some useful interdisciplinary opportunities here for involving computer science students for example in programming games.

Certainly for GEES, we have found that the historical domination of didacticism in HE learning is slowly becoming outdated. The kinds of innovative remote learning that have been catalysed by the COVID-19 pandemic highlight the intrinsic value of considering how teaching modules are initially designed to mitigate unforeseen circumstances, even before considering the way in which they will be delivered in situ, in the lecture theatre or seminar room. The case study discussed in this chapter has demonstrated that with remote teaching and learning it is possible to support active learning, involve external parties through outreach activity and still meet learning outcomes.

REFERENCES

Adedoyin, O. B., & Soykan, E. (2020). Covid-19 pandemic and online learning: the challenges and opportunities. In *Interactive Learning Environments*. Routledge. doi:10.1080/10494820.2020.1813180

Adkins, J. K. (2018). Active Learning and Formative Assessment in a User-Centered Design Course. *Information Systems Education Journal*, *16*(4), 34–40.

Baytak, A., & Land, S. (2011). An investigation of the artifacts and process of constructing computers games about environmental science in a fifth grade classroom. *Educational Technology Research and Development*, *59*(6), 765–782. doi:10.100711423-010-9184-z

Bell, D. M., & Pahl, K. (2018). Co-production: Towards a utopian approach. *International Journal of Social Research Methodology*, *21*(1), 105–117. doi:10.1080/13645579.2017.1348581

Bothwell, E. (2020). Blended research methods 'will become more prevalent' post-Covid | Times Higher Education (THE). *Times Higher Education*. https://www.timeshighereducation.com/news/blended-research-methods-will-become-more-prevalent-post-covid?utm_source=THE+Website+Users&utm_campaign=705eb14464-EMAIL_CAMPAIGN_2020_11_30_02_52&utm_medium=email&utm_term=0_daa7e51487-705eb14464-73364557

Bovill, C. (2020). Co-creation in learning and teaching: The case for a whole-class approach in higher education. *Higher Education*, *79*(6), 1023–1037. doi:10.100710734-019-00453-w

Bryson, J. R., & Andres, L. (2020). Covid-19 and rapid adoption and improvisation of online teaching: Curating resources for extensive versus intensive online learning experiences. *Journal of Geography in Higher Education*, *44*(4), 1–16. doi:10.1080/03098265.2020.1807478

Crawford, J. (2020). COVID-19: 20 countries' higher education intra-period digital pedagogy responses. *Journal of Applied Learning & Teaching, 3*(1), 09–28. doi:10.37074/jalt.2020.3.1.7

Deslauriers, L., McCarty, L. S., Miller, K., Callaghan, K., & Kestin, G. (2019). Measuring actual learning versus feeling of learning in response to being actively engaged in the classroom. *Proceedings of the National Academy of Sciences of the United States of America, 116*(39), 19251–19257. doi:10.1073/pnas.1821936116 PMID:31484770

Dhawan, S. (2020). Online Learning: A Panacea in the Time of COVID-19 Crisis. *Journal of Educational Technology Systems, 49*(1), 5–22. doi:10.1177/0047239520934018

Dolan, E. L., Elliott, S. L., Henderson, C., Curran-Everett, D., St. John, K., & Ortiz, P. A. (2018). Evaluating Discipline-Based Education Research for Promotion and Tenure. *Innovative Higher Education, 43*(1), 31–39. doi:10.100710755-017-9406-y

Dumford, A. D., & Miller, A. L. (2018). Online learning in higher education: Exploring advantages and disadvantages for engagement. *Journal of Computing in Higher Education, 30*(3), 452–465. doi:10.100712528-018-9179-z

Gibbs, B. (2020). What are we learning about assessment? In B. Gibbs & G. C. Wood (Eds.), *Emerging Stronger: Lasting Impact from Crisis Innovation* (pp. 3–18). Engineering Professors' Council.

Gibbs, B., & Wood, G. C. (2020). How do we partner with students in learning design at a distance? In B. Gibbs & G. C. Wood (Eds.), *Emerging Stronger: Lasting Impact from Crisis Innovation*. Engineering Professors' Council.

Gibbs, B., & Woods, G. (2020). Emerging Stronger: Lasting impact from crisis innovation. In B. Gibbs & G. C. Woods (Eds.), *Engineering Professors Council*. Engineering Professors' Council.

Healey, M., Flint, A., & Harrington, K. (2014). *Engagement through partnership: students as partners in learning and teaching in higher education*. https://www.advance-he.ac.uk/knowledge-hub/engagement-through-partnership-students-partners-learning-and-teaching-higher

Healey, M., Flint, A., & Harrington, K. (2016). View of Students as partners: Reflections on a conceptual model. *Teaching & Learning Inquiry, 4*(2), 8–20. doi:10.20343/teachlearninqu.4.2.3

Henderson, C., Connolly, M., Dolan, E. L., Finkelstein, N., Franklin, S., Malcom, S., Rasmussen, C., Redd, K., & St. John, K. (2017). Towards the STEM DBER Alliance: Why we Need a Discipline-Based STEM Education Research Community. *International Journal of Research in Undergraduate Mathematics Education, 3*(2), 247–254. doi:10.100740753-017-0056-3 PMID:30631670

Khuri, S., & Reed, M. (2020). *Tips and tools for making your online meetings and workshops more interactive.* https://www.fasttrackimpact.com/post/tips-and-tools-for-making-your-online-meetings-and-workshops-more-interactive

Kolb, A. Y., Kolb, D. A., Passarelli, A., & Sharma, G. (2014). On Becoming an Experiential Educator: The Educator Role Profile. *Simulation & Gaming, 45*(2), 204–234. doi:10.1177/1046878114534383

Kythreotis, A. P., & Mercer, T. G. (2020). Collaborative education as a new urban civil politics of climate change. In V. Castán Broto, E. Robin, & A. While (Eds.), *Climate urbanism: towards a critical research agenda* (pp. 195–210). doi:10.1007/978-3-030-53386-1_12 .

Kythreotis, A. P., Mercer, T. G., Howarth, C., Jonas, A., & Castree, N. (2019). *The 'New Civil Politics of Climate Change': Can schools, colleges and universities augment government policy action?* EAUC. https://www.eauc.org.uk/the_new_civil_politics

Lincolnshire County Council. (2021). Lincolnshire Climate Summit.

Littlejohn, A., Gourlay, L., Kennedy, E., Logan, K., Neumann, T., Oliver, M., Potter, J., & Rode, J. A. (2021). Moving teaching online: Cultural barriers experienced by university teachers during covid-19. *Journal of Interactive Media in Education, 2021*(1), 1–15. doi:10.5334/jime.631

Manduca, C. A., Iverson, E. R., Luxenberg, M., Heather Macdonald, R., McConnell, D. A., Mogk, D. W., & Tewksbury, B. J. (2017). Improving undergraduate STEM education: The efficacy of discipline-based professional development. *Science Advances, 3*(2), e1600193. doi:10.1126ciadv.1600193 PMID:28246629

Mercer, T. G., & Kythreotis, A. P. (2020). Towards Citizen Governance for Climate Change Education and Justice: A Science–Policy Perspective. In S. Myers, S. Hemstock, & E. Hanna (Eds.), *Science, Faith and the Climate Crisis* (pp. 79–92). Emerald Publishing Limited., doi:10.1108/978-1-83982-984-020201008

Mercer, T. G., Kythreotis, A. P., Robinson, Z., George, S., & Sands, D. (2019). Step out of your comfort zone, try discipline-based educational research. *Times Higher Education.* https://www.timeshighereducation.com/blog/step-out-your-comfort-zone-try-discipline-based-educational-research

Mercer, T. G., Kythreotis, A. P., Robinson, Z. P., Stolte, T., George, S. M., & Haywood, S. K. (2017). The use of educational game design and play in higher education to influence sustainable behaviour. *International Journal of Sustainability in Higher Education, 18*(3), 359–384. Advance online publication. doi:10.1108/IJSHE-03-2015-0064

Mintz, S. (2020). *Top five strategies for integrating active learning into virtual classes.* THE Campus. https://www.timeshighereducation.com/campus/top-five-strategies-integrating-active-learning-virtual-classes

Moore, J. L., Dickson-Deane, C., & Galyen, K. (2011). E-Learning, online learning, and distance learning environments: Are they the same? *Internet and Higher Education, 14*(2), 129–135. doi:10.1016/j.iheduc.2010.10.001

Müller, A. M., Goh, C., Lim, L. Z., & Gao, X. (2021). COVID-19 Emergency eLearning and Beyond: Experiences and Perspectives of University Educators. *Education Sciences, 11*(1), 19. doi:10.3390/educsci11010019

QAA. (2020a). *Guidance - Building a Taxonomy for Digital Learning*. https://www.qaa.ac.uk/docs/qaa/guidance/building-a-taxonomy-for-digital-learning.pdf

QAA. (2020b). Guidance - Preserving Quality and Standards Through a Time of Rapid Change: UK. *Higher Education.*

QAA. (2020c). *Guidance - Questions to Inform a Toolkit for Enhancing Quality in a Digital Environment.* https://www.qaa.ac.uk/docs/qaa/guidance/questions-to-inform-a-toolkit-for-enhancing-quality-in-a-digital-environment.pdf

Sands, D., Parker, M., Hedgeland, H., Jordan, S., & Galloway, R. (2018). *Using concept inventories to measure understanding.* doi:10.1080/23752696.2018.1433546

SEDA. (2020). *SEDA - Home.* https://www.seda.ac.uk/

Solent University. (2020). *Transforming Solent's learning delivery during Covid-19.* https://www.solent.ac.uk/news/transforming-solents-learning-delivery-during-covid-19

University of Kent. (2020). *CSHE Events 2020/21 - Centre for the Study of Higher Education - University of Kent.* https://www.kent.ac.uk/cshe/news-events.html#

University of Lincoln. (2020). *Teaching Remotely Essentials – Remote Teaching.* https://remoteteaching.blogs.lincoln.ac.uk/teaching-remotely-essentials/

Walkington, H. (2015). *Students as researchers: Supporting undergraduate research in the disciplines in higher education.* Academic Press.

Section 2

Resources in Geography Teaching

Chapter 6
Landscape and Heritage:
Didactic Resources for Sustainability Education

María Casas Jericó
https://orcid.org/0000-0003-1637-7606
University of Navarra, Spain

Luis Erneta-Altarriba
https://orcid.org/0000-0001-5932-9466
University of Navarra, Spain

Ana Villarroya Ballarín
University of Navarra, Spain

ABSTRACT

The growing technical, political, social, and cultural interconnectedness of our world helps advance societies. However, globalization also entails some negative environmental and cultural effects, which are usually intertwined. For example, the spread of Western consumption models accelerates ecological and social degradation worldwide. A sustainable development model is necessary to face these problems, which needs a conscious and competent society. This chapter explores how heritage and landscape are, at the same time, integrative concepts and didactic resources that promote interdisciplinary learning. Both aspects have great potential for advancing sustainability education. The chapter presents some didactic proposals that use diverse strategies and resources related to heritage and landscape, adopting a local focus whenever possible to make the learning experience more significant. Some of these activities address environmental and social challenges described in the United Nations Sustainable Development Goals.

DOI: 10.4018/978-1-7998-9598-5.ch006

INTRODUCTION

From its beginning in the mid-20th century, globalization has triggered environmental and cultural effects. While some of these are positive changes that contribute to the advancement of society —such as open access to information and collective building of knowledge (Rebolloso, 2008)—, others are negative outcomes that affect how that progress occurs.

Among the adverse effects of globalization, there is one that stands out: the homogenization and global spreading of a consumerist culture. The export of the so-called "Western model" to other countries has frequently entailed the degradation of the environment (UN Environment, 2019) and the impoverishment of the poorest communities on Earth (UN Development Programme, 2019). Such a model is rooted in an idea of progress linked to consumption and in a reductionist conception of nature as a mere source of material goods (Puig & Casas, 2017).

The spread of Western culture has also left its mark on the territory. "Aterritorial landscapes" are replicas of very similar landscapes that can be found anywhere in the world and built to consume the image they offer (Muñoz, 2007). Such landscapes have lost their identity, meanings, and values because they no longer reflect the characteristics of the land or the people that inhabit them. As Mitchell (2007, p.105, own translation) states: "(...) current landscapes do not represent a break with the modern past, but they rather materialize an atrocious continuation of a modern present (...) that is rooted in and shapes itself from globalizing capitalism".

In this line, cultural homogenization reveals itself in the loss of heritage, especially the immaterial assets, which are particularly vulnerable because they lack a material substrate. Language extinction is a paradigmatic example: 9% of the world's languages have disappeared since the 1950s, and 23% of the remaining ones are in a critical situation (Moseley, 2010).

In this context, adopting a sustainable development model is necessary for finding solutions to the ecological and social problems the environmental crisis entails. According to the UN (1987, p. 54), "sustainable development is a development that meets the needs of the present without compromising the ability of future generations to meet their own needs". To be effective, this model requires an informed, trained, aware, and committed society that is open to changing its habits and attitudes. To face this challenge, education is an essential tool.

Education for Sustainable Development (ESD) must be holistic, covering environmental and social matters, and attending to their material and immaterial (identity, cultural, spiritual) aspects. ESD must be transversal and interdisciplinary, joining efforts with other disciplines such as Environmental Education (EE), which has significant experience in citizen's awareness and training (Hume & Barry, 2015). Besides, ESD must respect local values and cultures (UNESCO, 2006) by educating individuals to live in a global society without forgetting their roots: meaning the identity and culture of the local community they belong to.

The Sustainable Development Goals (SDG) approved for the 2030 Agenda boost ESD and EE. SDG four ("Quality education") states in one of its goals the need that "(…) all learners acquire the knowledge and skills needed to promote sustainable development, including, among others, through education for sustainable development and sustainable lifestyles, human rights, gender equality, promotion of a culture of peace and non-violence, global citizenship, and appreciation of cultural diversity and of culture's contribution to sustainable development" (UN, 2015, p. 17).

Even if the SDGs inherit the spirit of previous international conferences and declarations, they "(…) put the concept of sustainability at the core of their discourse, and they do so from a 'glocal' perspective, which means global and local at the same time" (Bautista-Cerro, Murga-Menoyo, & Novo, 2019, p. 1103-11, own translation). In this sense, Geography didactics can significantly contribute to the achievement of the SDGs (Meadows, 2020; Yli-Panula & Jeronen, 2020).

This chapter explores how teaching geographical contents, especially heritage and landscape, can contribute to ESD by highlighting the importance of the local culture and environment. First, the authors describe the arguments that support the educational potential of landscape and heritage. Then, they present eight didactic proposals that use different resources and strategies to teach these concepts. These proposals are teaching guidelines that are open to the modifications that each teacher deems necessary. Similarly, although they all use landscape and heritage as a common thread, the proposals can be adapted to teaching any other aspect of Geography.

LANDSCAPE AND HERITAGE: CONCEPTS FOR EDUCATION FOR SUSTAINABILITY

The terms "heritage" and "landscape" refer to two complex and interrelated realities. Diverse disciplines —such as history, geography, ecology, architecture, or education— have studied them from different angles, giving these terms a high conceptual richness, as a result. This makes heritage and landscape excellent didactic resources for integral education.

Common Points of Landscape and Heritage

Despite their differences, heritage and landscape have several points in common. We will focus on four of them:

1. **Landscape elements can be heritage assets:** We can define landscape as the result of the interaction of natural (physical or biological) and human (material or immaterial) elements in the land under the look of an observer (Figure 1). On the other hand, heritage can be defined as the set of goods —of natural or human origin and tangible or intangible nature— that people deem valuable (from a historical, anthropological, scientific, or aesthetic point of view) and to which they attribute a sense of identity. For these reasons, heritage assets usually get protected and preserved for future generations' knowledge and enjoyment. For example, a centennial tree of high scientific value is, at the same time, a biological element of the landscape (vegetation) and a natural heritage asset. Similarly, a medieval bridge of high historical and architectural value can be considered a cultural heritage asset while being a human element (infrastructure) of the landscape.

The interrelation between these two concepts is such that there are landscapes that cannot be fully understood without the heritage they contain; and heritage assets that, if considered apart from their original environment, lose part of their sense. For example, an aqueduct built by the Romans may have become the key element in shaping the landscape around it, so we can only explain the latter because of the presence of an infrastructure that is actually a heritage asset. Similarly, a white marble sculpture originally built in a garden to stand out in the vegetation loses part of its meaning when placed in a museum.

2. **The landscape itself is also heritage:** In 2000, the European Council published the European Landscape Convention, which sets a roadmap for managing, regulating, and protecting European landscapes. The Convention recognizes the landscape as an asset: "the landscape contributes to the formation of local cultures and that it is a basic component of the European natural and cultural heritage, contributing to human well-being and consolidation of the European identity" (Council of Europe, 2000, p.1).
3. **There are specific landscape heritage protection figures:** It is the case of cultural landscapes, which are "(…) cultural properties and represent the 'combined works of natural and of man' (…). They are illustrative of the evolution of human society and settlement over time, under the influence of the physical constraints and/or opportunities presented by their natural environment and of successive social, economic and cultural forces, both external and internal." (UNESCO, 2021, p. 47). Although the original World Heritage Convention from 1972 only defined two kinds of heritage (natural and cultural), it included the term "mixed heritage", afterwards. This shares

Figure 1. Relation between the landscape elements and the types of heritage defined by UNESCO (1972). Source: own elaboration.

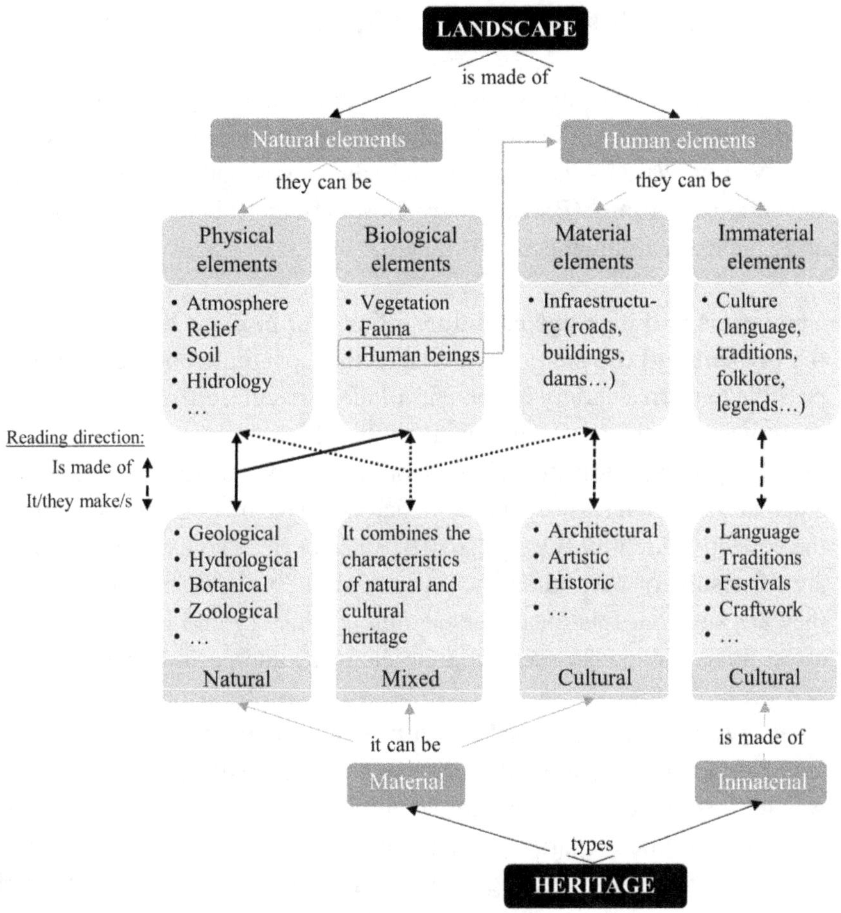

characteristics with the natural and cultural types, so it can also include certain landscapes.

4. **Heritage and landscape are social products:** Human beings have historically used the natural resources of the landscape to create goods that we now deem exceptionally valuable and therefore consider part of the material cultural heritage (castles, bridges, cathedrals, paintings, or sculptures). Besides human works, we can also recognize an exceptional value in nature; elements such as centennial trees, waterfalls, caves, or gorges can constitute natural heritage assets. Finally, due to living as communities in specific territories, human beings have also developed a rich immaterial heritage, which includes assets

such as language, folklore, legends, sayings, games, artisan knowledge and techniques, or plant medicine (Ortega, 2010).

The Educational Potential of Heritage and Landscape

Several authors have argued that studying heritage and landscape —either separately (Barbadillo, 2016; Calaf, 2009) or together as cultural landscapes (Cobas, 2016; Morón & Morón, 2012)— has educational advantages both in conventional and non-conventional settings (Baena, 2016; Busquets & Rubert, 2011). Below we synthesize the main arguments in this regard.

1. **Landscape and heritage contain explicit and implicit information about the present and the past:** The elements that constitute both realities make it possible to learn a variety of contents of the curricula, either related to natural components of the territory (climate, relief, hydrology, vegetation, fauna) or to the way humans transform it (economic activity, infrastructure building) or attribute value to it (historical, aesthetic, sacred) to make it an essential part of the collective identity (Ortega, 2010). Since current landscapes and heritage are the results of past human activities and the basis for building the future, they are key concepts for understanding our history and establishing cause-effect relationships between different elements and periods (Cuenca, Estepa & Martín, 2015).
2. **They allow studying specifics within an integrative context:** Because landscape and heritage are complex issues, they are usually taught in a fragmented way. For example, when learning about a landscape, climate and economic activity are studied separately, even if the latter depends on the first. Although this approach allows studying contents in deeper detail, it entails the risk of ignoring the interrelationships among natural elements and between those and human beings. Landscape and heritage provide an integrative framework for study (Fernández, 2005; Liceras, 2013).
3. **They offer an opportunity for interdisciplinary work:** Traditionally, the teaching of landscape and heritage has corresponded to the areas of Social Sciences, Geography, History, and Arts. However, since both realities give room for a wide range of contents, landscape and heritage allow studying them in an interdisciplinary way (Calbó, Juanola & Vallés, 2011). Working by projects can be a way of doing this.
4. **They allow using different resources and methodologies:** The diversity of contents that landscape and heritage allow covering is an opportunity for using varied resources and methodologies that will foster the development of different abilities. Some of these resources are photography, orthophotography,

miniatures, maps, Geographical Information Systems, and cartographic viewers (Fraile, Sánchez & Medina, 2016; Montsalvatge & Oller, 2014). Some possible methodologies are problem-based learning (García de la Vega, 2012), visits to facilities (museums, botanical gardens, protected areas) (Melgar & Silvio, 2011), didactic itineraries, and *in situ* learning (García de la Vega, 2004; Pérez & Morón, 2016). The Technologies of Information and Communication play an important role, as they allow accessing digital content and developing new teaching methods (Tourón & Santiago, 2015).

5. **They foster respectful values and behaviors towards the environment and society:** Since landscape and heritage are a consequence of human communities living in the territory, they can promote positive values and respectful attitudes towards close and distant environments and cultures. About landscapes, Busquets (2011, p. 80) says that "(...) they are the local expression of global phenomena, and their look and evolution depend on what happens in that specific place and also beyond its limits" (own translation). For this reason, its study may foster responsible attitudes towards the environment, may it be close or far away. Heritage, on its side, can play an essential role in educating for multiculturalism and developing critical thinking (Baena, 2016). These are key tools for facing the cultural homogenization processes caused by globalization.
6. **They foster the development of key competences:** Competence is defined as "(…) the combination of practical abilities, knowledge, motivation, ethical values, attitudes, emotions, and other social and behavioral components necessary for effective actions" (DeSeCo 2002 p.8, own translation). Studying landscape and heritage helps develop the key competences stated in the curriculum.

According to the framework for the international implementation of the UN Decade for ESD (2005-2014), these arguments significantly match the characteristics that the Education for Sustainable Development (ESD) should have. ESD should be holistic and interdisciplinary, promote value acquisition and development of critical thinking, focus on problem-solving and decision making, use different methodologies, allow students to participate in their learning process, apply to real life, and be relevant to local contexts by tackling local and global issues (UNESCO, 2006).

DIDACTIC RESOURCES AND STRATEGIES FOR LANDSCAPE AND HERITAGE TEACHING

This section describes eight educational proposals that use different resources for teaching about landscape and heritage. In each proposal, the authors detail the necessary resources, the didactic strategy, and the educational level the activities can be applied. They also include alternatives to adapt the proposal to different contexts or learning requirements.

Interpreting the Environment through Texts and Pictures

Ascertaining students' previous knowledge before studying new content is essential for sound educational practice. In this sense, introductory activities help detect, among other things, mistaken preconceptions that may nevertheless be quite widespread. One of these is the idea of the landscape as a natural, beautiful place that brings pleasure to our senses (Maderuelo, 2005). Something similar happens with the concept of heritage, which is usually linked to monuments and unique buildings (Marin-Cepeda & Fontal, 2020).

This proposal (Table 1) works as an introductory activity for knowing the students' notions about landscape or heritage. It allows working the guidelines for interpreting images, and it fosters creativity and written and artistic expression. Although it can fit any educational level, it is especially recommended for university students because higher education studies usually give less room for creative competence than in other contexts (Caballero, Sánchez & Belmonte, 2019).

Flipped Classroom

The Flipped classroom methodology, developed within the last decade, exchanges the roles of the teacher and the students (Tourón & Santiago, 2015; Ozdalmi & Asiksoy, 2016). In this model, the students perform the simplest cognitive tasks outside of the school and focus the in-person sessions on those most complex activities that need further discussion, explanation, or interaction. Social Sciences —and, more specifically, Geography and History— can take great advantage of this methodology.

In summary, the Flipped classroom methodology has two steps. First, the teacher explains the most theoretical and abstract contents of the subject, and the students analyze those on their own. Then, the second step focuses on reflecting, interacting, and applying what the students have learned (Akçayır & Akçayır, 2018). The aim is to offer the students the opportunity to apply their new knowledge in activities in the classroom. In addition, students can also use these sessions to solve the questions they have about the contents they worked on autonomously (Blair, Maharaj & Primus, 2016).

This proposal applies the Flipped classroom methodology using videos about landscape and heritage (Table 2). Although it is conceived for university students, it can also work for high secondary courses (10th-12th grade in the USA, 11th-13th grade in the UK).

Table 1. A detailed description of the didactic proposal: resources, didactic strategy, and alternatives

Resources	Didactic strategy	Alternatives
• Photographs of landscapes, with or without heritage assets. The authors recommend using images that show high contrasts: natural, rural, urban, industrial landscapes, artistic, architectural, natural, or immaterial heritage assets. • Paper, writing materials and colors.	1. Each student receives one photograph. 2. The teacher explains some basic guidelines for describing images (see Barbadillo, 2016). For example: • Observe the image and identify its elements • Classify the landscape or heritage asset according to its main elements or features • Describe each element in detail (color, shape, texture), indicating its location in the image in relation to the other components. • Personal opinion about the landscape or heritage asset (aesthetic valuation, sensations or memories it recalls). 3. Students write a detailed description of their pictures. 4. The students exchange their descriptions. It is important to make sure that each one receives a text corresponding to an image different from the one they got at the beginning. 5. Students draw a picture reflecting what they read in the description they just got. 6. Finally, they all share pictures and descriptions, comparing both. It is recommended to highlight the elements that appear in the pictures but were not present in the descriptions or photographs and the personal opinions about the observed reality. This information will be an indicator of the students' preconceptions about the subject of study.	This activity can be done *in situ*, directly observing the landscape or heritage asset instead of photographs. In this case, the authors recommend following the approach of Benayas and collaborators (1994).

Source: own elaboration

Comparative and Retrospective Analysis of Orthophotographs

Research and practice have proved the power of images as educational resources. Spatial photography or orthophotography (taken from flights or satellites) has traditionally been used to study natural and human elements' morphological and structural components (Álvarez & de Lázaro, 2017). Remote sensing, compared temporal analysis, and photography software give orthophotographs great potential as educational tools for Earth Sciences, Geography, and History (González, 2012).

This proposal focuses on a compared time series analysis of the territorial effects of climate change, one of the most challenging environmental problems that humanity faces (SDG number 13: Climate action) (Table 3). The prospective study of photography series of the same territory, taken in different moments, gives

Table 2. A detailed description of the didactic proposal: resources, didactic strategy, and alternatives

Resources	Didactic strategy	Alternatives
• Video explaining the contents of the subject. The authors recommend using short videos to simplify the message. Videos can be made by the teacher or downloaded from educational platforms. The authors propose some sources as an orientation: • UNESCO cultural heritage. [1] • Yann Arthus-Bertrand projects: "Home" (2009) and "Human" (2015). [2] • National Geographic Education Project (2020), available on National Geographic Foundation's YouTube channel. [3]. • Internet access is necessary for visualizing these contents.	1. Give the video to the students. 2. The teacher gives some guidelines for the students to focus on the key aspects of the video. These guidelines should include questions or activities about: • Contents explained in the video, including general and specific issues. • Reasons why the online platform, server, or organization has included this video in its catalogue. • Characterization of the contents: kind of heritage [1], kind of geopolitical conflict [2], or kind of environmental problem [3]. These questions can take different formats: a list of brief questions, a multiple option test, or a descriptive text. These activities or questions can appear linked to the video using specific software (Edpupzze, Vialogues, Videonot, HapYak, or Blubbr). The objective of these activities or questions is to revise, verify and reinforce the essential concepts that the video explains. 3. Share and comment on the questions and answers among the students. This should help the students understand the concepts and processes explained. These activities should help students synthesize, argue, and compare the previously analyzed data, theories, or opinions. Written reviews, reports, debates, summary charts or comparative tables are useful activities for exchanging views, fostering integrated and interactive learning.	The students may make and share the video instead of using an existing one. Using specific videos is usually time-consuming, but at the same time, it allows for specifying better the contents. This alternative requires a first step of preparing the resource (the video) that implies both the teacher and the student. However, this is a participative and proactive strategy that gives students the opportunity of getting involved in the teaching-learning process in a more significant way.

Source: own elaboration

a visual narration of some facts that leave traces in the land. The proposal aims to foster deductive reasoning, which can be implemented at any educational level.

Geographic Information Systems and Online Maps

Current digital platforms can growingly integrate different kinds of information and have a great capacity for updating —sometimes in real-time— the spatial data represented. Cell phones geopositioning, the accessibility of web browsers, the development and simplification of Geographic Information Systems (GIS) software, the availability of digital cartographic information, and the integration of cartography in online services or social networks have democratized cartography

Table 3. A detailed description of the didactic proposal: resources, didactic strategy, and alternatives

Resources	Didactic strategy	Alternatives
• Time sequences of spatial orthophotographs. Most public administrations have cartographic services and data infrastructure platforms that offer such information. • Some reference institutions, such as NASA and the European Spatial Planning Observation Network (ESPON), also provide this kind of graphical information (Moigne, Grubb & Milner, 2013). The authors remark on the project "Images of Change" (NASA), which displays a selection of spatial photographs of unique environmental and cultural interest places related to climate change.	Photography analysis needs a systematic approach that includes both general and deductive aspects. 1. The students describe the source of the images. This description should focus on the basic data: place, dates, technique. 2. The students describe the spatial phenomenon, focusing on the differences before and after a particular moment. The description must include structural and functional characteristics (texture, form, color), the distribution of the elements, and their spatial frequency. 3. The students describe the possible causes of the observed changes, identifying their dynamics. Also, they should describe change patterns that can be used for making predictions. This last aspect is the most abstract and deductive of all.	Oblique photographs can represent a more accessible alternative to orthophotographs. The authors suggest two repositories for this kind of resource: • The official Photos website of Yann Arthus-Bertrand • Geophotopedia: a database of geographical elements and phenomena of the library of the University of Sevilla.

Source: own elaboration

as a tool for an increasingly broad audience. It is also a great educational resource (Boix & Olivella, 2007; Wiegand, 2001).

This activity proposes using GIS in the classroom (Table 4). The complexity of the tasks carried out with this tool depends on the target educational level. Specialized GIS is usually a complex instrument and relies on complex concepts. In contrast, GIS included in multi-platforms or online browsers is usually intuitive and straightforward. For this reason, the following proposal relies on the latter.

Analysis of an Interview with an Expert

Expert interviews and conferences in different fields of knowledge have become popular in the last decade thanks to popular science, outreach, and education initiatives. Some well-known examples are "TED Talks" (internationally) or the "Learning together" ("Aprendemos juntos" in Spanish) initiative in Spain. These

Table 4. A detailed description of the didactic proposal: resources, didactic strategy, and alternatives

Resources	Didactic strategy	Alternatives
GIS software for creating and editing digital cartography. Some examples are: • Specialized GIS: (license required) Hexagon, Autocad's Autodesk Map 3D, ESRI's ArcGis and online ArcGis, or Bentley; (open software) Grass, GDAL, OGC, QGIS, OSGeo Project, gvSIG, PROJ.4, OpenJump. • Multiplatform GIS and online displayers: (license required) Power-Map, Excell; (open access) Open Street View, Google Maps, Google Earth. The authors advise open access multi-platform and online GIS for beginner levels.	The activity follows a four-step process: 1. Procedural: The students learn the essential characteristics of the GIS tool and how to use it: functions, procedures, visualizing and exporting options. 2. Demonstrative: The students learn about the methods for representing cartographic systems, considering the basic semiotics used in cartographic models: scale, projection, or legend. 3. Analytic: The students identify the spatial motives of the represented phenomenon through activities that allow integrating data from a territorial perspective. To this end, the teacher should propose questions about why, how, or where a specific process happens on the map. 4. Proactive: The students build new cartographic content, enriching the spatial representation's graphical side. For this, the teacher will propose activities focused on solving spatial problems that generate new variables as a result. Some examples are building themed choropleth maps, proportional symbols, or flow charts. The activities can also focus on building itineraries, calculating environmental parameters (altitude, slope, surface, volume), or determining cultural assets (administrative boundaries, value of particular social, territorial elements) (Manzanares & Quintana, 2019).	• Working with GIS may not be easy for all students, so the steps for developing the activity can be adapted to fit the group's needs. • Traditional (analogic) formats can work as a supporting resource at any proposed steps. They can show examples of types of symbols, causes of certain spatial distributions, or ways of representing new variables resulting from the cartographic analysis. Drawing on printed maps may be a good strategy for the latter.

Source: own elaboration

resources are growingly accessible thanks to Information Technologies and video platforms such as YouTube or Vimeo.

Oral sources have traditionally been an educational resource in Elementary Education, especially for teaching History and Social Sciences (Fuertes, 2014). Several authors consider that interviews also hold great potential for teaching Geography in higher education (Valero, 2016). The proposal below is based on interviews (Table 5). It is aimed at university, Baccalaureate, or high school students (10th-12th grade in the USA, 11th-13th grade in the UK). Although it takes the notion of landscape as an example, its strategy is applicable to contents of any discipline as long as there is an interview or a conference about them. The kind of questions that the teacher asks for analyzing the interview will determine the competencies to work through

Table 5. A detailed description of the didactic proposal: resources, didactic strategy, and alternatives

Resources	Didactic strategy	Alternatives
• Computer with Internet access. • Website or application to access interviews or conferences. The authors recommend using YouTube or Vimeo, but written interviews from newspapers, magazines, or blogs are appropriate. • List of questions for analyzing the interview or conference.	1. Choose the interview or conference. The authors recommend using short pieces of 5-10 minutes. When addressing non-university students, it will be necessary to adapt the language level, mainly if it uses technical terms. In this case, the teacher should view the interview in advance and write a summary or adapted transcription to give to the students. The following paragraphs show an adapted transcription of an interview with Dr Martínez de Pisón (Orte, 2007, 1:16:00-1:20:40) about the notion of the landscape for 14-15-year-old students: "The landscape is the sum of the components around you. These components are not divided into humanities and sciences; they are not Geology on one side and history on the other. The landscape is everything. In a mountain landscape, you can find rocks, creeks, forests, or meadows, and human beings who have adapted those environments for their survival for millennia. For example, they have built houses using mountain rocks, made roads that follow the natural contours, and constructed bridges that cross the rivers. They have usually –but not always– done this harmonically, integrating the new assets into the existing landscape. Thus, little by little, humans have progressively reconfigured the landscape. Therefore, the landscape is the sum of nature and culture. It includes the cultural assets we can see –such as buildings, meadows, and roads– and the ones we cannot see, but we can find in the library or the museum. The landscape is also the place of our experiences and feelings because we live in it. There are landscapes that we love because they contain our experiences: a buried family member, a good memory from our childhood. Therefore, landscapes are full of meanings." (Own adaptation and translation) 2. Provide the context of the interview/conference and its subject. Knowing the matter of the interview and the professional career of the speaker helps the students understand its contents. The teacher can assign the gathering of these data as a task for the students. 3. Watch or read the interview or conference. 4. Work individually or in small groups on the questions posed by the teacher to understand the interview or conference. For example, regarding the interview shown in point 1, the questions could be: • Define landscape in your own words. • What is the role of human beings in the landscape? • How can humans change the landscape? • Explain the following sentence: "landscapes are full of meanings". 5. Share the answers to the questions. The teacher will try to generate some debate by asking the students if they agree with some of the statements of the interview or conference.	Students can enrich the analysis by designing and carrying out the interviews (see proposal six about designing questionnaires). For example, they could interview a family member about their perception of changes in their close landscape or heritage assets. Fuertes (2014) adds some suggestions related to teaching about heritage.

Source: own elaboration

Table 6. A detailed description of the didactic proposal: resources, didactic strategy, and alternatives.

Resources	Didactic strategy	Alternatives
• Computer with Internet access. • Website or application to build questionnaires (Google Forms, SurveyMonkey, or Formsite). • (Optional) Website or application for generating QR codes (such as QR Code Generator).	1. The teacher assigns a subject to each group (4-5 people) to build three or four questions about it. The authors recommend dividing the questionnaire into sections (for example, natural elements of the landscape, human elements, impacts on landscape, landscape protection) and assigning one to each group. 2. The teacher explains some basic guidelines for designing questions (Bisquerra, 2014; Cohen, Manion & Morrison, 2011): kinds of questions, valuation scales, question order, question grouping, or socio-demographic variables (age, gender, place of residence). If the teacher deems it appropriate, the students can carry out this step as a research task. 3. Each group designs its questions. 4. The groups share their questions and exchange ideas. Together they decide the arrangement of the questions (order and grouping). 5. The groups design the final layout of the questionnaire using an online platform. The authors recommend Google Forms because it allows several people to work simultaneously on the same document. It is important to include an initial paragraph that briefly describes the goal of the questionnaire and names its authors. 6. The participants disseminate the questionnaire among the target audience through different channels (email, social media). Associating it to a QR code may make this step easier. 7. The questionnaire stays open for one or two weeks while the public is answering. The authors recommend closing it after that to avoid new data entering at the wrong time. 8. The students build a database for gathering the responses to the questionnaire. Google Forms allows downloading answers directly in .xls format. 9. The students carry out data treatment tasks. Each group will analyze the answers to their questions and the socio-demographic data, and they will divide the data according to socio-demographic variables and make comparisons. 10. The students will represent their information graphically (charts, maps, word clouds). 11. All the groups will share their results and draw together the study's conclusions.	• This proposal may work as a research introduction activity if the students first inquiry about the notion of landscape and then design questions based on their findings. This addition would significantly foster autonomous learning in them. • To strengthen the scientific soundness of the activity, the teacher may establish the respondents' profile from the beginning (for example, last year of the degree in Journalism) or fix the minimum responses necessary to get a statistically representative sample. S/he may also add a pilot study to check the quality of the questions. • The survey can be applied as personal interviews (in person or remotely) as an alternative to online platforms. This option entails higher costs and time but allows working communicative and social competencies. • The results of this activity can become the departing point for a Service-Learning proposal (see proposal eight).

Source: own elaboration

the activity. In general, this proposal allows working personal reflection, critical thinking, and linguistic competence.

Social Research in the Classroom: Design and Application of an Online Questionnaire

"A questionnaire is an instrument for gathering information that uses a limited set of questions that allow the [research] subject to provide information about his/herself or their environment" (Bisquerra, 2014, p. 240, own translation). Questionnaires are a widespread tool in social and educational research because they cover a variety of research objects and subjects. Regarding landscape, questionnaires have helped know different groups' social perceptions and preferences (Casas, Erneta & Puig, 2018; Górka, 2020; Zheng, Zhang & Chen, 2011). However, the potential of this tool as an educative resource remains largely unexplored.

This proposal is designed for university students, although it can also be appropriate for high secondary courses (10th-12th grade in the USA, 11th-13th grade in the UK) (Table 6). In small groups, students design a questionnaire to know their colleagues' perceptions of the landscape. Since they need certain notions of landscape to do this, the authors recommend implementing this activity after working on these concepts in class. After carrying out the survey, each group will analyze and represent the information, and they will share it with the rest of the students. This activity fosters collaborative and autonomous work and tackles digital and mathematical competencies. The proposal allows an interdisciplinary approach, encouraging collaboration among different disciplines.

Case Studies

Case studies are a widespread educational methodology that many disciplines apply. They help bridge the gap between theory and practice, a necessary step for fixing abstract concepts and acquiring new abilities. This may be especially relevant when dealing with assets whose scale or complexity may result challenging for students. Case studies constitute a valuable tool for exploring the human-environment interactions that configure landscape and heritage. In addition, case studies are a key tool for project-based learning methodologies, an approach that helps students develop skills for their professional future. This is especially significant in higher education systems, for institutions must transmit knowledge and train students to become professionals who can face real-world situations (Martinich, Solarz & Lyons, 2006).

The authors describe a proposal for using case studies when working on the landscape from the perspective of environmental impact assessment (EIA) (Table 7).

Table 7. A detailed description of the didactic proposal: resources, didactic strategy, and alternatives

Resources	Didactic strategy	Alternatives
Access to the documentation of an implemented or ongoing project subject to EIA. EIA procedures are accessible by the public in general terms. Preferably, the project should have an important social side.	1. The students review the existing information about the case. This is usually the technical report of the proposed project and the EIA report (if already existent) but may also include other sources such as previous environmental studies in the area or news reports. 2. In groups, students prepare an in-depth assessment of each environment component that is part of the EIA and explain it to the rest of the class in a field visit. These components are: • Abiotic environment: geology, landforms, climate, hydrology. • Biotic environment: vegetation, fauna. • Social environment: affected communities, economic situation, cultural assets. 3. Students elaborate an impact matrix (Leopold et al., 1971) to identify and value the positive and negative results of the interaction between the proposed project and each environmental component. 4. Based on the matrix results, students decide whether they would authorize or not the proposed project, and they explain their arguments to the teacher. 5. After finishing the course project, the students meet with the teacher to evaluate their performance and reflect on their learning process.	Different cases may work for this activity, and the authors recommend prioritizing those that allow frequent field visits to the area. This has the additional benefit of working at a local scale, which is a necessary step for eventually grasping the global picture. Students may rely on graphic material such as maps, photographs, and geographical information systems if that is not possible.

Source: own elaboration

Teaching this discipline poses the challenge of transmitting trainees' environmental knowledge and expertise to make decisions in various places, and case studies help cover both. Consequently, several authors recommend this teaching resource for EIA courses (Pope & Morrison-Saunders, 2018; Sánchez & Morrison-Saunders, 2010). The example below describes a course project for an EIA university course. Students work on an actual case assigned by the teacher during the semester, retracing the steps of actual EIA processes in the nearby environment. Besides the ecological side, the chosen cases always have a significant social component to show the interaction between both elements of the landscape.

For this reason, the following proposal allows an interdisciplinary approach and may be easily linked to the Sustainable Development Goals (SDG). Some examples are the construction of a breakwater that altered the ecology of an estuary and caused economic damage to a fish farm (see full description in Puig & Villarroya, 2013) or the extension of a quarry that would maintain jobs in a rural area at a high ecological cost. These kinds of projects help students focus on SDGs, mainly

Table 8. A detailed description of the didactic proposal: resources, didactic strategy, and alternatives

Resources	Didactic strategy	Alternatives
A close natural or naturalized environment. Students should be able to visit it regularly.	1. Identification of environmental restoration needs of the chosen area. Students carry out an environmental evaluation of the zone identifying its natural and social values and the negative impacts on them. These impacts will be the starting point for the restoration proposal. 2. Identification of the public potentially affected by the project. This includes the main actors involved (in the present and the past) in the area's development. For example, university students and teachers, campus managers, gardeners, or citizens that walk their dogs in the zone. These actors should have a voice in the project through the appropriate means (e.g., direct interviews, surveys, group seminars). 3. Elaboration of a restoration proposal to mitigate the negative impacts identified at the beginning of the process. Students apply their previous knowledge and the new concepts seen in the course to design a restoration project that is detailed enough to be applied. It shall include the following sections: • Introduction and justification • Description and valuation of the area • Description of the restoration activities proposed • Timeline • Budget 4. Public participation session/s. Students hold at least one seminar to inform and get feedback from the project's actors involved or potentially affected. This activity refines the proposal and helps students take the pulse of their performance before the final academic evaluation. 5. Final presentation seminar. At the end of the semester, the students present their final project in a public session. 6. Reflection on the learning process. The teacher holds individual and group sessions to talk with the students about the whole activity, and this helps them reflect on their learning process and gives feedback to the teacher to improve future courses.	The activity can occur in an urban setting if there is no natural or naturalized area available. In that case, the absence of natural components shall be considered a negative impact addressed through the restoration proposal.

Source: own elaboration

numbers 8 (Decent work and economic growth), 14 (Life below water), and 15 (Life on land). Such cases challenge the environmental knowledge of students and make them face difficult decisions (should we approve the extension of the quarry? could we have foreseen the impact of the breakwater on the fish farm?) that help develop attitudes and abilities necessary for professional practice allows an interdisciplinary approach, encouraging collaboration among different disciplines.

Service Learning

Service-learning is an educational methodology that links "community service and academic study so that each strengthens the other" (Ehrlich, 1996). Students participate in activities that aim to solve community needs, which require them to get new knowledge and apply it. Post-action reflection helps students gain a "further understanding of course content, a broader appreciation of the discipline, and an enhanced sense of personal values and civic responsibility" (Bringle, Hatcher & McIntosh, 2006, p.12). Landscape and heritage offer an excellent opportunity to apply this methodology, as they combine a material setting (natural environment or built assets) and an important social component. Therefore, students' actions on these frameworks can easily benefit the community. In this context, service-learning also helps students achieve the Sustainable Development Goals (SDGs) (García et al., 2019).

This proposal describes a course project about landscape restoration, which the authors have applied as the core activity of a university course on this subject (Table 8). Its main aim is to help students put their previous and new knowledge into practice to understand it better while also grasping new abilities and getting familiar with the challenges and opportunities of professional practice. In the example carried out by the authors, the students design a restoration project for a part of their university campus, which is at the same time a public park. Therefore, the potential beneficiaries are not just university students and workers but all the local citizens (Villarroya & Larrañeta, 2019). Besides the specific technical concepts of the course, this project helps students develop creativity, organizational skills, and commitment. Even if the proposal focuses on university students, it can easily be adapted to other educational levels to work on subjects related to the environment and human action.

CONCLUSION

Teaching about landscape and heritage is an educational opportunity to foster holistic and interdisciplinary education in direct contact with the studied object and students' local reality. It can be applied both to formal and non-formal educational contexts. Knowing and valuing the local environment can foster positive and responsible behavior towards nature and society, both at local and global scales.

There are many points in common between the educational possibilities that landscape and heritage offer and the characteristics of Education for Sustainable Development. Therefore, these concepts can significantly contribute to educating future citizens, especially by knowing and valuing the local environmental and cultural reality.

This chapter has explored this educational potential through eight proposals that use different resources and strategies, some of which aim to work on some of the Sustainable Development Goals. These proposals can act as guidelines for teaching and may be tailored to different contexts and educational levels.

Images of diverse kinds are the main educational resource of three of these proposals. The first describes an activity of image analysis and landscape description as a strategy to foster analytic abilities, reading comprehension, and oral expression. Drawing pictures from written descriptions also fosters creativity.

The third proposal describes the comparative analysis of a time series of orthophotographs to detect changes in the territory and investigate their causes. This strengthens analytic abilities and deductive reasoning. The fourth proposal follows a similar idea, describing a cartographic analysis using Geographic Information Systems. In this case, students can also represent the results of their analysis in maps.

Videos are other interesting educational resources for teaching geography. The second proposal uses videos as part of a Flipped classroom activity. Viewing short videos and completing questions and activities can foster students' autonomous learning, improving their involvement in the teaching-learning process. The proposal considers the option of students elaborating on educational videos, fostering their creativity.

The fifth proposal is based on oral resources as educational materials. Analyzing expert interviews or conferences and debating them develops personal reflection, critical thinking, and linguistic competences. The proposal includes the possibility of students designing and carrying out the interviews. This option would need them to build a questionnaire similar to the one described in the sixth proposal.

Designing and applying questionnaires is an educational strategy that allows students to work on geographic content autonomously and collaboratively. Designing and publishing the questionnaire boosts linguistic and digital skills while analyzing and representing results develops mathematical competence.

The last two proposals highlight the connection between society and the environment. The seventh proposal describes a methodology for case analysis based on the Environmental Impact Assessment actual procedure. It fosters autonomous learning, critical thinking, and complex decision-making. This is especially true when the teacher chooses cases that deal with challenging or controversial issues.

The last proposal materializes the educational action in the students' learning and some service to the community. It proposes a strategy based on the Service-

Learning methodology that helps students develop creativity, organizational skills, and commitment by designing and carrying out a project that responds to a specific social need.

All the above shows that geographical education in general —and landscape and heritage teaching more specifically— can significantly contribute to Education for Sustainable Development and, therefore, to achieving Sustainable Development Goals.

REFERENCES

Akçayır, G., & Akçayır, M. (2018). The flipped classroom: A review of its advantages and challenges. *Computers & Education*, *126*, 334–345. doi:10.1016/j.compedu.2018.07.021

Álvarez, J., & de Lázaro, M. L. (2017). Spatial Data Infrastructures and Geography learning. *European Journal of Geography*, *8*(3), 19–29. http://eurogeojournal.eu/articles/2.Spatial data infrastructures and Geography learning.pdf

Baena, J. M. (2016). Patrimonio y educación obligatoria. Deseos y realidades. *Revista UNES: Universidad, Escuela y Sociedad.*, *1*, 6–20. https://www.revistaunes.com/index.php/revistaunes/article/view/1/1

Barbadillo, F. (2016). *Manual para observar e interpretar paisajes*. Tundra.

Bautista-Cerro, M. J., Murga-Menoyo, M. A., & Novo, M. (2019). La Educación Ambiental en el S. XXI (página en construcción, disculpen las molestias). *Revista de Educación Ambiental y Sostenibilidad, 1*(1), 1103.

Benayas, J., Heras, F., De Lucio, J. V., Marcén, C., & Ruiz, J. P. (1994). *Viviendo el paisaje: guía didáctica para interpretar y actuar sobre el paisaje*. Fundación Natwest.

Bisquerra, R. (2014). *Metodología de la investigación educativa* (4ª). La Muralla.

Blair, E., Maharaj, C., & Primus, S. (2016). Performance and perception in the flipped classroom. *Education and Information Technologies*, *21*(6), 1465–1482. doi:10.100710639-015-9393-5

Boix, G., & Olivella, R. (2007). Los Sistemas de Información Geográfica (SIG) aplicados a la educación. El proyecto PESIG (Portal Educativo SIG). In M. J. Marrón, J. Salom, & X. M. Souto (Eds.), *Las competencias geográficas poara la educación ciudadana* (pp. 23–32). Grupo de Didáctica de la Geografía de la Asociación de Geógrafos Españoles.

Bringle, R., Hatcher, J., & McIntosh, R. (n.d.). Analyzing Morton's Typology of Service Paradigms and Integrity. *Michigan Journal of Community Service Learning, 2006*(13), 5–15.

Busquets, J. (2011). La importància de l'educació en paisatge. In J. Nogué, L. Puigbert, G. Bretcha, & À. Losantos (Eds.), *Paisatge i Educació. Plecs de Paisatge: reflexions 2* (pp. 69–88). Observatori del Paisatge de Catalunya. http://www.catpaisatge.net/fitxers/publicacions/paisatge_educacio/paisatge_i_educacio.pdf

Busquets, J., & Rubert, J. (2011). *La sensibilización en paisaje: un reto para el siglo XXI. Awareness-raising on landscape: a challenge for the 21st Century* (J. Busquets & J. Rubert, Eds.). Generalitat de Catalunya - Pays.Med.Urban. http://www.paysmed.net/pdf/paysurban_attivita_sensibilizzazione_es.pdf

Caballero, P. A., Sánchez, S., & Belmonte, M. L. (2019). Análisis de la creatividad de los estudiantes universitarios. Diferencias por género, edad y elección de estudios. *Educatio XX1, 22*(2), 213–234. doi:10.5944/educxx1.22552

Calaf, R. (2009). *Didáctica del patrimonio*. Trea.

Calbó, M., Juanola, R., & Vallés, J. (2011). *Visiones interdisciplinares en educación del patrimonio: Artes, Culturas, Ambiente*. Documenta Universitaria.

Casas, M., Erneta, L., & Puig, J. (2018). La encuesta como herramienta para conocer las ideas previas sobre paisaje del alumnado de ESO: una experiencia piloto en Navarra. *Didáctica Geográfica, 19*(2018), 47–76. https://www.age-geografia.es/didacticageografica/index.php/didacticageografica/article/view/416/389

Cobas, I. (2016). El concepto de paisaje cultural como recurso para la educación patrimonial en la educación secundaria. *Cadernos de Arqueoloxía e Patrimonio, 37*, 60. http://digital.csic.es/handle/10261/134694

Cohen, L., Manion, L., & Morrison, K. (2011). Surveys, longitudinal, cross-sectional and trend studies. In *Research Methods in Education* (6th ed., pp. 205–225). Routledge Falmer.

Council of Europe. (2000). *European Landscape Convention*. https://rm.coe.int/1680080621

Cuenca, J. M., Estepa, J., & Martín, M. (2015). La enseñanza de las sociedades actuales e históricas a través del patrimonio. *Íber. Didáctica de Las Ciencias Sociales. Geografía e Historia, 79*, 33–40.

DeSeCo. (2002). *Definition and selection of competences (DeSeCo): Theoretical and conceptual foundations. Strategy paper.* http://deseco.ch/bfs/deseco/en/index/02.parsys.34116.downloadList.87902

Ehrlich, T. (1996). Foreword. In B. Jacoby & ... (Eds.), *Service-Learning in Higher Education: Concepts and Practices* (pp. xi–xvi). Jossey-Bass.

Environment, U. N. (2019). *Global Environment Outlook - GEO6: healthy Planet.* Healthy People., doi:10.1017/9781108627146

Fernández, V. (2005). Finalidades del patrimonio en la educación. *Revista de Investigación En La Escuela, 5*, 7–18. http://www.investigacionenlaescuela.es/articulos/56/R56_1.pdf

Fraile, P., Sánchez, D., & Medina, B. (2016). El uso de la fotografía como recurso docente para la interpretación del territorio en el aula: las potencialidades de Geophotopedia. *Didáctica Geográfica, 17*, 63–78. https://www.age-geografia.es/didacticageografica/index.php/didacticageografica/article/view/376/349

Fuertes, C. (2014). ¿La historia oral en las aulas de infantil y primaria? Difusión, aportaciones y propuestas de aplicación. *CLÍO. History and Teaching, 10.* http://clio.rediris.es/n40/articulos/Fuertes2014.pdf

García, A., Martín, D. A., Costafreda, J. L., Nuñez, E., & Rodríguez, J. A. (2019). Aprendizaje-Servicio (ApS) como metodología para alcanzar los Objetivos de Desarrollo Sostenible (ODS). *V Congreso Internacional Sobre Aprendizaje, Innovación y Competitividad*, 441–446.

García de la Vega, A. (2004). El itinerario geográfico como recurso didáctico para la valoración del paisaje. *Didáctica Geográfica, 6*, 79–95. http://www.didacticageografica.es/index.php/didacticageografica/article/view/182/173

García de la Vega, A. (2012). El aprendizaje basado en problemas en los itinerarios vinculados al patrimonio. *Educación y Futuro, 27*(2012), 155–175. https://cesdonbosco.com/documentos/revistaeyf/EYF_27.pdf

González, M. E. (2012). *Las Infraestructuras de Datos Espaciales (IDE) como un recurso educativo TIC.* https://oa.upm.es/11015/1/MARIA_ESTER_GONZALEZ.pdf

Górka, A. (2020). Landscape perception and the teaching of it in Poland. *World Transactions on Engineering and Technology Education, 18*(2), 123–128. http://www.wiete.com.au/journals/WTE&TE/Pages/Vol.18,No.2 (2020)/07-Gorka-A.pdf%0A

Hume, T., & Barry, J. (2015). Environmental Education and Education for Sustainable Development. In J. D. Wright (Ed.), *International Encyclopedia of the Social & Behavioral Sciences* (Vol. 7, pp. 733–739). Elsevier. doi:10.1016/B978-0-08-097086-8.91081-X

Le Moigne, J., Grubb, T. G., & Milner, B. C. (2013). IMAGESEER: NASA IMAGEs for Science, Education, Experimentation and Research. *IEEE Geoscience and Remote Sensing Magazine*, *1*(1), 44–58. doi:10.1109/MGRS.2013.2244694

Leopold, L. B., Clarke, F. E., Hanshaw, B. B., & Balsley, J. R. (1971). A procedure for evaluating environmental impact. Geological survey circular. doi:10.3133/cir645

Liceras, A. (2013). Didáctica del paisaje. Lo que es, lo que representa, cómo se vive. *Íber. Didáctica de Las Ciencias Sociales. Geografía e Historia*, *74*, 85–93.

Maderuelo, J. (2005). La definición de paisaje. In *El paisaje. Génesis de un concepto* (pp. 15–40). Adaba Editores.

Manzanares, J. A., & Quintana, S. M. (2019). El uso de los itinerarios didácticos (SIG) en la educación ambiental. *Enseñanza de Las Ciencias: Revista de Investigación y Experinecias Didácticas*, *37*(2), 173–188.

Marín-Cepeda, S., & Fontal, O. (2020). Percepciones de estudiantes de Educación Secundaria Obligatoria en torno al Patrimonio. *Arte, Individuo y Sociedad, 32*(4), 917–933. https://revistas.ucm.es/index.php/ARIS/article/view/64657/4564456554158

Martinich, J. A., Solarz, S. L., & Lyons, J. R. (2006). Preparing students for conservation careers through project-based learning. *Conservation Biology*, *20*(6), 1579–1583. doi:10.1111/j.1523-1739.2006.00569.x PMID:17181792

Meadows, M. E. (2020). Geography Education for Sustainable Development. *Geography and Sustainability, 1*(1), 88–92.

Melgar, F., & Silvio, D. (2011). Salir del aula. Aprender de otros contextos: patrimonio natural, museos e Internet. *Revista Eureka Sobre Enseñanza y Divulgación de Las Ciencias, 8*(3), 323–333. http://revistas.uca.es/index.php/eureka/article/view/2720/2368

Mitchell, D. (2007). Muerte entre la abundancia: los paisajes como sistemas de reproducción social. In J. Nogué (Ed.), *La construcción social del paisaje* (pp. 85–110). Biblioteca Nueva.

Morón, H., & Morón, M. C. (2012). Paisaje y patrimonio como medio para enseñanza y aprendizaje de la geografía. *I Congreso Internacional "El Patrimonio Cultural y Natural Como Motor de Desarrollo: Investigación e Innovación,"* 1670–1681. https://dialnet.unirioja.es/servlet/articulo?codigo=4160136

Moseley, C. (Ed.). (2010). *Atlas of the World's Language in Danger* (3rd ed.). UNESCO Publishing. http://www.unesco.org/languages-atlas/en/atlasmap.html

Muñoz, F. (2007). Paisajes aterritoriales, paisajes en huelga. In J. Nogué (Ed.), *La construcción social del paisaje* (pp. 293–323). Biblioteca Nueva.

Orte, D. (2007). *Territorio de abejas, paisaje de hombres*. Imanat Documentales.

Ortega, N. (2010). Paisaje e identidad en la cultura española moderna. In E. Martínez de Pisón & N. Ortega (Eds.), *El paisaje: valores e identidades* (pp. 47–68). Universidad Autónoma de Madrid - Fundación Duques de Soria. https://libros.uam.es/uam/catalog/view/5/6/25

Ozdamli, F., & Asiksoy, G. (2016). Flipped classroom approach. *World Journal on Educational Technology: Current Issues*, *8*(2), 98–105.

Pérez, M., & Morón, H. (2016). El itinerario didáctico como herramienta para la activación patrimonial: una experiencia docente desde el parque de María Luisa. *Campo Abierto, 35*(2), 83–94. https://mascvuex.unex.es/revistas/index.php/campoabierto/article/view/2750/pdf_29

Pope, J., & Morrison-Saunders, A. (2018). Teaching impact assessment. In *Special Publication Series, 10*. International Association for Impact Assessment.

Puig, J., & Casas, M. (2017). El impacto ambiental: Un despertar ético valioso para la educación. *Teoría de la Educación*, *29*(1), 101–128. doi:10.14201/teoredu291101128

Puig, J., & Villarroya, A. (2013). Ecological quality loss and damage compensation in estuaries: Clues from a lawsuit in the Basque Country, Spain. *Ocean and Coastal Management*, *71*, 46–51. doi:10.1016/j.ocecoaman.2012.09.005

Rebolloso, R. (2008). El conocimiento en tiempos de globalizacion. *Ingenierías, 11*(38), 24–32. http://ingenierias2.uanl.mx/38/38_elconocimiento.pdf

Sánchez, L. E., & Morrison-Saunders, A. (2010). Teaching impact assessment: Results of an international survey. *Impact Assessment and Project Appraisal*, *28*(3), 245–250. doi:10.3152/146155110X12791029734641

Tourón, J., & Santiago, R. (2015). El modelo Flipped Learning y el desarrollo del talento en la escuela. *Review of Education, 368*, 196–231. doi:0.4438/1988-592X-RE-2015-368-288

UN. (1987). *Report of the World Commission on Environment and Development: Our Common Future*. http://www.un.org/en/ga/search/view_doc.asp?symbol=A/42/427

UN. (2015). *Transforming our world: the 2030 Agenda for Sustainable Development*. https://undocs.org/A/RES/70/1

UN Development Programme. (2020). *Human Development Report 2020 - The next frontier: Human development and the Anthropocene*. https://hdr.undp.org/sites/default/files/hdr2020.pdf

UNESCO. (1972). *Convention concerning the protection of the World Cultural and Natural Heritage*. https://whc.unesco.org/archive/convention-en.pdf

UNESCO. (2006). *Framework for the United Nations Decade of Education for Sustainable Development. International Implementation Scheme.* https://unesdoc.unesco.org/images/0014/001486/148650E.pdf

UNESCO. (2021). *Operational Guidelines for the Implementation of the World Heritage Convention*. https://whc.unesco.org/document/190976

Valero, J. R. (2016). La fuente oral en la enseñanz a y la investigación geográficas. In L. Alanís, J. Almuedo, G. De Oliveira, R. Iglesias, & B. Pedregal (Eds.), *Nativos digitales y Geografía en el siglo XXI: educación geográfica y estilos de aprendizaje* (pp. 65–79). Asociación de Geógrafos Españoles. Grupo de Didáctica de la Geografía - Universidad Pablo de Olavide - Universidad de Alicante.

Villarroya, A., & Larrañeta, M. (2019). La regeneración de un bosque en el campus de la Universidad de Navarra. In R. Española de Aprendizaje-Servicio (Ed.), XII Encuentro estatal de Aprendizaje-Servicio - I Jornada Internumiversitaria: "El ApS en las universidades navarras. Academic Press.

Wiegand, P. (2001). Geographical information systems (GIS) in education. *International Research in Geographical and Environmental Education, 10*(1), 68–71.

Yli-Panula, E., Jeronen, E., & Lemmetty, P. (2020). Teaching and Learning Methods in Geography Promoting Sustainability. In Education Sciences (Vol. 10, Issue 1). doi:10.3390/educsci10010005

Zheng, B., Zhang, Y., & Chen, J. (2011). Preference to home landscape: Wildness or neatness? *Landscape and Urban Planning, 99*(1), 1–8. doi:10.1016/j.landurbplan.2010.08.006

ADDITIONAL READING

Brabdt, C. B. (2013). Landscapes as contexts for learning. In R. B. Stevenson, M. Brody, J. Dillon, & A. E. J. Wals (Eds.), *International Handbook of Research on Environmental Education* (pp. 275–283). Routledge. doi:10.4324/9780203813331-44

Council of Europe. (2009). *5th Council of Europe Conference on the European Landscape Convention. Education on Landscape for Children.* https://rm.coe.int/16806b081b

Mendoza, R., Baldiris, S., & Fabregat, R. (2015). Framework to Heritage Education Using Emerging Technologies. *Procedia Computer Science*, *75*, 239–249. doi:10.1016/j.procs.2015.12.244

Pateraki, I., & Scimeca, S. (2018). *Learning from the past, designing our future: Europe's cultural heritage through eTwinning*. https://www.etwinning.net/downloads/Online – eTwinning publication 24.09.2018.pdf

Riesco-Chueca, P., & Gómez-Zotano, J. (2013). Landscape Fieldwork: Scientific, Educational and Awareness-Raising Requirements in the Context of the European Landscape Convention. *Landscape Research*, *38*(6), 695–706. doi:10.1080/01426397.2012.716028

UNESCO. (2014). *Aichi-Nagoya Declaration on Education for Sustainable Development.* https://unesdoc.unesco.org/images/0023/002310/231074e.pdf

UNESCO. (2015). *UNESCO science report: towards 2030.* https://unesdoc.unesco.org/ark:/48223/pf0000235406

UNESCO. (2021). *World Heritage Education Programme.* https://whc.unesco.org/en/wheducation/

KEY TERMS AND DEFINITIONS

Didactic Resources: Any material or tool used to foster the teaching-learning process.

Didactic Strategies: A set of planned actions or steps aimed at achieving specific didactic goals. They are based on the use of different didactic resources and methods.

Education for Sustainability: An educational approach that aims to foster the social commitment to build a sustainable future. It encompasses actions in formal, non-formal, and informal educational contexts.

Environment: The result of the interaction between nature (natural environment) and humanity (human environment).

Globalization: A political, economic, social, technological, and cultural phenomenon based on the interconnection increase worldwide.

Heritage: The set of goods that people deem valuable (from a historical, anthropological, scientific, or aesthetic point of view) and to which they attribute a sense of identity

Landscape: The result of the interaction of natural and human elements in the land under the look of an observer.

Service Learning: An educational approach that combines the learning of contents in the classroom and their application to real-life situations. The students provide community service as a part of their learning process.

Chapter 7
Identifying Challenges in Openstreetmap That Potentially Prevent New Users From Contributing:
Implications for Addressing OSM in Education

Robert Vogler
https://orcid.org/0000-0003-0249-9549
Paris-Lodron-University of Salzburg, Austria

Sabine Hennig
Paris-Lodron-University of Salzburg, Austria

ABSTRACT

OpenStreetMap (OSM) is known as the most successful application of VGI. The fact that it is based on user-generated content often leads to the common assumption that OSM data is neutral. These assumptions are legitimate, but the vast majority of OSM contributors are middle-aged, white, male users. Accordingly, OSM data represents the worldview of a delimited social group and therefore lags behind its actual potential regarding the richness and quality of data with regard to diversity. To overcome this, new and diverse user groups have to be recruited. It is therefore necessary to identify the challenges and barriers specific new user groups face. This chapter elucidates these mechanisms based on literature. Subsequently, the authors present a study with teacher training students that identifies challenges of new OSM users. The results indicate that there are three dimensions of potential barriers: the complexity of OSM, the user experience in OSM, and the user interface of OSM.

DOI: 10.4018/978-1-7998-9598-5.ch007

1.INTRODUCTION AND BACKGROUND: OSM AND ITS COMMUNITY

Web maps and web mapping applications are omnipresent and therefore a crucial part of the digitized world in the 21st century. Caused by the rapid technological advances, the use of geospatial applications dealing with digital spatial data have become an every-day technology (Thielmann, van der Velden, Fischer, & Vogler, 2012). Due to massive improvements in usability and the broad availability of necessary soft- and hardware, not only the use but also the production of spatial data has changed over the last decade. Besides automated data production mechanisms, especially crowdsourcing mechanisms known from the web 2.0 and turning passive users into active "produsers" (Bruns, 2008) changed the domain of geoinformation. Terms such as volunteered geographic information (VGI) (Michael F. Goodchild, 2007) and neogeography (Muki Haklay, Singleton, & Parker, 2008; Turner, 2006) evolved and are the blueprints for the new mechanisms in the production of geographic data based on bottom-up volunteered engagement (in contrast to top-down approaches by traditional, mainly state owned cartography authorities (M. F. Goodchild, 2008)). This development provides many advantages such as up-to-date data (due to flexibility) and first of all the openness and open access use (due to the crowdsourcing framework).

The world's largest collection and therefore the most successful initiative of crowdsourced open access spatial data is OpenStreetMap (OSM). Launched in 2004, OSM serves as a platform to produce user generated spatial data following the crowdsourcing approach and aims at involving the general public in data collection to close existing spatial data gaps. As a consequence, the frontend of OSM is a free editable and therefore constantly changing map of the world (Forghani & Delavar, 2014). As. Herfort, Eckle, de Albuquerque, & Zipf (Herfort, Eckle, Albuquerque, & Zipf, 2015) discuss, OSM data is a promising alternative for obtaining and maintaining administrative and authoritative data. Moreover, OSM geometric data quality is comparable to both authoritative and professional commercial spatial data in many cases since many years (see e.g. C.C. Fonte, basin, See, Foody, & Estima, 2015; Mordechai Haklay, 2010; Mahabir, Stefanidis, Croitoru, Crooks, & Agouris, 2017).

Exceeding the OSM map itself, OSM data is used for a huge variety of applications and domains, e.g. urban- and transportation planning (Tarapiah & Atalla, 2015), navigation- and routing systems (Amirian, Basiri, Gales, Winstanley, & McDonald, 2015), pedestrian navigation (Rousell & Zipf, 2017), user-centered applications for specific target groups (see e.g.Arsanjani, Zipf, Mooney, & Helbich, 2015) and even land cover mapping (C. C. Fonte et al., 2017), disaster management (Neis & Zielstra, 2014) or population distribution modelling (Rizqihandari & Indratmoko, 2016). While some of these domains primarily focus on geometry data, others specifically aim at the availability of detailed attribute data collected and provided by the OSM community. Hence, it is interesting to investigate the OSM community itself. Who are the main contributors to OSM? Does the composition of the community affect the type of data and its quality?

In general, everybody with internet access can contribute to OSM. However, just the possibility of open participation does not necessarily imply that OSM data is based on contributions of a broad public. Due to the structure of the OSM account system and the possibilities to participate anonymously (e.g. with anonymous avatar accounts) it is quite challenging to get real picture of the social structure of the OSM community. Nevertheless, there are several studies which at least deliver an idea, who is actively contributes to crowdsourced cartographic information (especially OSM), which motivations stand behind the participation and what kind of problems arise (e.g. (Brown & Kyttä, 2014; Budhathoki & Haythornthwaite, 2013; Das, Hecht, Gergle, & Assoc Comp, 2019; Steinmann, Häusler, Klettner, Schmidt, & Lin, 2013; Stephens, 2013; Vrbik, 2016)). Summing up these insights there are some characteristic common threads:

- The number of actually registered users is fewer than expected.
- The majority of registered users never mapped or stopped mapping before contributing the data.
- In conclusion, just a very small part of users (<10%) is actually mapping and thus contributing.
- The group of people contributing to OSM is socio-structurally very homogeneous:
    - the very vast majority of contributors is male
    - the vast majority of contributors is well educated (college degree or higher)
    - the majority of contributors is between age 20 and 50
    - the majority of contributors has a stable employment and income

The assumption that the crowdsourcing background of OSM ensures a broad and heterogeneous societal basement of its data, has to be disproved. Most of the OSM data is based on the contributions of a very small and very specific group in the context of the general public, namely: well-educated, well-situated middle-aged male users. Interestingly, this diagnosis is also true for other (non-spatial) crowdsourcing platforms, first of all Wikipedia (Ford & Wajcman, 2017), but especially in the case of OSM the numbers are dramatically imbalanced, with female contributors representing as little as 3-4% of the OSM community (Schmidt, Klettner, & Steinmann, 2013; Steinmann et al., 2013).

2. UNINTENTIONAL EXCLUSION MECHANISMS IN THE OSM COMMUNITY AND DATA

Having these findings in mind, the question arises if there is any direct link from the homogeneity of contributors to the quality and the structure of the OSM data. As "Critical Cartography" (Crampton, 2001; Crampton & Krygier, 2005) showed, every map is attached with intentions and further on a catalyst of power regarding the map authors' (e.g. professional cartographers, authorities or even lay mappers) interpretational sovereignty of spatial meaning (e.g. (D Wood, 1993; D. Wood, 2010)) and therefore the social construction of space through action (Werlen, 1993).

At first sight, OSM seems to be an exception from this effect: the exclusiveness of the interpretational sovereignty of spatial meaning seems to be mitigated by the crowdsourcing background itself (bottom-up vs. top-down). Investigations show that the main motives of OSM contributors root in an altruistic nature to provide a free digital map (Budhathoki & Haythornthwaite, 2013). It could therefore be assumed that OSM contributors have no intentions to include own interests in the map but 'just' want to help mapping 'ground truth'. This assumption may be reasonable when just focusing information like geometry data, land cover classes or physical objects. But OSM delivers more than that. One major benefit beyond the map itself is the attached information to each object, i.e. the attribute data (which is e.g. the basis for several third party applications using OSM data). In consequence, the semantic quality of the data is seen as an important issue as well (Vandecasteele & Devillers, 2015).

The attribute data in OSM is described with tags. Each tag consists of a key and a value (form "key=value"; e.g. "amenity=parking"). The OSM community agrees on certain key and value combinations regarding tags. These serve as informal standards. So-called primary tags are used to describe what the feature is. They define the feature type. Optional tags (secondary tags), describe further properties (for example, primary tag: "amenity=parking"; secondary tag: "capacity=200"). Since OSM uses a free tagging system, this permits the map/database to include an unlimited number of attributes to describe features.

The attached information via tags to each object is one crucial benefit of OSM data. This potential heterogeneity of attribute data also is one major advantage to authoritative or commercial data, which is characterized by strict specifications (Loidl, Krampe, Zagel, & Pucher, 2014). This is especially true for third party (target group oriented) applications using OSM data. One example is the application WheelMap (https://wheelmap.org/map), which provides a map visualizing public places with regard to accessibility (the information in this map is directly rooted in OSM data, querying public amenities based the secondary tag "wheelchair=yes/

no"). The success and the potential of such applications consequently rely on the variety and the quality of the attribute data in OSM.

However, when it comes to attaching attribute data to an object in OSM, the contributors' social background unavoidably influences his or her mapping action based on subjective interpretations and priorities, even if unintentionally so. In case of WheelMap: If data on accessibility is not of interest and therefore has no priority for a contributor, then he or she will comprehensibly not provide such kind of information. This would not cause any problems in single cases. But in combination with the very homogeneous socio-economic structure of the OSM community outlined above, such effects have consequences for the amount and quality of the OSM attribute data, as it is based on the worldview and the (non)priorities of a very small part of a heterogeneous society (in which e.g. wheelchair users are under-represented). Having a closer look on the above-mentioned WheelMap application, it impressively illustrates this phenomenon: the majority of the mapped features are grey, which means "no information". Of course, the example of data on accessibility also addresses just a small target group of society. But it illustrates the (unintentional) exclusion mechanisms caused by the (non)priorities of a homogeneous community and can be transferred to several other groups.

Traces of e.g. the under-representation of women (Budhathoki & Haythornthwaite, 2013; Das et al., 2019; Steinmann et al., 2013) in the contributing OSM community also can be found in the OSM data and its representation in the standard renderer Mapnik (https://wiki.openstreetmap.org/wiki/Mapnik). There are e.g. separate symbols for each cafes (N=465.357), bars (N=199.091), pubs (N=174.604) and nightclubs (N=21.671), which seem to be considered as important features by the majority of the OSM community (at least important enough to be each respectively represented with an own symbol in the Mapnik visualization). On the other hand, there are no symbols for e.g. schools (N=1.166.661), kindergartens (N=269.020) or social facilities (N=119.149) (numbers retrieved from https://taginfo.openstreetmap.org; February, 11th 2022). Another effect of the imbalance of priorities caused by the homogeneous structure of the contributing community has been (very controversially) discussed by Stephens (Stephens, 2013) on the example of secondary tags. This effect also has been discussed as "Self-Focus Bias" by Das et al. (Das et al., 2019). In sum, there is a relationship between socio-economic characteristics of the contributing users and their contributions (and in consequence the data itself) (Arsanjani & Bakillah, 2015; Bittner, 2017; Das et al., 2019; Duggan, 2019; Glasze & Perkins, 2015; Perkins, 2014).

Comparable effects show up having a look behind the mechanisms of a) setting and changing the informal standards of the tagging system or b) setting and changing the symbols of the most common OSM map renderer Mapnik. In both cases, everybody can provide a proposal to add or change the existing structures, but the

OSM community decides whether to accept the offer or not. In consequence, the homogeneous structure of the community accelerates the exclusion mechanisms based on the priorities of well-educated, well-situated middle-aged male users, even if this is not their primary intention. This also has been discussed in detail by Stephens (Stephens, 2013).

These outlined symptoms of exclusion mechanisms in OSM may not be critical for mapping the geometry and locations of features but at least cause some problems regarding the equality of the OSM attribute data (Das et al., 2019). This leads to the effect that OSM lags behind its actual potential based on crowdsourcing in the real sense of the word. Such effects are known in other crowdsourcing platforms as well. Wikipedia, for instance, also shows a significant imbalance in the contributing community regarding gender (Ford & Wajcman, 2017) and socio-cultural indicators (Panciera, Halfaker, & Terveen, 2009). It is also known that this has effects on the content generated, in this case: Wikipedia articles (Bear & Collier, 2016; Hargittai & Shaw, 2015; Panciera et al., 2009; Pfeil, Zaphiris, & Ang, 2006). In the case of OSM, this effect is even more dramatic due to its structure of information, which is not based on language (like Wikipedia articles) but on a formalized tagging system.

The solution to overcome these circumstances seems simple at first sight: The OSM community has to get more heterogeneous, especially by recruiting new community members, who are not male or younger than 20 or older than 50 or not highly educated. To answer the question why these groups are so dramatically under-represented in OSM, many different research initiatives are needed, as Neis & Zielstra (Neis & Zielstra, 2014) showed. One crucial question is to identify challenges new users face, which will be addressed in the following section.

OSM as such, its social and technological mechanisms and especially these findings have a very important means and show a lot of potential for (digital) (geography) education. Therefore, the study described and discussed in the following, is especially interesting for the (geography) education domain because its empirical basis is rooted in data collected from geography teacher training students (see section 3.2 for further details).

3. CHALLENGES AND PROBLEMS NEW USERS FACE WHEN TRYING TO CONTRIBUTE TO OSM

3.1. Research Question and Related Research

Among others, one question to overcome the situation sketched in the previous section is to identify the reasons, which discourage new users and user groups from joining the OSM community. One crucial core component of this issue and the central research question of this paper is to identify, examine and classify the concrete challenges new users face when trying to contribute to OSM.

There has occasionally been research related to this topic. While many studies had a deeper look at the motivations of (non)contributing (e.g. Budhathoki & Haythornthwaite, 2013; Hennig, 2020; Lin, 2011; Neis & Zielstra, 2014; Yang, Fan, & Jing, 2016), Schmidt et al. were the first to observe the general barriers for not contributing to VGI and OSM. In a survey-based study, they identified different reasons for not contributing to OSM. According to their results, the most prevalent barriers reported by more than a third of all survey respondents are: a) the time-consuming aspect of mapping, b) the completeness of maps, c) forgetting and d) the complexity of the editing process (Schmidt et al., 2013). Especially the issue of complexity appears to be critical for the topic addressed in this paper and has been pointed out in other contexts, as well. Carraro and Wissink, for instance, report linguistic and technical barriers of OSM in the framework of participation and marginalization in Jerusalem (Carraro & Wissink, 2018). However, while these insights are a good starting point for further research, such results are too general to answer the question of actual challenges new OSM users face.

On the contrary, several studies deliver deeper insights in concrete usability issues in the OSM editing process. Some compare the user interfaces of different VGI platforms including OSM (e.g. Picanco & Delazari, 2016), others directly focus on the usability of OSM editing tools. Jones and Weber presented a usability test of the Potlach 2 online editor and identified several usability problems like difficulties with the interface elements, user interactions or information provision (Jones & Weber, 2012). Behrens et al. investigated the usability of the other online editor directly embedded in the OSM website, namely iD. They identified concrete usability issues and concluded with several possible improvements, like more feedback, a better clarity and consistency of the editing process and options or more help options (Behrens, van Elzakker, & Schmidt, 2015). Such usability studies deliver rich information for further OSM improvement and usability research. But they are too specific to gain knowledge on concrete barriers and challenges new users face when trying to start to contribute to OSM, because the technical part of the editing process is certainly one core component, but not the only dimension to be observed.

In an exploratory micro-study we initially started to fill this gap (Vogler, Hennig, & Ferber, 2017). Based on documented observations made in two courses aiming at OSM basic knowledge and skills in the framework of a senior citizen education initiative, we recognized and described five concrete problems to be considered as crucial barriers for new OSM users (see Table 1).

With this exploratory approach, we initially started to identify the challenges for new OSM users, but the study has its limitations: (1) The observations were made in courses explicitly designed to serve as OSM introduction, where participants

Table 1. Barriers[1] for new OSM users on the example of senior citizens

Issues identified	Brief description
signup/ login	Based on several statements of course participants, the signup was seen as a first barrier: creating an online account is perceived critical due to personal data privacy. It is obvious that an account is required to contribute to OSM but this barrier should be taken into account when aiming to recruit new users. It is recommended to offensively provide short and clear information a) why an account is necessary and b) that OSM is a nonprofit organization and will not valorize or spread personal data.
GUI and usability	Several usability issues in the online editing tools iD and Potlach 2 have been observed, e.g.: the primary tag suggestions in iD, different feature symbols used in the editors and in Mapnik, the missing of an "stop editing" button. In conclusion, there should be some small interface edits, like additional buttons and/or small popups with short explanations, primary addressing new and unexperienced OSM users.
tag structure of the attribute data	The participants stated that the logic behind the tags is very confusing and challenging for lay people regarding databases and/or geoinformatics. Especially the language behind is the key factor for a better understanding and usability. Even if the needed information is very well documented in the OSM wiki, it is recommended to additionally provide the key information also during the editing process (e.g. via popups). It is also helpful to use a language suitable to lay people and beginners avoiding (or at least explaining) technical terms like key, value, primary and secondary tag.
Mapnik rendering	Immediately after editing or adding a feature, the participants expected it to be directly visible in the OSM map based on the Mapnik renderer. This is not the case, because the automatic rendering itself takes some time and not every feature is visualized by Mapnik. This circumstance caused uncertainty: most participants thought they did something wrong. To avoid such misunderstandings, it would help to offensively provide a direct feedback after the editing procedure a) confirming the success of editing and especially b) briefly explaining rendering background.
perceiving the community as a closed system	A very crucial barrier for the course participants was the OSM community itself, as they perceived it as an exclusive and closed system and reported difficulties to get in touch with other users (this is in line with findings on the very complex but somehow closed collaboration networks in the OSM community, see e.g. (Truong, de Runz, & Touya, 2019)). Of course, this is not a problem that can be solved structurally. But it is an issue established community members should be aware of and take into account

[1]documented as an exploratory micro study in two OSM courses for senior citizens (see (Vogler et al., 2017))

intentionally decided to take part and therefore can be considered as exceptionally motivated. (2) The courses were designated for senior citizens, so the findings have to be interpreted in the context that senior citizens are a very specific target group regarding the acceptance of computer-based information (Agudo, Pascual, & Fombona, 2012; Hernandez-Encuentra, Pousada, & Gomez-Zuniga, 2009; Phang et al., 2006). (3) The exploratory approach was based on documented observations and therefore lacks a deeper empirical fundament.

To sum up: While general survey-based empirical data are not able to show concrete barriers for new OSM users in detail and usability studies per definition just aim at the tool interaction process, additional rule-guided empirical research is needed to identify the concrete challenges new OSM users have to face and which potentially prevent them from contributing.

3.2 Data Collection

To examine the challenges and barriers for novice OSM users and potential contributors, it is necessary to gather authentic data. This is quite challenging due to at least two crucial restrictions. (1) It is hard to reach this very specific user group "new OSM users" because there is no stringent communication channel. For instance, there is good reason to posit that novice users are not familiar with the OSM community and therefore cannot be reached via OSM discussion forum platforms. In consequence, such data and information only can be gathered in an experimental setting with a fixed subject group. (2) It is crucial how the data on challenges and potential barriers is gathered and processed. While quantitative surveys are very fruitful to deliver information about concrete pre-defined issues on a large scale (e.g. assessing the difficulty of a concrete task), they are inadequate to survey open items. Per definition, items in a survey generate a framing effect (Stalans, 2012) and therefore prime respondents' answers. Directly asking respondents for concrete experienced challenges would imply that there are such and therefore distort the information collected.

To minimize such framing and priming effects and to get authentic material, it is necessary to very openly ask the respondents for general experiences made in the first OSM contribution and then analyze the data collected with regard to potentially reported problems. In order to gather authentic data, the study described and discussed in the following addresses the two above-mentioned restrictions by specifically analyzing 41 feedback documents written by university students on their first OSM experiences.

These documents were collected in the framework of two university courses in autumn 2020. The courses were embedded in the basic curriculum (Bachelor) of the initial teacher-training program at the University of Salzburg (Austria) with regard to the subject of "geography and economics education", i.e.: the course participants were prospective high school teachers. The courses aimed at geomedia didactics and education and introduced geomedia use in the classroom. One course module was a short overview of OSM and its potential. All participants were OSM novices, i.e. they never contributed to OSM before. After introducing OSM as a VGI platform and compactly demonstrating the possibilities to contribute online (iD and Potlach 2), the students had the task to contribute at least one OSM edit at home. As a proof of fulfilling this task, they had to post the changelog-ID to an learning management system.

After that, students were asked to write and submit a short report (approximately one page), in which they had to reflect the experiences made during their mapping contribution. The concrete task was "Write a short reflection (approx. 1 page) about the experiences made during your first contribution to Openstreetmap! Please do not refer to the importance and potential of OSM itself, but only to working with OSM (mapping/ editing)!" (translation, the original task was given in German language). This approach delivered 41 written texts each reflecting the procedure of contributing to OSM for the first time. 16 students were male, 25 female. The students' age range was 20-28 years.

3.3 Methodology

As outlined above, this kind of data (i.e. freely written reflections on experiences) shows a much higher level of authenticity than directly surveyed problems and is highly valuable to examine potential barriers or concretely mentioned challenges. Due to the nature of the material (qualitative data, i.e. text), a suitable method is needed to analyze these texts regarding the research question. One widely recognized and accepted method to extract concrete information from qualitative material in a rule-guided and therefore objective reliable way, is Qualitative Content Analysis (QCA) (Mayring, 2014). QCA provides a bundle of techniques for systematic text analysis, while its main idea is to preserve the advantages of quantitative content analysis and to transfer them to qualitative steps of analysis (Mayring, 2000). One QCA technique is the so-called inductive category formation. Its aim is to summarize qualitative categories with a strict focus on a pre-defined question directly from the material and not from theoretical concerns and therefore to authentically describe the material without biases from preconceptions of the analyzing researcher (Mayring, 2014). Despite its openness, inductive category formation QCA follows a very strict process model and a coding ruleset that has to be pre-defined before analyzing the material to ensure a rule-guided, reliable and objective analysis.

Hence, the approach is considered a very fruitful method to analyze the 41 reflection texts gathered in the above-mentioned setting. Thus, it has been applied for the further analysis based on the pre-defined rule-set listed in Table 2.

Based on the rules listed in Table 2, the material was analyzed line by line. The first time, a text phrase fitting the category definition was found, the concrete phrase was marked and a category was constructed. A term, which characterizes the passage based on the defined level of abstraction, served as category label. The next time a text passage fitting the category definition was found, it was checked, whether it falls under the previous category. If so, it was subsumed under this category. If not, a new category was formulated. Following this procedure, the whole material (i.e.

Table 2. Pre-defined rules for material analysis applying inductive category formation QCA[2]

Rule type	Rule definition	Comment
Research question	Which elements of the OSM contribution process are being perceived as challenges?	This is the research question to be answered while analyzing the material.
Coding unit	Phrase or clause (word sequences)	This defines the smallest component of the material that can be coded.
Context unit	The respective reflection report	This defines the background for the respective coding decision.
Recording unit	All documents (i.e. all reflection reports)	This defines the text portion confronted with the developed category system.
Category definition	OSM elements, which are evaluated negatively and/or perceived as challenge; clearly denominated issues; issues directly derived from the context of the respective text passage exclusion: statements that are verifiably wrong	This category definition sets the criteria, which have to be met to determine the relevant material from the text
Level of abstraction	Concretely denominated issues and issues directly derived from the context	This defines how specific (or general) the categories have to be formulated.

[2]following the inductive category formation QCA criteria provided by Mayring (Mayring, 2014).

all 41 reflection reports) was scanned, interpreted and coded and, consequently, the category system with each category labelling one concrete issue was built.

The coding itself was conducted using QCAmap (https://qcamap.org/). QCAmap is a freely available open access online tool to analyze text material, which is specifically designed for the techniques of QCA. Its advantage is the exact suitability to the strict procedures of QCA. Beyond that, the coding results and category counts can be easily exported to Excel files for further processing and analysis.

To illustrate this coding procedure, Table 3 shows some selected examples of phrases in the text and the extracted categories.

3.4 Results

After working through the whole text material and coding each text phrase fitting the rules defined in Table 2, a total of 94 passages in the text were identified and extracted in a category or alternatively subsumed under an existing category. This procedure resulted in 22 different categories, i.e. 22 identified challenges and potential barriers:

- confusion, how to exit the editing mode

Table 3. Selected text phrase examples and extracted categories to illustrate the coding procedure

Text example[3]	Extracted category	Comment
"Without an a priori explanation of the tagging process, it would not have been clear to me how this system works."	Difficulties to understand the tagging system	The text passage states difficulties to understand the tagging system
"During the editing, it was good for me to have received an instruction in advance. Otherwise I would have had great difficulties."	Difficulties to acquire the basics (help of an expert or introduction needed)	The text passage states the need of an expert introduction to understand the basics and predicts problems, if not.
"In addition, exiting the editing mode is inappropriate. I don't think anyone intuitively suspects that the OSM logo ends the editing function."	Confusion, how to exit the editing mode	The text passage states a confusion how to exit the editing mode.
"Because the map does not immediately show what you have produced, you think here that you have made a mistake."	Confusion, because edits appear with a delay (afraid to have made a mistake)	The text passage states the apprehension to have made a mistake while editing, because the map does not immediately show the changes.
"I was a little confused by the tagging scheme and thought that it is almost necessary to take a designated course on the different primary and secondary tags."	Difficulties to understand the tagging system	The text passage states difficulties to understand the tagging system and therefore can be subsumed under this already existing category.
"I just had a look at my edits in OSM again and saw that nothing had changed. That astonishes me a little. The changes I made are not on the map."	Confusion, because edits appear with a delay (afraid to have made a mistake)	The text passage states an irritation because the map does not immediately show the changes. Therefore, it can be subsumed under this already existing category.

[3]Translation; the original text was written in German

- too many details to deal with at the beginning
- difficulties to understand the tagging system
- difficulties to acquire the basics (help of an expert or introduction needed)
- difficulties to acquire the basics (too time consuming)
- slow site navigation in editing mode (performance)
- confusion, because edits appear with a delay (afraid to have made a mistake)
- for beginners very complicated user interface
- different layouts are confusing
- confusion, which editor 'has to' be used
- help page does not open in new browser tab
- confusion how to finish a polygon (double click)
- uncertainty, that and how one have to save edits
- partly no up-to-date orthophotos

- found no mapping instruction
- registration / account necessary
- not very pleasing design
- too many colors in map design
- labels are hard to read
- too many labels in map
- difficulties to find the right tag
- OSM Wiki at the beginning to overloaded

These identified categories demonstrate the diversity regarding the stated challenges and deliver a first insight with regard to the research question. Otherwise, this first overview is initially very confusing on one hand and does not explain any significance of each issue on the other.

Therefore – following the QCA procedure – the next step is to construct main categories on a higher abstraction level subsuming these 22 issues (Mayring, 2014). After a having a deeper look at the identified categories, three main category domains could be identified:

- challenges caused by complexity
- challenges caused by the user experience
- challenges caused by the user interface

After subsuming all 22 categories under these three domains, the respective frequencies can be counted to conclude the significance and importance of each item. In this case, the crucial index is – obviously – not the total amount of each category's mentions, but the number of documents, in which it occurs. Table 4 shows all 22 items subsumed to the three main categories and with their respective frequencies.

This overview delivers a way more nuanced picture of the analysis results and allows to draw some first conclusions. Regarding the numbers, more than 75% of the students reported challenges caused by the complexity of the editing procedure in OSM, more than a third stated challenges caused by the user experience and almost 30% were challenged by the user interface in one way or the other. The item with the largest frequency derives from mentions which stated that it is very difficult to acquire the basic knowledge to contribute to OSM and that there is a need for an introduction provided by an expert. This issue has been reported by more than 50% of the students. Another critical issue is the complexity of the OSM tagging system. 16 of 41 students reported difficulties to understand the it.

Table 4. All identified challenges subsumed to main domains and arranged according to frequency

Category	occurs in % of all documents (N=41)
Challenges caused by complexity	**75,6**
difficulties to acquire the basics (help of an expert or introduction needed)	53,7
difficulties to understand the tagging system	39,0
difficulties to acquire the basics (too time consuming)	14,6
too many details to deal with at the beginning	12,2
difficulties to find the right tag	4,9
OSM Wiki at the beginning to overloaded	2,4
Challenges caused by the user experience	**36,6**
slow site navigation in editing mode (performance)	12,2
confusion, because edits appear delayed (afraid to have made a mistake)	12,2
confusion, which editor 'has to' be used	7,3
confusion how to finish a polygon (double click)	2,4
uncertainty, that and how one have to save edits	2,4
found no mapping instruction	2,4
registration / account necessary	2,4
Challenges caused by the user interface	**29,8**
confusion, how to exit the editing mode	14,6
for beginners very complicated user interface	7,3
different layouts are confusing	2,4
help page does not open in new browser tab	2,4
partly no up-to-date orthophotos	2,4
not very pleasing design	2,4
too many colors in map design	2,4
labels are hard to read	2,4
too many labels in map	2,4

Even if the other issues have significantly lower frequencies, one should put these numbers in context. Having a closer look at the individual documents and the corresponding distribution of the identified main category domains (complexity, user experience and user interface) can help to evaluate this:

- 37 of the 41 students' reports (90,3%) contain at least 1 of the 3 main categories,

- 16 of the 41 students' reports (39,0%) contain at least 2 of the 3 main categories,
- 5 of the 41 students' reports (12,2%) contain all 3 main categories,
- 4 of the 41 students' reports (09,8%) contain no reported challenges

Having in mind that the analyzed material consists of freely written reflection reports and no student was explicitly asked to describe challenges or problems, these numbers are quite impressive. Only four reports did not mention any issue perceived as a potential challenge. On the other hand, slightly more than 90% of all documents reported problems. Due to this outcome, each category listed in Table 4 can and should be taken seriously.

Another insight is reflected in the category frequencies in the context of gender. This is especially of interest because women are dramatically underrepresented in the OSM community as described and discussed in the beginning of this paper. Therefore, Table 5 shows the distribution of the category frequencies divided by gender affiliation. In this context, all categories mentioned just once in total have been excluded because per definition they would have been mentioned in only one gender domain, anyway.

Of course and as it is not the primary goal of qualitative research, these numbers and differences show no statistical significance due to the small sample size of N=41. However, the tendencies in the differences are quite interesting. The only challenge mentioned distinctly more frequent by female students is the difficulty to acquire the basic knowledge in combination with the necessity of an introduction by an expert. On the other hand, male students report way more problems and challenges regarding the user interface of OSM.

3.5 Discussion

Based on the information contained in the data and extracted in the analysis, we were able to gain nuanced knowledge with regard to the research question to identify challenges and potential barriers for new OSM users (see Table 4). These are divided in three sub-domains: complexity, user experience and user interface. The frequency analysis allows to evaluate the significance of each issue identified and extract the most important indications. Each issue mentioned by at least two test subjects is listed in Table 5. In addition, we could identify slight differences with regard to gender: male participants complained more about issues emerging in the context of the user experience and the user interface, female participants more often stated the necessity of an introduction or the need for expert help (see Table 5).

Of course, these results have to be interpreted with regard to the limitations of this study. Due to the sample size of N=41, the numbers obviously have no statistical

Table 5. Gender differences in the category frequencies[4]

Category	**occurs in % of all documents** *male* **(N=16)**	**occurs in % of all documents** *female* **(N=25)**
Challenges caused by complexity	75,00	76,00
difficulties to acquire the basics (help of an expert or introduction needed)	43,75	**60,00**
difficulties to understand the tagging system	**50,00**	32,00
difficulties to acquire the basics (too time consuming)	**25,00**	8,00
too many details to deal with at the beginning	12,50	12,00
difficulties to find the right tag	--	8,00
Challenges caused by the user experience	**43,75**	32,00
slow site navigation in editing mode (performance)	**18,75**	8,00
confusion, because edits appear delayed (afraid to have made a mistake)	12,50	12,00
confusion, which editor 'has to' be used	6,25	8,00
Challenges caused by the user interface	**56,25**	12,00
confusion, how to exit the editing mode	**31,25**	4,00
for beginners very complicated user interface	12,50	4,00

[4]Category frequencies with a double-digit gender difference are highlighted in bold

significance and have to be evaluated carefully. On the other hand, the primary goal of this study was to authentically identify challenges, which occur to new users. To accomplish this, such a qualitative approach has some substantial advantages as discussed in sections 3.2 and 3.3. Nevertheless and even if not statistically significant, the frequencies give a first hint on the importance of each issue.

Another limitation that has to be taken into account is the sample composition of the subject group delivering the data. All reflection reports were written by university students, so it has to be expected that additional issues occur when surveying a different socio-economic target group. However, when having a closer look on the nature of the identified challenges, there is no reason to expect these issues to be irrelevant for other socio-economic groups. To verify this assumption, it is worthwhile to compare the results of the study described and discussed in this paper with the outcomes of the exploratory study referred to in section 3.1 and Table 1. Even if these two studies observed two completely different target groups (university students mainly in their early twenties on one hand and senior citizens on the other), there is a huge overlap in the concrete results. 8 of the 10 most important issues identified in this study surveying university students (Table 5) also were carried out in the exploratory approach observing senior citizens (Vogler et al., 2017). Just the two categories "difficulties to acquire the basics (too time consuming)" and "confusion, which editor 'has to' be used" did not occur in the previous study. This can be reasoned with the specific situation of a very intense interaction between instructors and course participants, which made these two items obsolete in the first exploratory setting.

In this context, the results presented here serve as a basis for further research. There are at least two dimensions to be addressed. (1) The challenges presented in this paper were identified qualitatively, but the QCA techniques applied here do not allow for a deeper understanding of each issue. Therefore, there is a need for follow-up studies using methods for a deeper qualitative analysis (e.g. Grounded Theory (Corbin & Strauss, 1990)) based on semi-structured interviews with novice OSM users or concrete observations specifically focussing the nature of these problems. (2) The issues identified in this study can serve as basis for the item creation in further quantitative research based on surveys to deliver a statistically significant picture of the frequency distribution of each challenge. However, the results presented here deliver both a first concrete insight regarding the research question and a starting point for further research (and in consequence a considerable contribution to the recent state of knowledge).

4 CONCLUSION

This paper initially discussed the exclusiveness of the OSM attribute data reasoned by the socio-economically very homogeneous composition of its contributing community. This has effects on the quantity (Arsanjani & Bakillah, 2015) and especially on the quality and diversity of the data (Glasze & Perkins, 2015; Perkins, 2014; Stephens, 2013). Consequently and despite the already huge success of OSM in general, its data richness lags behind its actual potential. To overcome this, there is a need for a socio-economically more diverse community. With regard to this goal, several issues need to be addressed and knowledge needs to be gained. Among others, one crucial part is to identify concrete challenges new users have to face when trying to contribute to OSM for the first time, because these challenges (when perceived as frustrating experience) are potential barriers, which prevent them to (furtherly) contribute to OSM. This research question was addressed in an empirical study based on 41 qualitatively gathered and analyzed reflection reports of novice OSM users.

Despite the limitations previously discussed, the results of the study presented in this paper deliver first empirically driven insights and therefore allow drawing some first conclusions for measures needed. Many challenges identified (especially the ones subsumed under the category "caused by the user interface", see Table 4) can be minimized by e.g. some minimal adjustments in the design and user interface of

OSM and its integrated editors iD and Potlach 2 as already concluded in the previous exploratory study (Vogler et al., 2017). In addition, some challenges subsumed under the category "caused by user experience" (e.g. confusion, because of edits appearing delayed, confusion, which editor 'has to' be used or confusion how to finish a polygon) can be addressed by delivering small on-site hints via popup or mouse-over tooltip. One crucial issue for novice OSM users is its tagging system. Despite the problems caused by its informal nature in general (like discussed in e.g. Hennig, 2017; Mooney & Corcoran, 2012), results show that many new OSM users have problems to understand the logic behind the tags and/or to find the right tags. This problem – by definition – cannot be solved with simple interface adjustments, because it is rooted in the system itself. Therefore, the tagging system needs to be explained to novice users in a more stringent way. How this can be achieved exactly, must be addressed in further research.

The complexity of the tagging system is directly related to the most important issue detected in this study, i.e. the need for an instruction and/or support by an expert (or experienced OSM user). This issue is crucial from two points of view. (1) It has been stated by more than half of the students and therefore is the most reported concern. (2) This aspect has been reported significantly more often by female students, which should be of particular interest due to the substantial underrepresentation of women in the OSM community discussed above. The core of this issue leads to the conclusion that providing education material and activities (courses, lectures, presentations, e-tutorials etc.) on OSM and its benefits is essential to face this major challenge. Additionally, such measures allow to raise the public's awareness of OSM and to motivate new users and user groups. One domain that can be directly addressed in this respect is secondary school education. This would not just generate a surplus value from an educational point of view, as e.g. Bartoschek & Keßler (Bartoschek & Keßler, 2013) discuss, but also could serve as a powerful multiplier to engage new users and user groups. Another domain, which constantly gained importance in recent years, is informal adult education. Initiatives in this field (like e.g. (Vogler & Hennig, 2013)) are considered as powerful means to provide and distribute the knowledge needed to contribute to OSM.

Beyond secondary education serving as a powerful framework to promote OSM as such, the findings discussed in this paper are also very important for (geography) education from a pedagogical point of view. First of all, OSM has the potential to supplement geography classes by a) teaching the basics and technological backgrounds of digital cartography, b) modern VGI mechanisms and c) identifying and de-construct the exclusion effects described at the beginning of this paper to foster critical thinking and therefore serve a critical cartography approach in school – as e.g. part of concepts such as Spatial Citizenship (Gryl & Jekel, 2012). To show the full potential of OSM in the geography classroom, it is – not surprisingly –

important that the respective teachers are competent in OSM handling. From this point of view, the findings of this paper are even more substantial based on the fact that they are rooted in a study, in which geography teacher training students have been observed and interviewed.

REFERENCES

Agudo, S., Pascual, M. A., & Fombona, J. (2012). Uses of Digital Tools among the Elderly. *Comunicar, 20*(39), 193–201. doi:10.3916/C39-2012-03-10

Amirian, P., Basiri, A., Gales, G., Winstanley, A., & McDonald, J. (2015). The Next Generation of Navigational Services Using OpenStreetMap Data: The Integration of Augmented Reality and Graph Databases. In J. J. Arsanjani, A. Zipf, P. Mooney, & M. Helbich (Eds.), *Openstreetmap in Giscience: Experiences, Research, and Applications* (pp. 211–228). Springer-Verlag Berlin. doi:10.1007/978-3-319-14280-7_11

Arsanjani, J. J., & Bakillah, M. (2015). Understanding the potential relationship between the socio-economic variables and contributions to OpenStreetMap. *International Journal of Digital Earth, 8*(11), 861–876. doi:10.1080/17538947.2014.951081

Arsanjani, J. J., Zipf, A., Mooney, P., & Helbich, M. (Eds.). (2015). *OpenStreetMap in GIScience: Experiences, Research, and Applications*. Springer-Verlag Berlin. doi:10.1007/978-3-319-14280-7

Bartoschek, T., & Keßler, C. (2013). VGI in Education: From K-12 to Graduate Studies. In D. Sui, S. Elwood, & M. Goodchild (Eds.), *Crowdsourcing Geographic Knowledge: Volunteered Geographic Information (VGI) in Theory and Practice* (pp. 341–360). Springer Netherlands. doi:10.1007/978-94-007-4587-2_19

Bear, J. B., & Collier, B. (2016). Where are the Women in Wikipedia? Understanding the Different Psychological Experiences of Men and Women in Wikipedia. *Sex Roles, 74*(5), 254–265. doi:10.100711199-015-0573-y

Behrens, J., van Elzakker, C. P. J. M., & Schmidt, M. (2015). Testing the Usability of OpenStreetMap's iD Tool. *The Cartographic Journal, 52*(2), 177–184. doi:10.1080/00087041.2015.1119457

Bittner, C. (2017). OpenStreetMap in Israel and Palestine - 'Game changer' or reproducer of contested cartographies? *Political Geography, 57*, 34–48. doi:10.1016/j.polgeo.2016.11.010

Brown, G., & Kyttä, M. (2014). Key issues and research priorities for public participation GIS (PPGIS): A synthesis based on empirical research. *Applied Geography (Sevenoaks, England)*, *46*, 122–136. doi:10.1016/j.apgeog.2013.11.004

Bruns, A. (2008). *Blogs, Wikipedia, Second Life, and Beyond: From Production to Produsage*. Peter Lang.

Budhathoki, N. R., & Haythornthwaite, C. (2013). Motivation for Open Collaboration: Crowd and Community Models and the Case of OpenStreetMap. *The American Behavioral Scientist*, *57*(5), 548–575. doi:10.1177/0002764212469364

Carraro, V., & Wissink, B. (2018). Participation and marginality on the geoweb: The politics of non-mapping on OpenStreetMap Jerusalem. *Geoforum*, *90*, 64–73. doi:10.1016/j.geoforum.2018.02.001

Corbin, J., & Strauss, A. (1990). Grounded Theory Research: Procedures, Canons and Evaluative Criteria. *Zeitschrift für Soziologie*, *19*(6), 418–427. doi:10.1515/zfsoz-1990-0602

Crampton, J. W. (2001). Maps as social constructions: Power, communication and visualization. *Progress in Human Geography*, *25*(2), 235–252. doi:10.1191/030913201678580494

Crampton, J. W., & Krygier, J. (2005). An Introduction to Critical Cartography. *Acme-an International E-Journal for Critical Geographies*, *4*(1), 11-33.

Das, M., Hecht, B., Gergle, D., & Assoc Comp, M. (2019). The Gendered Geography of Contributions to OpenStreetMap: Complexities in Self-Focus Bias. *Proceedings of the 2019 Chi Conference on Human Factors in Computing Systems*. 10.1145/3290605.3300793

Duggan, M. (2019). Cultures of Enthusiasm: An Ethnographic Study of Amateur Map-Maker Communities. *Cartographica*, *54*(3), 217–229. doi:10.3138/cart.54.3.2018-0002

Fonte, C. C., Basin, L., See, L., Foody, G., & Estima, J. (2015). *Good practice guidelines for assessing VGI data quality*. Paper presented at the AGILE 2015, Lisbon, Portugal. https://agile-online.org/conference_paper/cds/agile_2015/posters/95/95_Paper_in_PDF.pdf

Fonte, C. C., Patriarca, J. A., Minghini, M., Antoniou, V., See, L., & Brovelli, M. A. (2017). Using OpenStreetMap to Create Land Use and Land Cover Maps: Development of an Application. In Volunteered Geographic Information and the Future of Geospatial Data (pp. 113-137). IGI Global.

Ford, H., & Wajcman, J. (2017). Anyone can edit', not everyone does: Wikipedia's infrastructure and the gender gap. *Social Studies of Science*, *47*(4), 511–527. doi:10.1177/0306312717692172 PMID:28791929

Forghani, M., & Delavar, M. R. (2014). A Quality Study of the OpenStreetMap Dataset for Tehran. *ISPRS International Journal of Geo-Information*, *3*(2), 750–763. doi:10.3390/ijgi3020750

Glasze, G., & Perkins, C. (2015). Social and Political Dimensions of the OpenStreetMap Project: Towards a Critical Geographical Research Agenda. In J. J. Arsanjani, A. Zipf, P. Mooney, & M. Helbich (Eds.), *Openstreetmap in Giscience: Experiences, Research, and Applications* (pp. 143–166). Springer-Verlag Berlin. doi:10.1007/978-3-319-14280-7_8

Goodchild, M. F. (2007). Citizens as sensors: The world of volunteered geography. *GeoJournal*, *69*(4), 211–221. doi:10.100710708-007-9111-y

Goodchild, M. F. (2008). Assertion and authority: the science of user-generated geographic content. In G. Navratil (Ed.), *Proceedings of the Colloquium for Andrew U. Frank's 60th Birthday* (pp. 1-18). Vienna: Department of Geoinformation and Cartography, Vienna University of Technology.

Gryl, I., & Jekel, T. (2012). Re-centring Geoinformation in Secondary Education: Toward a Spatial Citizenship Approach. *Cartographica: The International Journal for Geographic Information and Geovisualization*, *47*(1), 18–28. doi:10.3138/carto.47.1.18

Haklay, M. (2010). How Good is Volunteered Geographical Information? A Comparative Study of OpenStreetMap and Ordnance Survey Datasets. *Environment and Planning. B, Planning & Design*, *37*(4), 682–703. doi:10.1068/b35097

Haklay, M., Singleton, A., & Parker, C. (2008). Web Mapping 2.0: The Neogeography of the GeoWeb. *Geography Compass*, *2*(6), 2011–2039. doi:10.1111/j.1749-8198.2008.00167.x

Hargittai, E., & Shaw, A. (2015). Mind the skills gap: The role of Internet know-how and gender in differentiated contributions to Wikipedia. *Information Communication and Society*, *18*(4), 424–442. doi:10.1080/1369118X.2014.957711

Hennig, S. (2017). OpenStreetMap used in protected area management. The example of the recreational infrastructure in Berchtesgaden National Park. *Eco Mont-Journal on Protected Mountain Areas Research*, *9*(2), 30–41. doi:10.1553/eco.mont-9-2s30

Hennig, S. (2020). Motivation and Its Consideration in Participatory Spatial Data Contribution. *The Professional Geographer, 72*(2), 238–252. doi:10.1080/00330124.2019.1676799

Herfort, B., Eckle, M., Albuquerque, J. P. d., & Zipf, A. (2015). *Towards assessing the quality of volunteered geographic information from OpenStreetMap for identifying critical infrastructures.* Paper presented at the ISCRAM.

Hernandez-Encuentra, E., Pousada, M., & Gomez-Zuniga, B. (2009). ICT and older people: Beyond usability. *Educational Gerontology, 35*(3), 226–245. doi:10.1080/03601270802466934

Jones, C. E., & Weber, P. (2012). Towards Usability Engineering for Online Editors of Volunteered Geographic Information: A Perspective on Learnability. *Transactions in GIS, 16*(4), 523–544. doi:10.1111/j.1467-9671.2012.01319.x

Lin, Y.-W. (2011). A qualitative enquiry into OpenStreetMap making. *New Review of Hypermedia and Multimedia, 17*(1), 53–71. doi:10.1080/13614568.2011.552647

Loidl, M., Krampe, S., Zagel, B., & Pucher, G. (2014). *Aufbereitung von Open Street Map Daten für GIS-Modellierungen und Analysen.* Paper presented at the AGIT 2014, Salzburg. http://gispoint.de/fileadmin/user_upload/paper_gis_open/537543048.pdf

Mahabir, R., Stefanidis, A., Croitoru, A., Crooks, A. T., & Agouris, P. (2017). Authoritative and Volunteered Geographical Information in a Developing Country: A Comparative Case Study of Road Datasets in Nairobi, Kenya. *ISPRS International Journal of Geo-Information, 6*(1), 25. doi:10.3390/ijgi6010024

Mayring, P. (2000). Qualitative Content Analysis. *Forum Qualitative Social Research, 1*(2). Advance online publication. doi:10.17169/fqs-1.2.1089

Mayring, P. (2014). *Qualitative content analysis: theoretical foundation, basic procedures and software solution.* Beltz.

Mooney, P., & Corcoran, P. (2012). The Annotation Process in OpenStreetMap. *Transactions in GIS, 16*(4), 561–579. doi:10.1111/j.1467-9671.2012.01306.x

Neis, P., & Zielstra, D. (2014). Recent Developments and Future Trends in Volunteered Geographic Information Research: The Case of OpenStreetMap. *Future Internet, 6*(1), 76–106. Advance online publication. doi:10.3390/fi6010076

Panciera, K., Halfaker, A., & Terveen, L. (2009). Wikipedians are born, not made: a study of power editors on Wikipedia. *Proceedings of the ACM 2009 international conference on Supporting group work.* 10.1145/1531674.1531682

Perkins, C. (2014). Plotting practices and politics: (im) mutable narratives in OpenStreetMap. *Transactions of the Institute of British Geographers*, *39*(2), 304–317. doi:10.1111/tran.12022

Pfeil, U., Zaphiris, P., & Ang, C. S. (2006). Cultural Differences in Collaborative Authoring of Wikipedia. *Journal of Computer-Mediated Communication*, *12*(1), 88–113. doi:10.1111/j.1083-6101.2006.00316.x

Phang, C. W., Sutanto, J., Kankanhalli, A., Li, Y., Tan, B. C. Y., & Teo, H. H. (2006). Senior citizens' acceptance of information systems: A study in the context of e-government services. *IEEE Transactions on Engineering Management*, *53*(4), 555–569. doi:10.1109/TEM.2006.883710

Picanco, P. L., & Delazari, L. S. (2016). Usability Assessment of VGI Systems Interfaces in Feature Insert Task. *Boletim de Ciências Geodésicas*, *22*(3), 492–510. doi:10.1590/S1982-21702016000300028

Rizqihandari, N., & Indratmoko, S. (2016). Using OpenStreetMap Data for Population Distribution Model. In B. Handoyo (Ed.), *Proceedings of the Lst International Conference on Geography and Education* (Vol. 79, pp. 244-247). Paris: Atlantis Press.

Rousell, A., & Zipf, A. (2017). Towards a Landmark-Based Pedestrian Navigation Service Using OSM Data. *ISPRS International Journal of Geo-Information*, *6*(3), 22. doi:10.3390/ijgi6030064

Schmidt, M., Klettner, S., & Steinmann, R. (2013). *Barriers for contributing to VGI projects*. Paper presented at the 26th International Cartographic Conference, Dresden.

Stalans, L. J. (2012). Frames, Framing Effects, and Survey Responses. In L. Gideon (Ed.), *Handbook of Survey Methodology for the Social Sciences* (pp. 75–90). Springer New York. doi:10.1007/978-1-4614-3876-2_6

Steinmann, R., Häusler, E., Klettner, S., Schmidt, M., & Lin, Y. (2013). *Gender Dimensions in UGC and VGI: A Desk-Based Study*. Paper presented at the GI_Forum 2013, Wien.

Stephens, M. (2013). Gender and the GeoWeb: Divisions in the production of user-generated cartographic information. *GeoJournal*, *78*(6), 981–996. doi:10.100710708-013-9492-z

Tarapiah, S., & Atalla, S. (2015). Public Transportation Management System based on GPS/WiFi and Open Street Maps. *International Journal of Advanced Computer Science and Applications*, *6*(1), 189-194.

Thielmann, T., van der Velden, L., Fischer, F., & Vogler, R. (2012). Dwelling in the web. Towards a googlization of space. *HIIG Discussion Paper Series, 2012*(3). doi:10.2139/ssrn.2151949

Truong, Q. T., de Runz, C., & Touya, G. (2019). Analysis of collaboration networks in OpenStreetMap through weighted social multigraph mining. *International Journal of Geographical Information Science, 33*(8), 1651–1682. doi:10.1080/13658816.2018.1556395

Turner, A. J. (2006). *Introduction to neogeography*. O'Reilly.

Vandecasteele, A., & Devillers, R. (2015). Improving Volunteered Geographic Information Quality Using a Tag Recommender System: The Case of OpenStreetMap. In J. J. Arsanjani, A. Zipf, P. Mooney, & M. Helbich (Eds.), *Openstreetmap in Giscience: Experiences, Research, and Applications* (pp. 59–80). Springer-Verlag Berlin. doi:10.1007/978-3-319-14280-7_4

Vogler, R., & Hennig, S. (2013). Providing Geomedia Skills beyond (Post)Secondary Education. *GI_FORUM - Journal for Geographic Information Science, 2013*(1), 317-327. doi:10.1553/giscience2013s317

Vogler, R., Hennig, S., & Ferber, N. (2017). Redressing the Exclusiveness. Challenges which Prevent New Users from Contributing to OSM. *GI_FORUM - Journal for Geographic Information Science, 2017*(2), 294-299. doi:10.1553/giscience2017_01_s294

Vrbik, D. (2016). Crowdsourcing of Local Spatial and Historical Knowledge. *GI_Forum, 2*, 109-122. doi:10.1553/giscience2016_02_s109

Werlen, B. (1993). *Society action and space: an alternative human geography*. Routledge.

Wood, D. (1993). *The Power of Maps*. Routledge. doi:10.1038cientificamerican0593-88

Wood, D. (2010). *Rethinking the Power of Maps*. Guilford Press.

Yang, A. R., Fan, H. C., & Jing, N. (2016). Amateur or Professional: Assessing the Expertise of Major Contributors in OpenStreetMap Based on Contributing Behaviors. *ISPRS International Journal of Geo-Information, 5*(2), 15. doi:10.3390/ijgi5020021

Chapter 8
A Didactical Approach of Rhumb Line vs. Great Circle in Web Mercator Projection for 6th Grade Pupils

Nikolaos Lambrinos
https://orcid.org/0000-0002-5670-0909
Aristotle University of Thessaloniki, Greece

Ioanna Repanidou
Aristotle University of Thessaloniki, Greece

Georgia Intzidou
https://orcid.org/0000-0002-2880-1972
Mandoulides Schools, Greece

ABSTRACT

The chapter presents two studies on the integration of map projections in Greek geography curriculum. Perceptions and misconceptions about world maps of 6th grade elementary school students were investigated and analyzed by qualitative and quantitative methods. Through inquiry-based methods, students learnt about the distortions in distance, shape, and size of areas that are caused by transformation of the spherical Earth into a flat surface. Didactic interventions were carried out where students attained knowledge about 1) world map projections and 2) the overseas flightpaths and the measurements of the distance between two points in the Mercator projection. Through guided research, comparisons, experimentation on digital world maps, and experiential actions, students found that 1) all world maps are wrong and 2) the shortest distance between two points in Mercator projection can be a curved line (great circle) and not a straight line (rhumb line).

DOI: 10.4018/978-1-7998-9598-5.ch008

INTRODUCTION

In this chapter the two researches that are presented, examine beliefs of students as far as world maps are concerned (distortions in distance, shape and size). According to Crone (1968); Colledge & Zannaras (1973), from past centuries there was a need to map the world, in order to survive and communicate with other people (as mentioned in Tzotzis, 2007). In the 21st century, a map is primarily used as a tool to provide information and knowledge on the subject of Geography at schools. Despite the evolution of maps and the efforts to map the Earth's surface accurately, it is known that there are issues in small-scale maps. Earth´s ellipsoid shape is unable to be properly represented on a flat map surface (Stevens & Coupe, 1978). Thus, there are several map projections. Distortions on map projections include shape, area, distance and angles. No map projection can preserve all geometric properties (area, distance, direction, angle, shape) at every location when projecting terrestrial features from globe to plane. Such problems should be taken into consideration by the teacher and pointed out to students in order to avoid misconceptions while dealing with maps in classroom. The purpose of the study is primarily to promote student´s ideas and secondary to eliminate their misconceptions regarding:

- Distances measured in world maps. Students are going to attain knowledge in map projections and understand that the shortest route between two points at the Mercator projection could be a curved line (orthodrome) and not a straight line (loxodrome). Literature review showed no similar researches on both Elementary and Secondary Education.
- The shape and size of countries in different map projections. The students are expected to understand that world maps are different and show the countries in different ways.

Another purpose of the present study is to highlight the importance of Geography course and in particular the importance of map knowledge.

This chapter is composed of six sections:

- The theoretical background, where several issues are analyzed, such as map projections, great circle and rhumb line, map projections in Education, map projections and spatial thinking, GIS and Geography, pedagogical and didactic methods, TPACK and G-PCK.
- The methodology of the researches: research questions, participants, data collection and online software.
- The didactical interventions and the nine activities that were conducted in the classroom.

- The results.
- The conclusions.
- Future suggestions.

Two researches were conducted to 669 sixth-grade students in 39 schools in Thessaloniki, Greece. The students became aware of the map projections and the distortions in world maps. Through experiential activities, inquiry based lessons and comparative methods, they understood that a sphere cannot be transformed into a flat surface (a flat map) and therefore, maps are distorted as far as the distances (great circle and rhumb line), the angles, the shapes and sizes of the regions are concerned. It should be pointed out that there is no other related research regarding loxodrome and orthodrome in Elementary and Secondary Schools.

BACKGROUND

Map Projections

During the Renaissance, there was a wealth of investigations and inventions. For the first time in history, there was an increasing need to find a way to depict the whole world – the globe - into a flat surface. There were made globes and world maps where geographers were trying constantly to depict the true size, shape, distances and direction of the areas of the Earth (Keuning, 1955). From the sixteenth century to today, there are made several projections, but no one of them shows exactly how the Earth is and there are many distortions, because it is mathematically impossible to transfer the three-dimensional spherical shape of the Earth geometrically exactly to a two-dimensional map (Schommer, 2019).

Map projection is a method that allows spherical Earth to be represented on a flat map. A large number of map projections are developed by mathematicians or cartographers to represent (a part of) the globe on a flat surface (Lapon,Ooms & DeMaeyer, 2020). Map projections are designed with a particular purpose. However, distortions are an inevitable characteristic of every map projection. No map projection can preserve all geometric properties (area, distance, direction/angle, shape, and specific curves) at every location when projecting terrestrial features from globe to plane (Lambrinos, 2009, Basaraner&Cetinkaya, 2019). Teachers and students, as map users, need to recognize what is and what is not distorted on projections to be able to use maps better in their professional and personal lives (Olson, 2006).Teachers and students need to be able to select maps that are appropriate for their needs and they would benefit by knowing how to look at a map projection intelligently. Some basic understanding of projections can help them to be good map readers, aware of what can and cannot be concluded from a map (Olson, 2006).

Mercator Map Projection

The Mercator projection was designed in the 16th century by the Flemish Gerardus Mercator. It is a cylindrical conformal map projection, preserving the angles as they are on the globe. The projection method was especially valuable for maritime navigating, because the direction of a line between two locations on a map is the same direction that has to be followed while navigating (Monmonier, 2004). However, the Mercator projection is much debated because of its particular distortions that inflate the areas toward the poles. Due to the widespread and longtime use of this map projection, it is argued that the cognitive map of students might be most similar to this distorted map projection, which is referred to as the Mercator effect (Lapon,Ooms &DeMaeyer, 2020). The reason that Mercator map projection is selected for the present research is that it distorts the distances (great circle and rhumb line) and the size of continents.

Robinson Map Projection

The Robinson projection was especially designed for general-purpose world maps and made an attempt to find a compromise for distortions. Robinson map projection has a balance of size and shape of high latitude lands than Mercator. This is the reason that Robinson map projection is selected for the present research.

Peters Map Projection

The Peters projection world map is an area-accurate map that was first introduced by historian Arno Peters in 1974. The Peters map shows all areas, whether countries, continents, or oceans, according to their actual size but distorts the shape of continents. Arno Peters changed the world (Ashmore, 2003). Through his cartographic activity and his advocacy of the use of his map Peters engaged a wide range of people in an exploration of the way that flat, two-dimensional representations of the world distort not only the image of the world but also people's perceptions of it (Ashmore, 2003). He effectively changed millions of people's perception of the world. The reason that Peters map projection selected for the present research is that it shows all continents and countries in their actual size.

Great Circle vs. Rhumb Line

The prevailing perception is that the shortest distance between two points is always a straight line rather than a curved line. Moreover, geometry defines that a distance that unites two points is the minimum length of a straight section. People believe that the same happens with maps. But, in reality, there are few ways to construct a world map with a straight line as the shortest route between two points (Gilmartin, 2017). That happens because the Earth's shape is spherical (3D dimension), but the map's is flat (2D dimension). Therefore, since the Earth cannot be converted with absolute accuracy into a flat map, it is possible that the shortest distance between two points appears with a curved line. The projection of the spherical Earth into a flat piece of paper causes serious distortions (Gilmartin, 2017).

Mercator cylindrical projection displays a property: the straight line between two points, which is called rhumb line or loxodrome, is the longest route that an airplane is able to cross. A rhumb line or a loxodrome is called a line on the surface of the Earth that intersects the meridians at a constant angle (General Navigation, n.d.). On the contrary, the great circle or orthodrome appears in the Mercator projection as a curved line. This curved line is the shortest distance between two points that an airplane can cross. The great circle is a line on the surface of the Earth, a trace of a plane that passes through its center and divides it into two parts (General Navigation, n.d.).

In Figure 1 it is depicted a great circle and a rhumb line. At the globe a great circle appears as a straight line and a rhumb line as a curved line, on the contrary to the Mercator projection where a great circle is shown as a curved line. This happens because the Earth has transformed into a flat surface, resulting in the distortion of distances. Figure 2 shows the comparison of the representation of a rhumb line and a great circle into the Earth and the Mercator world map.

Map Projections in Education

Map projection is a current and timeless issue in the field of Geography Education. For example, in the field of education new didactical approaches in map reading are observed, such as Boston Public Schools that announced the switch to the Gall-Peters projection for all classroom maps showing the entire world (Šavrič,Patterson & Jenny, 2019).At the same time, a number of new applications and videos have recently appeared on the internet that approach the subject of map projections in a remarkable, simple, understandable and playful way. However, despite the timeless and importance of map projections in cartographic education, the number of research outcomes is still limited and there is much left to investigate regarding the map reader

Figure 1. Representation of great circle and rhumb line on the surface of the Earth (Tseng, 2014).

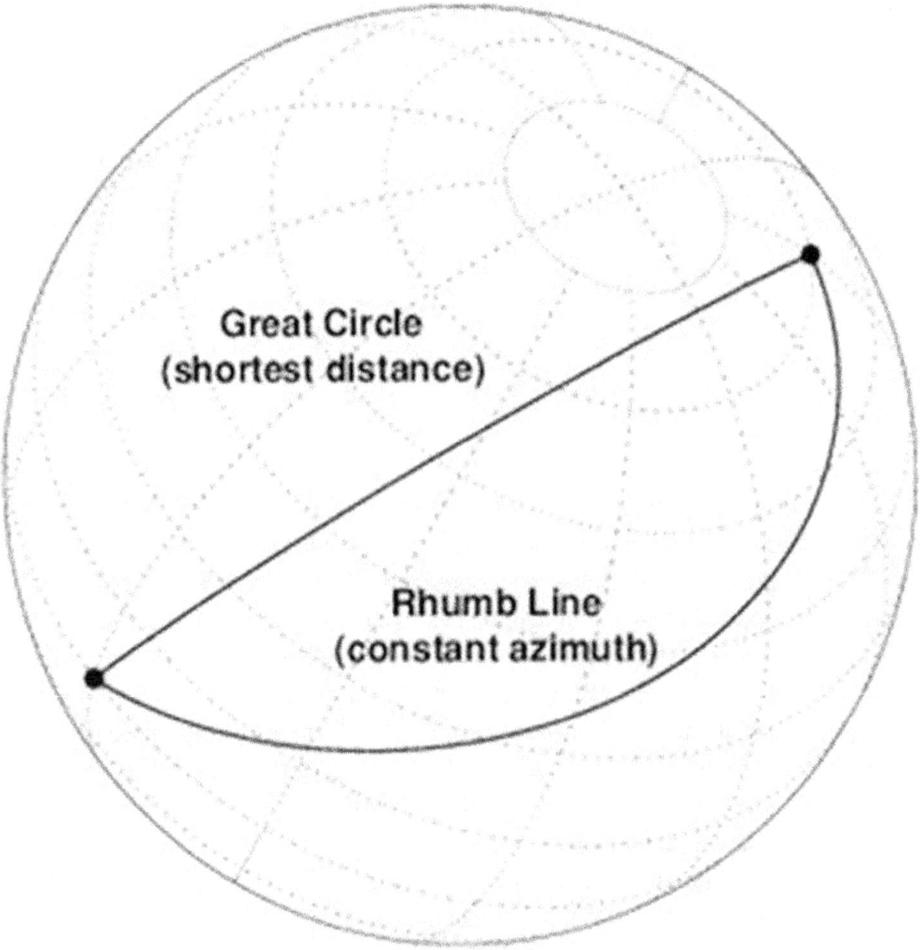

and the map projection (Kessler et al., 2017) and the skills and understandings in reading digital maps (Catling, 2018).

Some researchers have presented map projections to their students using the method of comparison (Gutstein, 2013, Valovicová,Vallo& Kramáreková, 2019; Schommer, 2019, Intzidou et al., 2021). Moreover, Catling (2018) suggests that children 5-8 years old can compare a globe and world maps and look for the same features on each. Children 9-12 years old can compare how parts of the world are

Figure 2.Representation of rhumb line and great circle on the surface of the Earth and a Mercator world map (https://www.coursera.org/lecture/introduction-gis-mapping/great-circles-and-rhumb-lines-jHUtd).

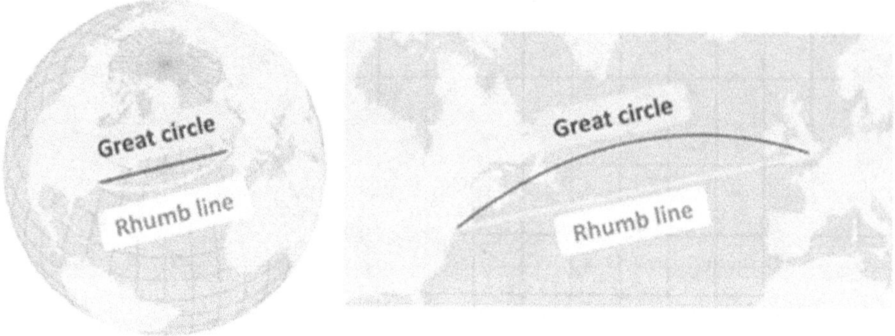

shown on globes, wall maps of the world and what is similar and different. Therefore, map projection can be taught in Elementary School via comparative method.

The map projection discipline is recognized as one of the most challenging in Geography (Olson, 2006). Evidence suggests that many children older than 11/12 years are not knowledgeable and do not understand what atlas maps really present or what their limitations are (Wiegand, 2006). Primary children encounter a variety of world map projections and have to grapple with the relationship between a globe and a flat map, as well as with satellite photographs of the Earth (Catling, 2018). The lack of knowledge of map projections leads to a number of difficulties in working with maps and often promotes misconceptions about the distances between two points, the shape of countries and continents and their spatial relationships to one another (Schommer, 2019). A typical example of a map projection that leads students to misconceptions regarding the true area size of countries is the cartographic representation of the world using the Mercator projection.

Unfortunately, in the Greek educational system, the teaching of maps is characterized poor. In school textbooks the extent given to map reading, construction and interpretation skills is minor, compared to other 26 geographical issues that are given. The maps in the 5th and 6th grades of the Primary School constitute only 10% of the Geography curriculum. In fact, from the school year 2021-2022 the subject of Geography was decreased to one teaching hour per week, contrary to the previous year which was two teaching hours per week (Government Gazzette, 3791-13/8/21).

Map Projections and Spatial Thinking

Spatial thinking is defined as knowledge, skills and thinking habits to use the concepts of space such as distance, orientation, distribution, and association; representational tools such as map, graph, and diagram; and reasoning processes such as cognitive strategy to facilitate problem solving and decision making to structure problems, find answers, and express solutions to problems (Collins, 2018). Concepts of space are divided into four subcategories: non-spatial, spatial primitives, simple spatial and complex spatial (Jo &Bednarz, 2009; Scholz et al., 2014). Map projections are part of complex spatial concepts and develop spatial thinking (Scholz et al., 2014).

GIS and Geography

Geographic Information System is a discipline for capturing, storing, analyzing, managing and presenting data and associated attributes which are spatially referenced to Earth. GIS is an effective and powerful tool in teaching Geography because it makes the presentation of data more attractive (Attfield et. al, 2002). GIS improves critical spatial thinking (Walshe, 2017), enhances the interdisciplinary approach (Rød, Larsen & Nilsen, 2010; Baker et al., 2012, Lambrinos & Asiklari, 2014), helps students answer geographical questions (Peterson et. al, 2020), help students ask geospatial questions and search for answers by querying the data (Roberts, 2013; Goldstein &Alibrandi, 2013; Fargher, 2017), leads students to solve real problems (Baker et al., 2012; Roberts, 2013; 2017; Lambrinos & Asiklari, 2014; Jadallah et al., 2017, Walshe, 2017), increases students' global and local spatial understanding by using real-world tools (Bednarz & Van der Schee, 2006) and, finally, is a powerful tool for teaching World Geography (Jo, Hong, & Verma, 2016). GIS is a helpful tool for teaching map projections in Elementary School students.

Pedagogical and Didactic Methods

1. Inquiry Method

The concept of IBE constitutes an activating interactive teaching and learning method. Although Inquiry-Based Education (IBE) has been developed theoretically and practically for several years, the use of inquiry methodological procedures in routine teaching practice of Geography is not widespread in Primary Schools (Karvánková & Popjaková, 2018). Although enquiry skills can be developed in other subjects, school Geography can make a major contribution to students' development because of the range of questions it addresses, its varied sources of information presented in different formats, sometimes from different viewpoints, and its use of different analytical and interpretive techniques (Roberts, 2017).

2. Comparative Method

Comparison is an ordinary thinking process that makes it easier for people to sharpen perceptions, practice categorization, and solve problems in their everyday lives. Comparison is a reflective and argumentative process. Based on a geographical question, a number of units are selected, whose similarities and differences are identified along determined variables in order to work out interrelations. Based on this, justified explanations are formulated to answer a specific question, or further questions are raised, which need to be considered to develop a final answer (Wilcke & Budke, 2019). Wilke & Budle (2019) suggest a systematic comparative method for Geography Education (Figure 3), which is used in the present research.

Technological Pedagogical Content Knowledge (TPACK)

Teachers should be specialized in GIS, in order to include them into their classroom. Teachers should be aware of how they choose and use the appropriate software that brings extra pedagogical value. The TPACK theoryhas a great impact as far as the researches in Education are concerned (Mishra, 2019). According to Pamuk, (2012); Brinkley-Etzkorn, (2018); Mishra, (2019); Purwanto et al. (2020) there are many effective ways that a teacher can use technology into a learning environment. TPACK theory contains three main structural elements such as: Technological Knowledge (TK), Pedagogical Knowledge (PK) and Content Knowledge (CK) (Figure 4). All the structural elements are interacting with each other during the lesson. If teacher can appropriately use these three structural elements, it means that the lesson has been conducted and completed efficiently.

As far as the Geography curriculum is concerned, it is necessary for teachers to have Geographical Technological Pedagogical Content Knowledge (G-TPCK) (Figure 5) (Doering and Veletsianos, 2008; Blankman et al., 2016). Teachers should be aware of geospatial technologies and know how to integrate GIS into their classroom. Moreover,they should be able to know how to teach Geography and find out the appropriate teaching methods. According to Hicks et al. (2014) an effective lesson inquires the appropriate pedagogical methods, in combination with the use of technological tools. Overall, teachers play a significant role in the success or failure of integrating new concepts in the learning process.

Figure 3. Method of comparison step by step (by Wilcke & Budke, 2019).

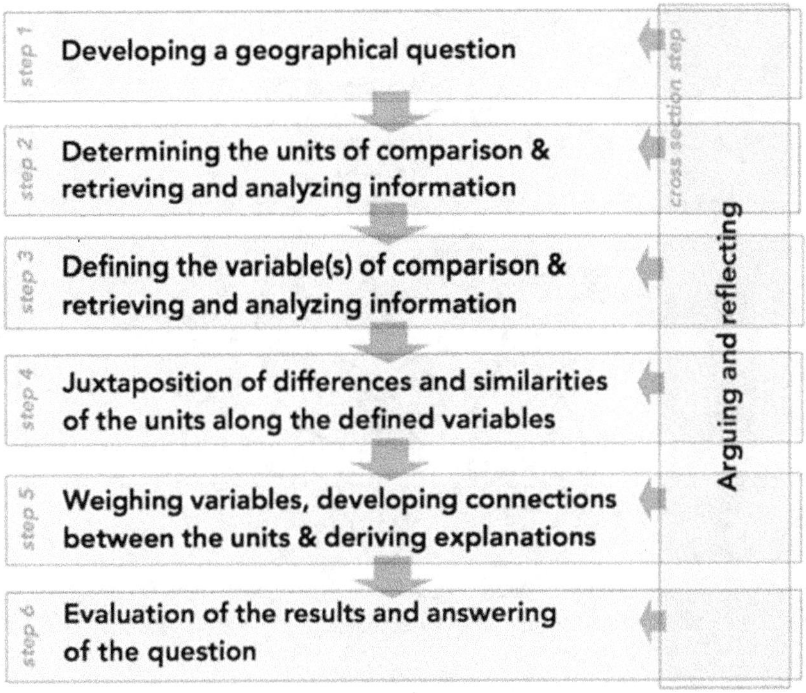

METHODOLOGY

Research Questions

This chapter presents the research questions, the tools of the research and the tools that were used in the classrooms.

Q1: Can students find out that all map projections contain errors and cannot imprint the Earth´s surface with total accuracy?

Q2: Can students' misconceptions about distance, size and shape be eliminated by experimentation with Geographical Information Systems (GIS) and experiential activities?

Participants

Two researches were conducted by the authors. The first one is a case study, that included 14 (9 boys and 5 girls)6th-grade Elementary School students from the city

Figure 4. Technological pedagogical content knowledge – TPACK) (Mishra, 2019).

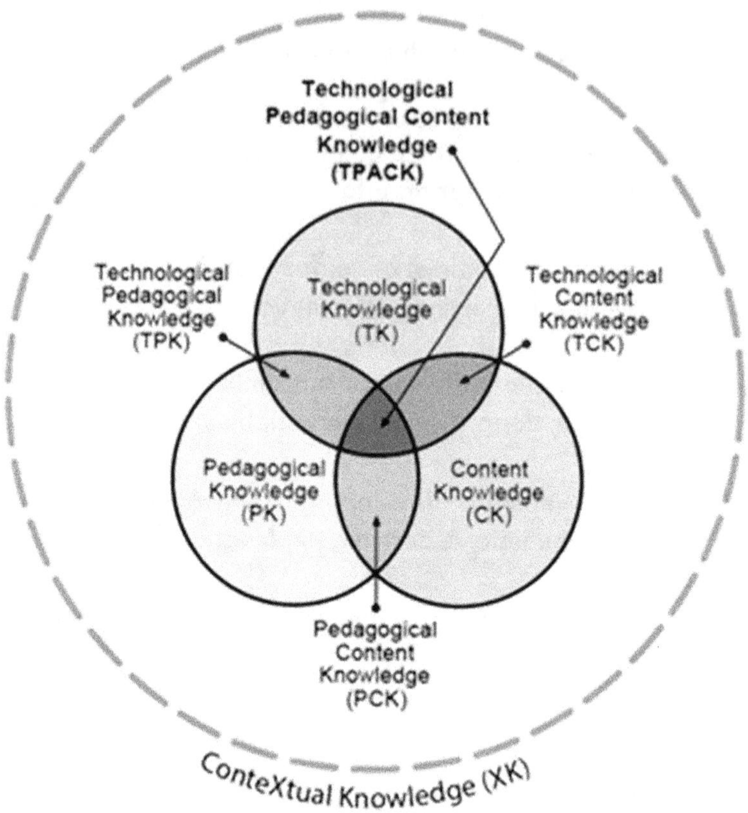

of Thessaloniki, Nea Apollonia region, Greece. The Elementary School of Nea Apollonia is located in the western regional unit of Thessaloniki. This school was selected as a convenient sample selection. It should be emphasized that 6th- grade students were selected, because at this specific grade they first learn about Earth, North and South Pole, meridians, parallels and get acquainted with world maps. In the second research six hundred fifty-five students participated ($N = 655$) from thirty-eight (38) sixth-grade classes in Thessaloniki city, Greece. The Urban Complex of Thessaloniki city was divided into 2 strata: the Directorate of Primary Education of Eastern Thessaloniki and the Directorate of Primary Education of Western Thessaloniki. Then, 10% of schools of each Directorate were selected.

Figure 5. Geographical pedagogical content knowledge (G-TPCK) (Doering & Veletsianos, 2008).

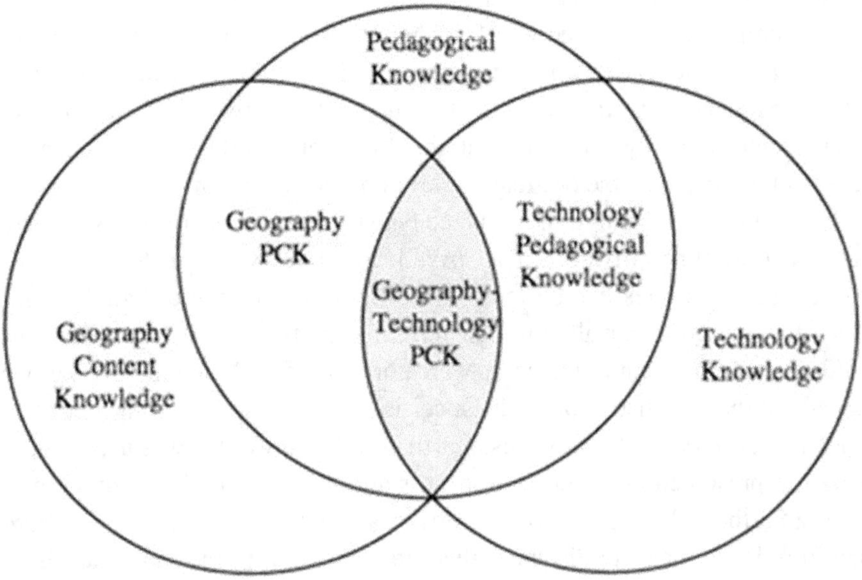

Data Collection

In the first research an open-ended questionnaire was used for data collection. It was given to students, before researchers intervened, as a pre-test, to record student's preconceived ideas and beliefs. After the completion of the intervention, the same questionnaire was given again to students, as a post-test, to record the knowledge they gained and to detect the change in their ideas and beliefs. The answers were analyzed qualitatively.

In the second research a questionnaire was used for data collection. Data from students' answers were coded and imported in IBM SPSS Statistics, version 25 (IBM Corp., Armonk, N.Y., USA) where they were statistically processed. The descriptive statistics were presented as frequencies (*N*) and percentages (%).

The tools that were used in the classroom were: a) video projectors, b) computers, c) rulers, e) multimedia and interactive material created by researchers, f) plasticine, g) two globes with large dimensions (one inflatable and one deflatable), h) balloons

Online Software

The first one was *ArcGIS StoryMaps* (https://arcg.is/1i1Xeb), where researchers created a multimedia material in order for the children to be introduced to the subject. Another online software used was *Scribble Maps*(https://www.scribblemaps.com/tools/distance-calculator), where students experimented by drawing in the Mercator projection transatlantic travel routes. The routes between two points (regions) should be represented not only with a straight line (loxodrome or rhumb line), but also with a curved line (orthodrome or great circle) in order to compare distances.

Also, students measured the distance between two points (airports) from many regions at the online software *Great Circle Map*(https://www.greatcirclemap.com/). In this software routes can be represented not only through a digital world map (Mercator projection) but also through 3D representation of the globe. Thus, students could compare how curved lines in Mercator projection look like to the globe and understand that the distortion in distances is due to the spherical shape of Earth.

Moreover, *Google Earth* (https://earth.google.com/web/) was used, where the globe is represented in 3D projection. It is an interactive tool, where the user can observe Earth, look for points of interest, measure distances etc. *Google Earth* was used in order to measure distances that are positioned to the same meridian. The aim was for the students to understand that when distances are measured in the same meridian, distortions are minor and in Mercator projection the great circle and the rhumb line have the same distances.

The final tool was three world maps with different map projection (Mercator, Robinson, Peters) from *ArcGIS Online* and *the website*www.thetruesize.com. This website includes a digital interactive Mercator map and gives the user the opportunity to move any country to any place on the map, observing its increase or decrease in size, as it moves away or closer to Equator, respectively (Intzidou et al., 2021).

Didactical Interventions

First Research

- **First activity: Pre-test (questionnaires) and exploration of prior knowledge of students. Duration 1 hour and 30 seconds.** The aim was to highlight student's ideas and misconceptions about distortions in world maps, especially as far as distances are concerned. Each student completed the pre-test questionnaire on their own. There were 9 open-ended questions.
- **Second activity: Introduction to map projections and brainstorming. Duration 1 hour.**The purpose of this activity was to introduce students to different map projections and to understand their importance. Targeted

questions were asked in order to highlight student´s ideas and develop their critical and spatial thinking. In the beginning, researchers showed a powerpoint with map projections and posed questions to help students think in a different way. These questions were:

- *"What shape does Earth have?",*
- *"What shape do maps have?",*
- *"What does Earth look like through maps?".*

In the latter question, students answered: *"it looks way bigger", "we can detect the places, the sea..", "it has different colors, terrain, shapes", "Earth is spherical, but maps are rectangles"*. At this stage it is obvious that students already knew that Earth looks different through the maps. It is noted that students said: *"Earth could also have been flat, because we see it in that way through the maps".*

Then, another question was posed to the classroom:

- *"Can the Earth be presented as it really is through the maps?".*

Students answered that: *"in maps the islands and the cities look smaller in contrast to the Earth", "maps are made of paper, but the Earth not", "maps can show us the airports, the names", "maps do not rotate, but the Earth does", "maps are smaller than the Earth".*

Analyzing the answers, it is clear that students have a general knowledge of maps compared to the globe. Only one student, who had some geographical knowledge, answered that: *"maps show us smaller dimensions and distances. Also through maps the places do not look like reality e.g. Russia looks bigger and different"*. Then, students came up with different map projections. They were impressed and they said that: *"these maps look like they are devastated", "I can't find the countries", "do that kind of maps really exist?", they are different from the maps that we see in our everyday lives".*

At the end of the presentation, students had already changed their way of thinking about map projections. Posing the final question:

- *"Are there finally any distortions in maps? What distortions are they?".*

Students answered that: *"some of them do not show accurately the shapes", "after all, there are problems through maps", "we did not know that there are some issues in maps, we thought that the countries look the same as on the earth".*

Then,the whole classroom visited the website: https://www.leventhalmap.org/digital-exhibitions/bending-lines/interactives/projection-face/. In this website, a face is distorted, when different map projections are chosen. Using this example, students detected differences in a playful manner.

To a similar question in the second research, students were asked to answer the open-ended question:

- *"Why do you think that on some world maps the size of the areas is displayed correctly while on others it is incorrect?".*

Two cases emerged. The first case attributed the different representations of the area´s size to reasons related to the *construction of the map* (errors in the construction process of the map, expediency, deception, date of construction of the map, purpose of construction of the map, Earth shape, problem of two-dimensional projection), while the second case was attributed to *map elements* (shape, scale, size, type).

- **Third activity: Printed maps with different map projections – comparison of distances. Duration 1 hour.** The aim of this activity was for students to understand that different map projections distort distances between two points. Each student individually measured distances in printed maps with a ruler and understood that distances are not the same.

Students used four printed world maps with different map projection, but in the same scale. Map projections were: Peters, Miller, Gall, Mercator. These four map projections were selected because of their similarity, but, in reality, there are differences in distance. In every map there were two points on the right side perpendicular to each other and two points on the center perpendicular to each other, as well. Students were asked to unite the points perpendicular to each other and measure with a ruler the distance between them. Afterwards, they wrote down their results and noticed that maps were not the same, as they initially thought. They noted that even though the points were the same, distances varied in different map projections. A discussion took place in the classroom, where students found out that, apart from distortions in area, shape and size, there are distortions in distances, too.

- **Fourth activity: Experiential activity with plasticine – converting a sphere into a flat surface. Duration 1 hour.** The aim of this activity was for students to understand via experiential way that the conversion of the Earth into a flat surface causes distortions in distances, sizes, shapes and direction. Students used two plastiscines in different colors of their choice. In the beginning, they were asked to form a sphere with a single color and imagine that it is a globe. Then, with the different color plasticine, they were asked to form a circle and place it on the top of the sphere (North Pole) and another circle at the center of the sphere. Circles were placed in specific areas deliberately, in order to help students understand that the circle on the top of the sphere is being distorted the most, as in the Mercator projection. Then, they were asked to reshape the sphere into a rectangle flat surface, like a world map, but the circles should not be distorted at all. Students believed that this

Figure 6. Children measure the distances between points in different cartographic projections.

would be easy, but then they realized that the sphere cannot be transformed into a flat map without distorting the circles. They understood also that the circle on the top of the sphere was distorted the most, as the regions/areas that are placed near the North Pole in Mercator projection. This activity was an occasion for discussion and students were introduced into a big problem that all cartographers face: The globe cannot be represented into a flat map without a distortion.

Figure 7. Transformation of the globe into a flat rectangle surface.

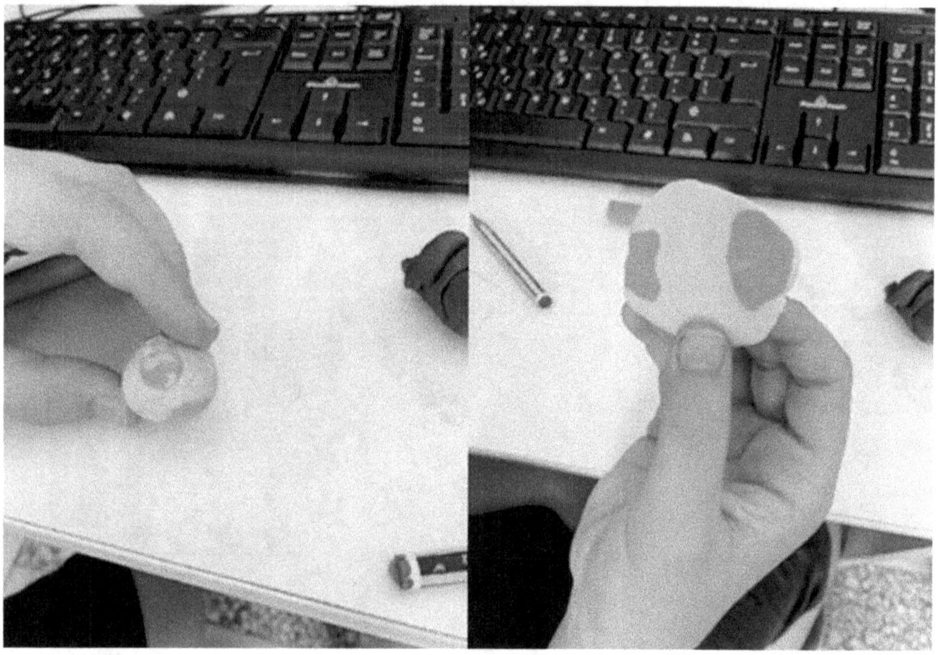

- **Fifth activity: Experimentation in digital world maps. Duration 3 hours.** The aim was to experiment in groups of two students and draw distances in digital interactive world maps in Mercator projection. Students should be able to understand that in Mercator projection a great circle (a curved line) and not a rhumb line (a straight line)is often the shortest distance that an airplane can travel. Online software that students used were: ArcGIS StoryMaps, Scribble Maps and Great Circle Map.

Students browsed at the online software ArcGIS StoryMaps, where researchers had created an interactive presentation (https://arcg.is/1i1Xeb) with questions related to different map projections and distances in Mercator digital world map (great circle and rhumb line). Students initially chose, in pairs, transatlantic travels that they want to go on in the future. Answers were recorded on worksheets. Other online software used by students include:ScribbleMaps and Great Circle Map, as shown in Figure 7. Through these software students experimented and measured distances, that they had chosen before (transatlantic travels), designing curved lines (great circle) and straight lines (rhumb line). They detected that regarding long distances in Mercator projection, the shortest path is a great circle. Initially, students were wondering how that was possible. Then, through guided investigation and experimentation, they understood that the curved line, that was shown into the digital world map, was due to the Earth´s spherical shape. It is necessary to point out that the points on digital world maps were different each time, in order for students to become aware of different routes. For example, in one case the shortest route could be plotted with a curved line to the North Pole while in another case to the South Pole. Routes on the same meridian were also chosen in order to understand that the difference as far as kilometers are concerned is negligible on both loxodrome and orthodrome.

- **Sixth activity: Google Earth and worksheet completion. Duration 2 hours.** The aim of this activity was the measurement of distances between two points that are located on the same meridian. Students were asked to measure distances between two pairs of cities that were located on the same meridian in Google Earth. Students discovered that when measuring distances between two points on the same meridian, distortions are minor and the loxodrome and orthodrome coincide with each other.
- **Seventh activity: Experiential activity with an inflatable and deflated globe/ deepening and consolidation. Duration 2 hours.** In the last activity, the experiential approach was chosen, with the ultimate goal of summarizing, practicing and evaluating. The purpose of this activity was for students to see alterations of distances again, by turning the sphere into a flat surface through experiential learning. Students worked into two groups. The first group had an inflatable globe and the second a deflated globe, which was represented by a world flat map. Each group had a worksheet with pairs of cities. They were asked to draw at the inflatable and the deflated globe the shortest route between two points. Researchers through ScribbleMaps online software, showed the points of the cities that students had to measure each time. Students found the points on the globe and then were asked to draw with a marker the shortest route that they thought a plane would take. At the end, they compared and measured the routes and noted that differences

were substantial, especially in long distances. Distances at the same meridian were deliberately put, so that students would observe that there is almost no difference in kilometers whether measuring in the inflatable or the deflated globe.

Figure 8. Activities with online software: Introduction to loxodrome and orthodrome.

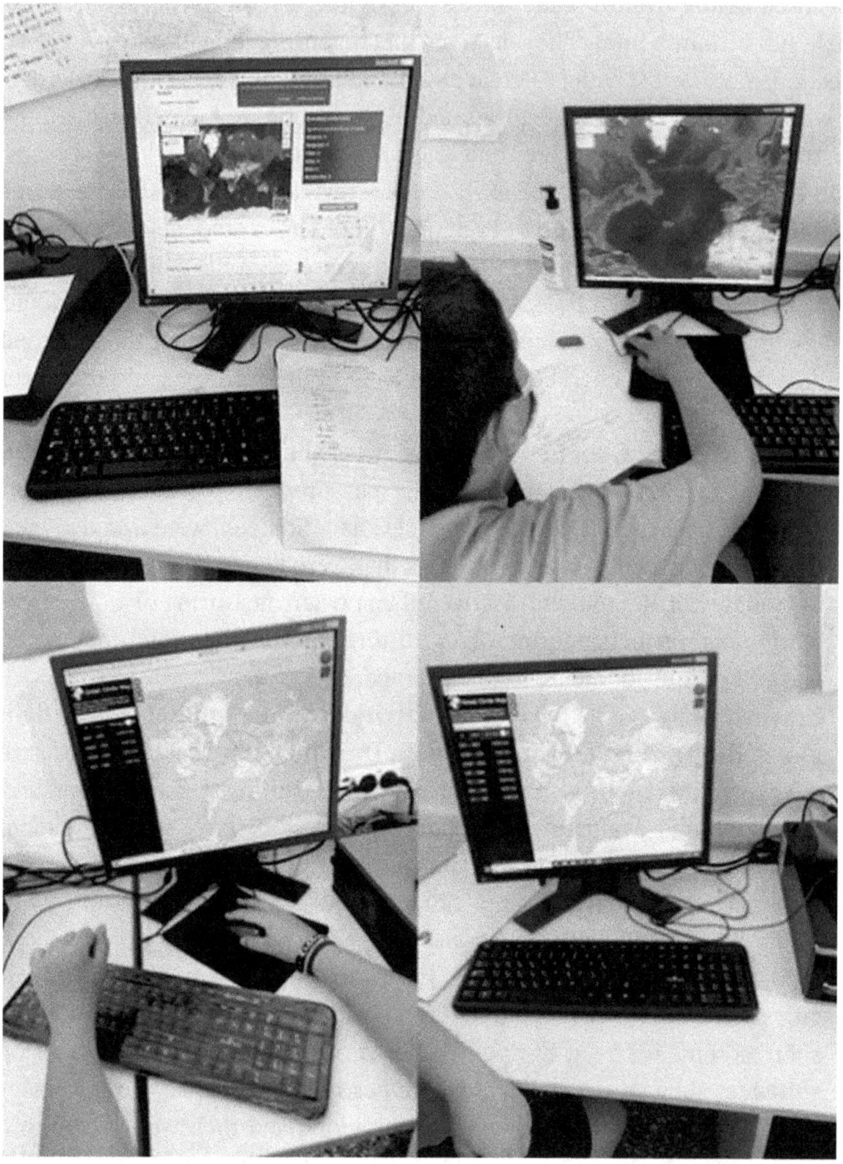

Figure 9. Students measure the distances at the same meridian at the online software Google Earth.

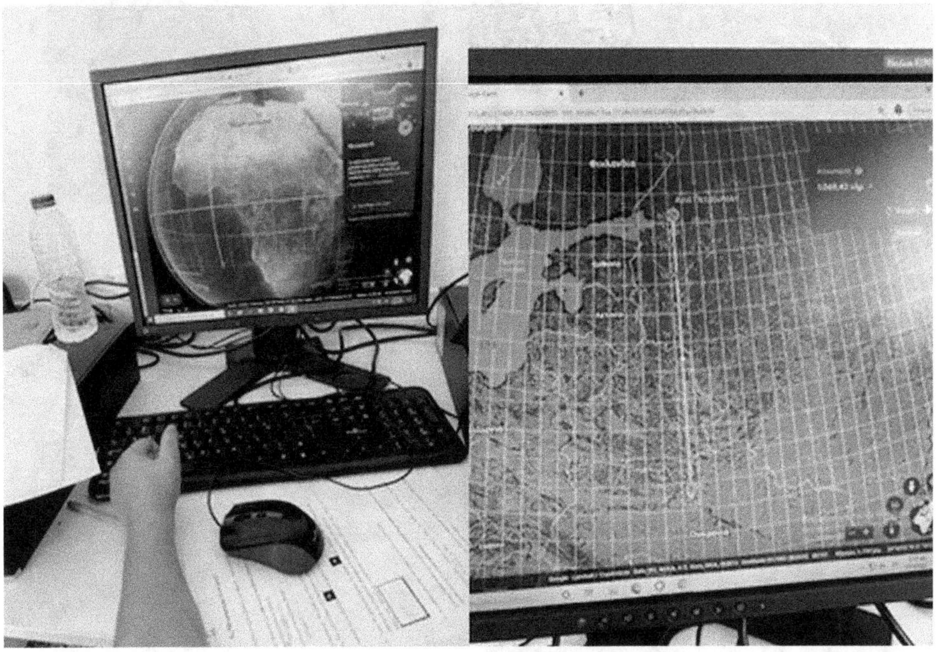

- **Eighth activity: Post-tests (questionnaires). Duration 1 hour and 20 minutes.** Finally, after two weeks of teaching interventions, the same questionnaires/post-tests were given to students.
- **Ninth activity (Second research):** In the second research students compared in pairs the size of particular countries in 4 different representation of the Earth: a) a 3d globe sphere, b) Mercator map projection, c) Robinson map projection and d) Peters map projection. They compared the pairs "Brazil and Greenland", "Spain and Iceland", "Mexico and Finland" in size and wrote their findings. For this activity students used ArcGIS Online and the website https://thetruesize.com Students had to answer the questions
    - *"Which of the 4 representations of the Earth perfectly reflects the actual size of the countries?".*
    - *"Which map shows the size of the countries approximately the same as the 3d globe sphere?".*
    - *"Why are the maps different?".*

Figure 10. Activity with the inflatable and deflated globe into the classroom.

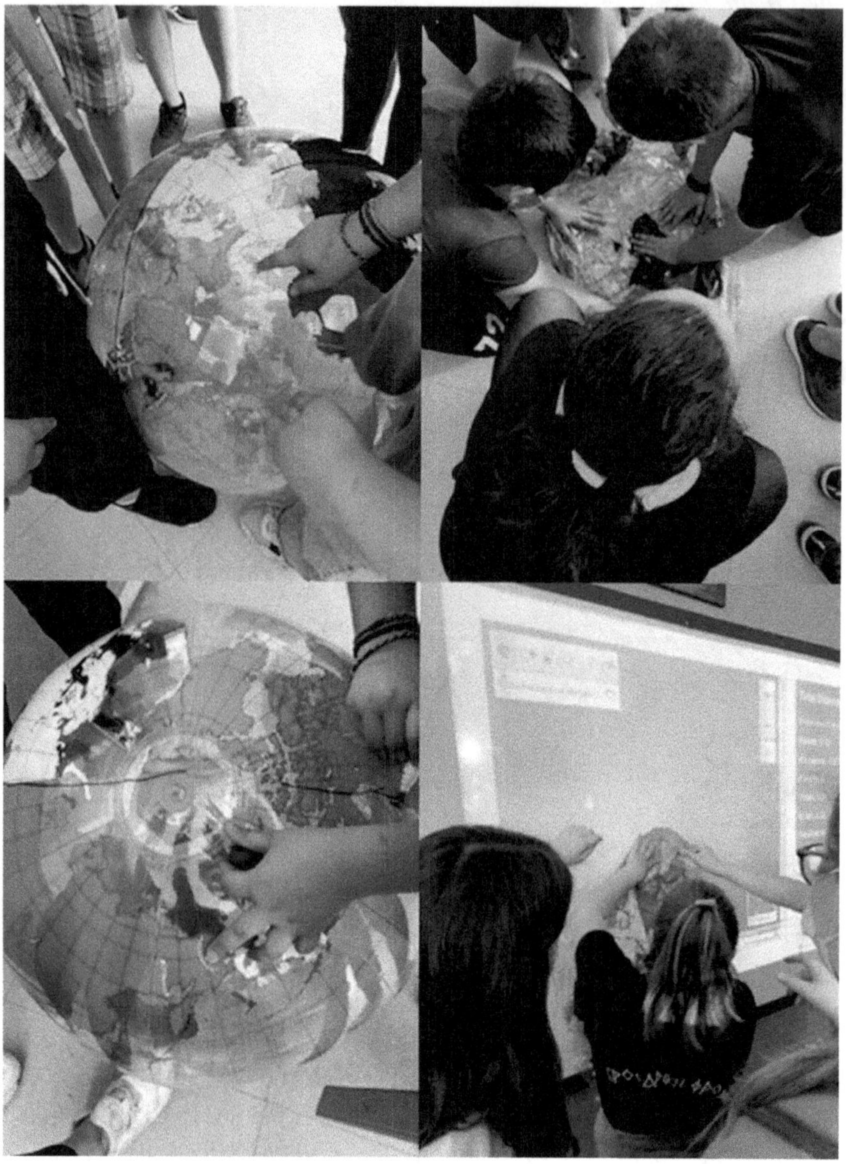

Then, students were asked to do an experiment: They inflated a balloon, and drew a face on it. Afterwards, they popped the balloon and tried to answer the question

- *"What does the face look like, when you stretch the balloon?"*.

The purpose of the experiment was for students to understand that it is impossible to transfer the three-dimensional spherical shape of the Earth exactly to a two-dimensional map without a distortion. Finally, students saw an educational video related to map projections (https://www.youtube.com/watch?v=kIID5FDi2JQ).

RESULTS

At Table 1student´s correct answers at pre-tests are presented. On the contrary, at Table 2 the correct answers at post-tests are presented.

Table 1. Correct answers pre-tests/ questionnaires.

Correct answers pre-tests/ questionnaires	
Question 1	4/13
Question 2	2/13
Question 3	1/13
Question 4	0/13
Question 5	2/13
Question 6	0/13
Question 7	0/13
Question 8	1/13
Question 9	0/13
Total: 10/117 correct answers (8,5%)	

The metacognition and the knowledge that students gained is obvious, not only from the pre and post-tests results, but also from their answers during the didactical interventions. The students' ideas were transforming constantly, which is clear from the recordings in the classroom.

Initially, students could not understand how it is possible for a curved line to be the shortest route in a world map. Some examples from the students' perceptions are given: *"Why does the plane not go straight?", "It is clear that the shortest distance is the straight line", "It is not possible that the shortest distance is a huge curved line than a straight line", "How is that possible?", "It is unbelievable. It is like you are telling us that if I want to go to the square the shortest distance is to go around the village rather than just go straight. It is not possible. Why can we not just go straight?".*

Table 2. Correct answers post-tests/ questionnaires.

Correct answers post-tests/ questionnaires	
Question 1	10/13
Question 2	8/13
Question 3	2/13 (11/13 half correct)
Question 4	5/13 (6/13 half correct)
Question 5	5/13
Question 6	7/13
Question 7	5/13
Question 8	9/13
Question 9	7/13
Total: 58/117 correct answers (49,6%)	

Through brainstorming it was found that students do not learn about map projections at school. Students were confused with different geographical terms and ended up thinking that the types of maps were the same with map projections. When showing the students different world map projections they started to realize that all maps have distortions. They said that *"some maps do not depict exactly the shapes", "eventually, there are problems in maps", "we did not know that there are problems in maps, we thought that the areas are like this on the Earth, also"*. In the second research, students were asked to explain why world maps are different. Many of the students' answers were related to mistakes made in *construction of the map* (errors in the construction process of the map, expediency, deception, date of construction of the map) or *to map elements* (shape, scale, size, type).

In the next activity, students had to draw and measure distances between two points in different printed world map projections. Through experimentation and measurements, they understood that there is one more distortion in world maps that they had not noticed earlier: the distortion in distances. Students said: *"we noticed that the distances in all these maps are different"*. Then, the researchers posed the following question: *"How is it possible to have different results on every map?"*. Students said that: *"it is happening because the map projections are different", "the distances are changing in every map", "the squares change", "at some projections the areas look bigger"*. As a result, students gained further knowledge and discovered alterations of distances on maps beyond the shape and size that they had perceived in the previous activity.

In the fourth activity, students tried to convert the sphere into a rectangle by trying not to distort the size, the shape and the distances of the circles. Initially, they thought that this would be easy, but they ended up saying that: *"I cannot change the shape of the sphere without distorting the circles on it", "we realized that the circles are distorted", "the circle on the top of the sphere looks way bigger than before".* Then, the researchers asked students: *"so, could a sphere be imprinted exactly as it is into a world map?".* Students answered: *"no, the areas are not the same as they are in reality", "in the end, there are not right maps", "we understood that as the circles deformed on the plasticine, the same is to be when we try to transform a sphere into a flat map".*

Continuously, students were troubled by the curved line that appeared in Mercator projection. In Figure 11 there is an example of the great circle and rhumb line between two points. Students had to think about which route was the shortest one.

Figure 11.Great circle vs. rhumb line.

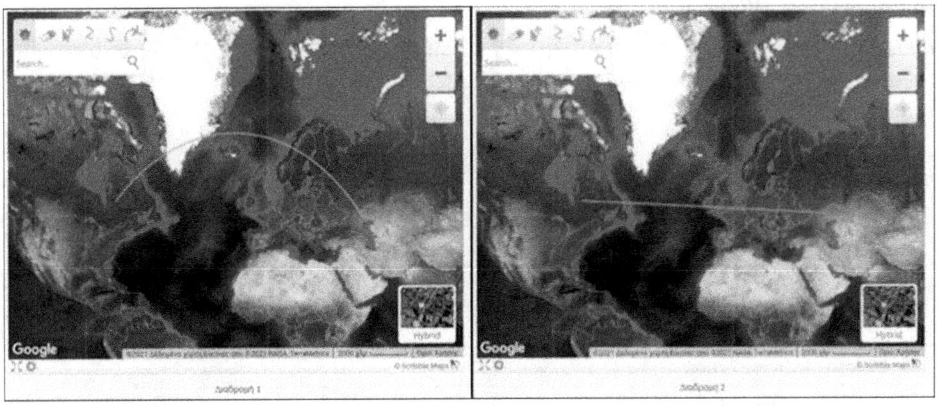

A student reported: *"why is there a curved line? Is it because the Earth is a sphere, so the airplane is going to fly around the Earth?".* Another student stated: *"the airplane should follow the shape of the Earth", "we cannot cross the Earth by going straight".*

Experimentation with the great circle and rhumb line in combination with online software (ScribbleMaps, Great Circle Map, Google Earth) led students to different paths of knowledge. Thus, most students ended up saying that: *"the curved line is the shortest path".* This point of view is totally different from their initial beliefs and thoughts.

Table 3 presents the frequencies and percentages of students' answers regarding the question *"Which of the four representations of the Earth perfectly reflects the actual size of the countries?".* The majority of students chose correctly the 3d global sphere (N = 408, 62.3%).

Table 3. Which of the 4 representations of the Earth perfectly reflects the actual size of the countries?

Map	*N*	%
3D Globe Sphere	408	62.3
Mercator	26	4.0
Robinson Map Projection	83	12.7
Peters Map Projection	122	18.6
None	11	1.7
All of them	5	0.8
TOTAL	655	100.0

Table 4 presents the frequencies and percentages of students' answers regarding the question *"Which map shows the size of the countries approximately the same as the 3d globe sphere?"*. The majority of students chose correctly the Peters Map Projection (N = 60.6%) and the Robinson Map Projection (N = 29.3%).

Table 4. Which map shows the size of the countries approximately the same as the 3d globe sphere?

Map	*N*	%
Mercator	37	5.6
Robinson Map Projection	192	29.3
Peters Map Projection	397	60.6
None	23	3.5
All of them	6	0.9
TOTAL	655	100.0

Table 5 presents the frequencies and percentages of students' answers regarding the question *"What does the face look like when you stretch the balloon?"*. The majority of students chose correctly that the face seems distorted (99.2%).

Table 5.What does the face look like when you stretch the balloon?

The face on balloon: *N*		%
The face seems the same as the spherical balloon	5	0.8
The face seems distorted.	650	99.2
Total	655	100.0

CONCLUSION

An alternative way of learning through maps was attempted in both researches. Students came up with plenty and different map projections, experimented by measuring distances in world maps especially in Mercator projection, compared the loxodrome and orthodrome, drew routes on inflatable and deflatable globes and compared results with each other. Moreover, students experimented by comparing the size of countries in world maps especially in Mercator, Robinson and Peters map projection. They found out by themselves that the spherical shape of the Earth cannot be transformed into a flat world map without any distortions. Through activities with different levels of difficulty, students developed critical and spatial skills, as well as communicative skills due to the fact that cooperative learning and discussion were promoted into the classroom. Students cultivated their spatial thinking, as they came into contact with complex spatial concepts, map projections, through activities that lead them to a high level of reasoning and justification, as they are asked to compare and contrast different map projections (Intzidou et al., 2021).

Before the conduction of didactic interventions, it was observed that students did not have any knowledge as far as cartography is concerned. This is highlighted when analyzing the pre-tests, where student´s primary ideas are documented. None of the students knew what a map projection was. Moreover, most of them had the same misconception that a map projection is the same as a map kind (political, geomorphological, etc.). Although the majority could understand the basic difference between the globe and a map (the globe is a sphere and maps are rectangular), they could not think that this difference is leading to distortions in distances, shapes and sizes of areas. Furthermore, it was difficult for them to accept that in Mercator projection a curved line is the shortest route that an airplane could travel.

Overall, common student misconceptions could be summarized in the following:

- Map projections are the same as the political or geomorphological maps.

- Mercator, Robinson, Peters and Mollweide projections are correct. All these map projections are common in student´s school life, so they believe that these maps do not have any distortions.
- There are no distortions in world maps.
- In Mercator projection the shortest distance between two points is a straight line and a curved line is the longest.
- World maps are different for reasons related to the *construction of the map* (errors in the construction process of the map, expediency, deception, date of construction of the map) or due to *map elements* (shape, scale, size, type).

At this point, it is vital to remember the research questions and present the interpretation of research findings.

Research question 1: Can students find out that all the map projections contain errors and cannot imprint the Earth´s surface with complete accuracy?

As it was mentioned before, students did not know that there are several map projections and they believed that the correct ones were the projections that they meet in their everyday school life, like Mercator or Robinson projection. It has been proven by researchers that students prefer the projections that are used in school or the ones that look like a sphere (Tzotzis & Lamprinos, 2004). Generally, in the present research it was found that the majority of the students had not cultivated their critical and spatial knowledge through the Geography lesson. The lack of the promotion of critical thinking skills could be due to the way of teaching Geography, where sterile knowledge predominates (Leinhardt, Stainton, & Bausmith, 1998; Wakabayashi & Matsui, 2013; Likouri, Klonari, & Flouris, 2017).

Through guided investigation and experimentation in digital world maps and experiential activities, student's beliefs changed on an important level. They comprehended that all maps are wrong. Specifically, students learned that:

- Maps cannot capture the Earth's surface with absolute accuracy.
- There are plenty of map projections, because scientists and researchers have not found so far a totally effective way of presenting the Earth into a flat surface.
- There are many distortions in maps as far as distance, size and shape is concerned.
- In Mercator projection the great circle (curved line) is a shorter route than a rhumb line (straight line).
- When measuring distances at the same meridian, the great circle and rhumb line in Mercator projection are almost the same or the same.

- 3d globe sphere perfectly reflects the actual size of countries.
- Robinson and Peters projections show the size of countries approximately the same as a 3d globe sphere.

Research question 2: Can students' misconceptions about distances, sizes and shapes be eliminated by experimentation with Geographical Information Systems (GIS) and experiential activities?

When students are involved with GIS and experiential activities (like the conversion of a sphere into a flat surface and/or the measurement in inflatable and deflatable globes), they understand more efficiently the representation issues of mapping.

The majority of the children stated that they did not learn Geography by practicing in such interesting activities using digital world maps. Particularly, students in Greece are taught about map projections once during in their school life at the age of 13-14 (Intzidou & Lambrinos, 2019). As a result, students cannot comprehend the distortions in world maps. The lack of interaction with different world maps at the Geography lesson is also mentioned by Amosun (2016).

Therefore, Schomer's (2019) view that the use of globes, internet mapping tools and maps with different map projection can prevent the manifestation of misconceptions in Elementary Schools, is confirmed. Comparing world maps is the right way to introduce the concept of map projections into the classroom (Lambrinos, 2009, Intzidou et al., 2021). It is necessary for students to study different types of maps, to record information that maps offer and to identify their differences. The students should know how to accurately interpret maps and acknowledge their limitations. Working on cartographic representation helps to show students that the map is constructed and therefore not neutral and objective. One of the challenges of education today is learning how to read critically. It is necessary, to incorporate techniques to teach young students how to evaluate images (Intzidou et al., 2021).

FUTURE RESEARCH DIRECTIONS

A suggestion for further research could be a replication study on younger students, such as less than 8-year-old, to confirm or disprove the view of Catling (2018), who stated that children between 5 and 8 years old can compare a globe and world maps and look for the same or different features on each, introducing them indirectly not only to the subject of map projections, but also to the problem of mapping the Earth's surface into a two-dimensional map.

ACKNOWLEDGMENT

We are grateful for the support we received from the Hellenic digital-earth Center of Excellence (www.digital-earth.edu.gr). We would also like to thank all the students who participated in this research.

REFERENCES

Amosun, P. A. (2016). Why Nigerian Geography Teachers Scarcely and Scantly Teach Map Reading and Why Students Are Scared of It. *American Educational Research Journal*, *4*(2), 42–48.

Ashmore, B. (2003). Arno Peters changed the world! Development education and the Peters' projection. *The Cartographic Journal*, *40*(1), 57–5. doi:10.1179/000870403235002097

Attfield, I., Tamiru, M., Parolin, B., & DeGrauwe, A. (2002). *Improving Micro Planning in Education Through a GIS: Studies on Ethiophia and Palestine*. UNESCO.

Baker, T., Kerski, J., Huynh, N., Viehrig, K., & Bednarz, S. (2012). Call for an agenda and center for GIS education research. *Review of International Geographical Education Online*, *2*(3), 254–288.

Basaraner, M., & Cetinkaya, S. (2019). New measures for analysis and comparison of shape distortion in world map projections. *Cartography and Geographic Information Science*, *46*(6), 518–531. doi:10.1080/15230406.2019.1567394

Bednarz, S. W., & Van Der Schee, J. (2006). Europe and the United States: The implementation of geographic information systems in secondary education in two contexts. *Technology, Pedagogy and Education*, *15*(2), 191–205. doi:10.1080/14759390600769573

Blankman, M., Schoonenboom, J., Van der Schee, J., Boogaard, M., & Volman, M. (2016). Learning to teach geography for primary education: Results of an experimental programme. *Journal of Geography in Higher Education*, *40*(3), 425–441. doi:10.1080/03098265.2016.1144731

Brinkley-Etzkorn, K. E. (2018). Learning to teach online: Measuring the influence of faculty development training on teaching effectiveness through a TPACK lens. *Internet and Higher Education*, *38*, 28–35. doi:10.1016/j.iheduc.2018.04.004

Catling, S. (2018). To know maps: Primary school children and contextualized map learning. *BoletimPaulista de Geografia*, *99*, 268–290.

Collins, L. (2018). The impact of paper versus digital map technology on students' spatial thinking skill acquisition. *The Journal of Geography, 117*(4), 137–152. doi:10.1080/00221341.2017.1374990

Doering, A., & Veletsianos, G. (2008). An investigation of the use of real-time, authentic geospatial data in the K-12 classroom. *The Journal of Geography, 106*(6), 217–225. doi:10.1080/00221340701845219

Fargher, M. (2017). GIS and Other Geospatial Technologies. In M. Jones (Ed.), *Handbook of Secondary Geography* (pp. 244–259). Geographical Association.

General Navigation. (n.d.). Retrieved August 20, 2021 from https://www.nordian.net/REPOSITORY/110_easa_general_navigation_demo.pdf

Gilmartin, P. P. (2017). Showing the Shortest Routes-Great Circles. In Handbook of Choosing a Map Projection. Lecture Notes in Geoinformation and Cartography. Subseries: Publications of the International Cartographic Association (ICA) (pp. 1-48). Springer.

Goldstein, D., & Alibrandi, M. (2013). Integrating GIS in the middle school curriculum: Impacts on diverse students' standardized test scores. *The Journal of Geography, 112*(2), 68–74. doi:10.1080/00221341.2012.692703

Gutstein, E. (2013). Math, Maps, and Misrepresentation. In E. Gutstein & B. Peterson (Eds.), *Rethinking Mathematics Teaching Social Justice by the Numbers* (2nd ed., pp. 187–200). Rethinking Schools.

Hicks, D., Lee, J., Berson, M., Bolick, C., & Diem, R. (2014). Guidelines for using technology to prepare social studies teachers. *Contemporary Issues in Technology & Teacher Education, 14*(4), 433–450.

Intzidou, G., & Lambrinos, N. (2019). Investigation of the knowledge of the students of Elementary Schools in digital cartography. *Proceedings of 12th International Conference of Greek Geographical Company "INNOVATIVE GEOGRAPHIES: Mapping and Modeling our World"*, 1-4.

Intzidou, G., Lambrinos, N., Tourtouras, Ch., & Seroglou, F. (2021). Metadata: A pedagogical tool for the teaching of map projections in elementary schools. *European Journal of Geography, 12*(3), 56–59. doi:10.48088/ejg.g.int.12.3.56.69

Jadallah, M., Hund, A. M., Thayn, J., Studebaker, J. G., Roman, Z. J., & Kirby, E. (2017). Integrating geospatial technologies in fifth-grade curriculum: Impact on spatial ability and map-analysis skills. *The Journal of Geography, 116*(4), 139–151. doi:10.1080/00221341.2017.1285339

Jo, I., & Bednarz, S. W. (2009). Evaluating geography textbook questions from a spatial perspective: Using concepts of space, tools of representation, and cognitive processes to evaluate spatiality. *The Journal of Geography*, *108*(1), 1, 4–13. doi:10.1080/00221340902758401

Jo, I., Hong, J. E., & Verma, K. (2016). Facilitating spatial thinking in world geography using Web-based GIS. *Journal of Geography in Higher Education*, *40*(3), 442–459. doi:10.1080/03098265.2016.1150439

Karvánková, P., & Popjaková, D. (2018). How to link geography, cross-curricular approach and inquiry in science education at the primary schools. *International Journal of Science Education*, *40*(7), 707–722. doi:10.1080/09500693.2018.1442598

Kessler, F. C., Battersby, S. E., Finn, M. P., & Clarke, K. C. (2017). Map Projections and the Internet. In M. Lapaine& E. Usery (Eds.), Choosing a Map Projection. Lecture Notes in Geoinformation and Cartography (pp. 117-202). Cham: Springer. doi:10.1007/978-3-319-51835-0_4

Keuning, J. (1955). The history of geographical map projections until 1600. *Imago Mundi*, *12*(1), 1–24. doi:10.1080/03085695508592085

Lambrinos, N. (2009). *Teaching about school geography*. GRAFIMA Press.

Lambrinos, N., & Asiklari, F. (2014). The introduction of GIS and GPS through local history teaching in primary school. *European Journal of Geography*, *5*(1), 32–47.

Lapon, L., Ooms, K., DeMaeyer, P. (2020). The Influence of map projections on people's global-scale cognitive map: A worldwide study. *ISPRS International Journal of Geo-Information*, *9*(4), 196. doi:10.3390/ijgi9040196

Leinhardt, G., Stainton, C., & Bausmith, J. M. (1998). Constructing maps collaboratively. *The Journal of Geography*, *97*(1), 19–30. doi:10.1080/00221349808978821

Likouri, A. A., Klonari, A., & Flouris, G. (2017). Relationship of Pupils' Spatial Perception and Ability with Their Performance in Geography. *Review of International Geographical Education Online*, *7*(2), 154–170.

Mishra, P. (2019). Considering Contextual Knowledge: The TPACK Diagram Gets an Upgrade. *Journal of Digital Learning in Teacher Education*, *35*(2), 76–78. doi: 10.1080/21532974.2019.1588611

Monmonier, M. S. (2004). *Rhumb Lines and Map Wars: A Social History of the Mercator Projection*. University of Chicago Press. doi:10.7208/chicago/9780226534329.001.0001

Olson, J. (2006). Map projections and the visual detective: How to tell if a map is equal-area, conformal, or neither. *The Journal of Geography*, *105*(1), 13–32. doi:10.1080/00221340608978655

Pamuk, S. (2012). Understanding pre-service teachers' technology use through TPACK framework. *Journal of Computer Assisted Learning*, *28*(5), 425–439. doi:10.1111/j.1365-2729.2011.00447.x

Peterson, E. G., Kolvoord, B., Uttal, D. H., & Green, A. E. (2020). High school students' experiences with geographic information systems and factors predicting enrollment in the geospatial semester. *The Journal of Geography*, *119*(6), 238–247. doi:10.1080/00221341.2020.1824009

Purwanto, U., Utaya, S., Handoyo, B., & Bachri, S. (2020). Transformation of Geospatial Technology Knowledge in Pre-service and Experienced Geography Teachers as Pedagogical Tools in the Technological-Pedagogical-Content Knowledge Framework. *International Journal of Learning, Teaching and Educational Research*, *19*(9), 58–76. doi:10.26803/ijlter.19.9.4

Roberts, M. (2013). *Geography through Enquiry*. Geographical Association.

Roberts, M. (2017). Geography education is powerful if … *Teaching Geography, 42*(1), 6-9.

Rød, J., Larsen, W., & Nilsen, E. (2010). Learning geography with GIS: Integrating GIS into upper secondary school geography curricula. *NorskGeografiskTidsskrift–Norwegian Journal of Geography*, *64*(1), 21–35. doi:10.1080/00291950903561250

Šavrič, B., Patterson, T., & Jenny, B. (2019). The Equal Earth map projection. *International Journal of Geographical Information Science*, *33*(3), 454–465. doi: 10.1080/13658816.2018.1504949

Scholz, M. A., Huynh, N. T., Brysch, C. P., & Scholz, R. W. (2014). An evaluation of university world geography textbook questions for components of spatial thinking. *The Journal of Geography*, *113*(5), 208–219. doi:10.1080/00221341.2013.872692

Schommer, M. (2019). Spatial orientation - competence expectations and common misconceptions based on map projections. In M. Gröger, C. Prust, A. Flügel (Eds.), Cultural appropriation of spaces and things, 20 – 30 October 2019 (pp. 231 – 244). Siegen: Universi.

Stevens, A., & Coupe, P. (1978). Distortions in Judged Spatial Relations. *Cognitive Psychology*, *10*(4), 422–437. doi:10.1016/0010-0285(78)90006-3 PMID:699514

Tseng, W. K. (2014). The shortest overall distance of two piecewise rhumb-lines. *10th International Conference on Natural Computation (ICNC)*, 1153-1157. 10.1109/ICNC.2014.6976004

Tzotzis, I. (2007). *The practical importance of map projections in the teaching of maps in school* [Doctoral Thesis]. Faculty of Primary Education, A.U.T.H., Thessaloniki, Greece.

Tzotzis, I., & Lambrinos, N. (2004). The cognitive competence of first grade of Gymnasium in the subject of mathematical geography. *Proceedings of 7th Pan-Hellenic Geographical Conference*, 2, 399-406.

Valovicová, L., Vallo, D., & Kramáreková, H. (2019). Mercator projection as a tool for development of interdisciplinary reasons among physics, mathematics and geography. In *21st International Conference DIDFYZ, 9 – 12 October 2019*. Terchova: AIP.

Wakabayashi, Y., & Matsui, Y. (2013). Variation of geospatial thinking in answering geography questions based on topographic maps. *Proceedings of the 26th International Cartographic Conference*.

Walshe, N. (2017). Developing trainee teacher practice with geographical information systems (GIS). *Journal of Geography in Higher Education*, *41*(4), 608–628. doi:10.1080/03098265.2017.1331209

Wiegand, P. (2006). *Learning and Teaching with Maps*. Routledge. doi:10.4324/9780203477793

Wilcke, H., & Budke, A. (2019). Comparison as a method for geography education. *Education in Science*, *9*(3), 225–240. doi:10.3390/educsci9030225

ADDITIONAL READING

Gosling, P. C., & Symeonakis, E. (2020). Automated map projection selection for GIS. *Cartography and Geographic Information Science*, *47*(3), 261–276. doi:10.1080/15230406.2020.1717379

Grafarend, E. W., You, R. J., & Syffus, R. (2014). *Map Projections: Cartographic Information Systems* (2nd ed.). Springer.

Greco, S. (2018). Seven possible states of geospatial data with respect to map projection and definition: A novel pedagogical device for GIS education. *Geo-Spatial Information Science*, *21*(4), 288–293. doi:10.1080/10095020.2018.1536406

Hruby, F., & Riedl, A. (2018). 2000 years of 'globes vs. maps' – Lessons (to be) learned. *International Journal of Cartography*, *4*(2), 186–200. doi:10.1080/23729333.2017.1422097

Kerkovits, K. (2020). Quadrature rules to calculate distortions of map projections. *The Cartographic Journal*, *57*(3), 249–260. Advance online publication. doi:10.1080/00087041.2020.1714278

Leitão, H., & Gaspar, J. A. (2014). Globes, rhumb tables, and the pre-history of the Mercator projection. *Imago Mundi*, *66*(2), 180–195. doi:10.1080/03085694.2014.902580

Monmonier, M. S. (2004). *Rhumb Lines and Map Wars: A Social History of the Mercator Projection*. University of Chicago Press. doi:10.7208/chicago/9780226534329.001.0001

Šavrič, B., Jenny, B., White, D., & Strebe, D. (2015). User preferences for world map projections. *Cartography and Geographic Information Science*, *42*(5), 398–409. doi:10.1080/15230406.2015.1014425

Sun, S. (2016). Symbolize map distortion with inscribed circles in polygons. *International Journal of Cartography*, *2*(2), 166–185. doi:10.1080/23729333.2016.1179863

Wiegand, P. (2006). *Learning and Teaching with Maps*. Routledge. doi:10.4324/9780203477793

KEY TERMS AND DEFINITIONS

Distortion: The misrepresentation of a feature on a map. There are no map projections that can maintain a perfect scale throughout the entire projection, because they are taking a spheroid and forcing it onto a flat surface.

Geographic Information Systems (GIS): A system that creates, manages, analyzes, and maps all types of data. GIS helps users understand patterns, relationships, and geographic context. The benefits include improved communication and efficiency as well as better management and decision making.

Great Circle: The largest possible circle that can be drawn around a sphere.

Map Projection: One of many methods used to represent the 3-dimensional surface of the earth on a two-dimensional plane in cartography (mapmaking) calledcartographic projection.

Misconception: A conclusion that's wrong because it's based on faulty thinking or facts that are wrong.

Rhumb Line: A line on the surface of the earth that follows a single compass bearing and makes equal oblique angles with all meridians also called loxodrome.

Spatial Thinking: A distinctive, universal, and powerful form of thinking used in multiple disciplines and in real-world activities. Space, representation, and reasoning are inseparably integrated in spatial thinking and therefore it is, and has always been, central to the teaching and practice of Geography in schools.

Chapter 9
Reading and Spatial Analysis Through Mental Maps in School Geography

Denis Richter
https://orcid.org/0000-0001-7133-5279
Universidade Federal de Goiás, Brazil

ABSTRACT

This chapter aims to highlight the mental map as a relevant pedagogical proposal for the work of school geography linked to the process of construction of the students' geographic thinking. To develop this idea, at first, the relationship of geographic science with the issue of spatiality as a way of thinking is highlighted, and, subsequently, the contribution of cartographic language as a means to systematize reading and spatial analysis. In this sense, the mental map is required as a proposal for a cartographic representation that enables the subject who builds it to highlight different contexts, interpretations, and perceptions about a given space or place. For this, some examples of the construction of mental maps from the perspective of school activities will be presented and discussed, highlighting the potential of this material for the development of a more critical reading and analysis of the geographic space.

INTRODUCTION

Teaching Geography is a task that requires the teacher a set of knowledge, expertise, and specific languages that contribute to the development of students' spatial reading and analysis of a given reality. For this to happen in the school context, the teacher

DOI: 10.4018/978-1-7998-9598-5.ch009

must understand the importance of articulating the concept of geographic space with the daily practices experienced by students. In this sense, one of the possible methodological paths to accomplish this demand is the use of mental maps linked to the work of school Geography.

This didactic-pedagogical approach is based on the recognition of Cartography as a powerful language for the development of the teaching-learning process in Geography, mainly for its major purpose which is to represent the space. It is understood that this function of the map, one of the main products of Cartography, makes it possible to promote the construction of questions and spatial reflections to the subject, either by the reading activity or by the construction of this cartographic representation. This quality of the map meets one of the specificities of Geography, which refers to its scope of organizing and problematizing spatial thinking. But, for these ideas to become effective in the school environment, it is essential to structure some methodological paths, such as the use of the mind map in Geography classes integrated with practices that value the reading and analysis of space.

Thus, this article aims to present how the mind map qualifies as a pedagogical proposal relevant to the teaching of Geography, from the perspective of developing the students' geographic thinking. For this, the first part of this text will discuss the existing link between geographic science and the conception of spatiality, from the idea that thinking about space is a specific way to reason through Geography. This debate will be articulated in the teaching work, taking as a parameter the challenges present in this practice. In this sense, to develop this proposal in the classroom context, it is essential to establish some methodological guidelines and orientations that strengthen the students' cognition, that is, to indicate a possible path for learning. Thus, the second part of this article will present some guidelines that enable the integration of the construction of mental maps in Geography classes, having as references research that has contributed to the advancement of this theme in the field of geography education.

Teaching Geography: Spatial Thinking in Focus

Geographical science has been concerned for a long time with recognizing that spatial thinking has a prominent place in the reading and analysis of phenomena and facts that occur in reality. Authors such as Moreira (2007) and Gomes (2017b) add to this debate by highlighting in their studies the appreciation of the spatial perspective concerning this science. This theme has been fostering the current theoretical discussions of Geography and, consequently, increasingly consolidating one of the characteristics that identify this knowledge, i.e., that developing this scientific knowledge means thinking about a certain space and problematizing the possible contexts that justify and define its construction process. It is also important to highlight that this kind of thinking is not restricted to the academic field; it is known that Geography can be practiced at all levels and, therefore, has this attribute of reflecting on the production of space. For Cavalcanti (2019, p. 83, author's translation),

The understanding of Geography, [...] is that there is a complex spatiality to be known and analyzed. It is always in motion and in dialectical tension between several of its dimensions. Thus, in order to reach its understanding, it is necessary to develop the capacity provided by a type of thinking - theoretical and conceptual, which goes beyond facts and phenomena in their empirical description and characterization. Finally, this thinking is capable of producing analyses that articulate subjective, imponderable elements that cannot be measured (the senses and meanings of the locations in their relational dimension).

Therefore, the concern with space is not a minor task of Geography, either in the context of academic production or in the school environment. For, it is understood that this demand is among the central objectives of this science and that its contribution is in enabling the subject to a more attentive spatial reading and analysis for the processes and situations that are part of their daily lives, at different scales.

In this sense, taking the presence of Geography in school as a reference, one can understand that the development of students' spatial thinking becomes one of the primary actions of this discipline, by recognizing its importance both for everyday practices and for more complex readings about places. It is assumed that having greater clarity about this focus helps to better define strategies and teaching methodologies that give meaning to the didactic work of Geography. In fact, lately, it has been common for teachers of Geography in Basic Education to have a certain doubt about the definition or understanding of what spatial thinking is and how we can integrate it into school practices. This context is observed from the experiences in initial and continuing teacher education courses, as well as in the development of scientific research that approaches the school.

By being aware of this scenario, it is important to highlight that the debate on spatial thinking has been increasingly recurrent in Geography studies and linked with greater strength to the approaches of the teaching-learning process. However, the construction of ideas about this theme is not recent. Castellar and De Paula (2020) highlight this historical process and the contribution of the development about spatial thinking from the perspectives of the cognitive psychology field and geographic education, especially in Anglo-Saxon studies. In this path, it is notorious the relevance that the publication of the National Research Council (NRC, 2006) document presented for the dissemination and formulation of more structuring concepts about spatial thinking. It can be said that this material has become one of the major references when dealing with the respective theme, regardless of the fact that there is no consensus among all researchers about its theoretical and methodological conceptions. For Silva [et al] (2018), despite identifying valuable contributions in the NRC document, some issues need to be analyzed more carefully, especially in relation to Brazilian research developed on geographic education.

On the other hand, even though there is some criticism about this material, it is possible to recognize a significant advance in the ideas regarding the understanding of what spatial thinking is. Based on the NRC document, it is evident the importance and the strong presence of spatial intelligence in the daily practices of individuals and its proximity to the focus of the study of Geography – spatial reading and analysis. It is in this context that the reflections and ideas produced so far should be highlighted in relation to spatial thinking. For, even acknowledging its limitation in what is understood as geographic, which goes beyond the Cartesian basis or only the logic of location, there is a fundamental reference in the development of spatial thinking that is found in what is specific to think about a given place - the where? This question can be taken as a central issue for the construction of geographical reading about a given context, fact, or situation. That is, it leads to the formulation of a cognitive structure of thinking through Geography.

If the goal of geographic education is the formation of geographic thoughts or reasoning, as advocated by several authors (CAVALCANTI, 2000 and 2002; CASTELLAR, 2005) and if we want the map to stop being used only for the location of phenomena and that, in addition to this, students/citizens are enabled to advance beyond the elementary level of map reading (SIMIELLI, 2007), we need to base our School Cartography methodology on the comprehension of cognitive operations that involve the understanding of socio-spatial phenomena with strong geographic content. (DUARTE, 2017, p. 199, author's translation)

It is pertinent to recognize that the ideas presented by Duarte (2017) about the development of spatial thinking have been occupying other spaces beyond the academic, such as curriculum documents. It has been increasingly recurrent the highlight of this type of cognition in the texts that guide and orient the geographic contents in Basic Education. This context can be evidenced in the United States, with the National Geographic Standards (DUARTE, 2017), in Portugal, with the National Curriculum for Basic Education (FREITAS and ALMEIDA, 2014), and in Brazil, with the National Common Curricular Base (RICHTER and MORAES, 2020), to name a few cases. These examples reveal a certain movement, theoretical and curricular, to bring the principles of geographic science closer to the teaching-learning process of this discipline in school, reinforcing the conception that the development of spatial thinking, even though it is a type of intelligence that is linked to different fields of knowledge, has a centrality for the studies of geographic science. Therefore, it is considered of utmost importance to think about strategies and methodologies that enhance the development of this type of thinking in school and the place of Geography in this process.

Concerning this debate, the contribution of Gomes (2017b) is pertinent when saying that geographic science is structured under three domains, namely: the spatial dimension, spatial intelligence, and the logic of locations. For the said author, "Geography is a field of study that interprets the reasons why diverse things are situated in different positions or why diverse spatial situations can explain different qualities of objects, things people, and phenomena" (id, p. 20). This perspective highlights the strong relationship and association that Geography has with the spatial issue since it is not just a nominal complement to this field of knowledge, but a specificity of this science. Therefore, developing a geographical reading on a given object requires a methodological approach that subsidizes the construction of a spatial perspective, as Castellar and De Paula (2020, p. 298, author's translation) point out:

[...] spatial thinking is a procedural content, an action directed to an end, we understand that it consists of three fields of knowledge 'that are aligned', associated with a geographic situation that will stimulate the student to argue with consistency through the geographic vocabulary, and that becomes a procedure that stimulates geographic reasoning that will deepen and make sense of the contents.

This discussion presented by the authors indicates the need for articulation of what has been advanced in the theoretical studies of Geography with its methodological approach, that is, how the spatial thinking is an important cognition for the teaching-learning process of Geography in school (GERSMEHL, 2008). Thus, it is understood the need to approach this discussion with the teaching practice, because the teacher's work requires an articulation with the theoretical and methodological bases of geographic science. These ideas are structured from the demands of actions of the teaching work because it is considered that one of the main concerns of the Geography teacher is to recognize what a student actually needs to know about this discipline in the development of school practices.

In a way, it is understood that for Geography the word "space" has great importance, both in terms of the idea of concept and of representation. For this, the term space receives a specific nomenclature - the geographic space. In other words, it is the reading, analysis, and interpretation of Geography about the spatial arrangements produced by society in a given area, territory, or scale.

To help the teacher in this process of better understanding the geographic space, it is essential the contribution of other concepts, such as place, region, scale, landscape, territory, nature, among others, which give meaning and, at the same time, articulate different perspectives on space. Of course, in school Geography, these concepts should not be treated directly, as if the purpose of the classes of this discipline was only to explain these concepts to the students. On the contrary, these concepts are configured as basic elements to help define geographic contents and themes, that is, these concepts are used to better define what will be effectively taught in Geography classes, such as the city, environmental issues, landforms, economic development, population growth, social inequality, globalization, etc.

Taking this context as a reference, it can be said that regardless of the content to be worked on and taught in the Geography class, some elements and concepts that give meaning to the geographic science. In this sense, one of the points of great articulation to foster this relationship is in the recognition that teaching Geography is teaching to think about space – to think geographically. For this, in addition to the content and knowledge that help detail each situation and phenomenon, some specific languages are used, as is the case of the map, an important product of cartographic language.

In this way, the map is highlighted in Geography classes because it provides an opportunity, through its own representation, to see how certain facts and processes are spatialized, that is, how they occupy, distribute, configure, and organize themselves in space. At first, these spatial readings are more related to the physical and metric dimension of the terrain, of how each phenomenon occupies space. In a way, it is initially more concerned with spatiality itself, that is, with the elements that give meaning to the spatial configuration than with the content represented on the map. This relates to spatial practices, concerning the order and position of things in space. Therefore, this type of intelligence - spatial intelligence - is used to perform numerous tasks and actions that are present in everyday life. For example, one can think of a hypothetical situation in this context as follows: when looking at a city map it is possible to identify an environmental preservation area located in the western region, this area is ten thousand square meters and in its northern portion there is a lake. From this short report it is understood that to describe this space, resources are validated that give dimension to its spatiality, its physical and locational configuration, which are very present in social and spatial practices.

However, it is important to consider that even though spatial thinking is strongly linked to human action, often innate, its development does not occur randomly or spontaneously. For this, according to Golledge (1995) and Jo and Bednarz (2009), spatial thinking requires different concepts, such as primitive and complex ones. Recognizing this difference between the types of spatial concepts present in the spectrum of spatiality helps to understand that there are readings about space, ranging from the simplest to others that are more elaborate. Therefore, having an understanding of this structure becomes fundamental so that school practices can effectively contribute to the development of students' spatial thinking. About this debate, Soares (2017, p. 24, author's translation) collaborates by saying that,

[...] for an evolution in students' spatial and geographic thinking to occur, teachers should ask questions that stimulate students to use spatial concepts. The first questions should be simple and require concepts according to the level of the question asked, and the following questions should be a little more complex and require concepts of a higher level.

This means that the organization of the teaching work needs articulated planning with the different stages linked to spatial thinking so that it can effectively contribute to geographic learning. However, this requires the use of teaching materials that make this process possible. Therefore, the spatial elements are used to give meaning to the configuration of space, as a first approach to recognizing a given content. In this work, one can point out the contribution of the map, from its representation, by allowing the visualization of these spatial elements through the cartographic language.

However, it should be noted that if this reading is restricted to spatial description alone, it is likely to limit the interpretations that can be made about space. To this end, we integrate in this intellectual effort the geographic questions, for example, questions that help to tense reflections beyond the spatial. Taking the hypothetical situation presented above as a reference, the questions articulated with the geographical issues would be constituted as follows: how is this environmental preservation area maintained until today? What is its importance for the city? What are the problems that can harm the environmental quality of this area etc. From these questions, it is possible to observe that ways of thinking about this space are brought up based on geographical principles, in which some reasoning such as extension, differentiation, connection, analogy are taken as a reference to better understand this place, its contexts, and the processes of transformation and permanence.

In other words, a geographic sense is highlighted at this moment for the analysis of this place, developing geographic thinking. This intellectual exercise also draws on the importance of the map to build such reasoning that helps to better organize questions or even answer them, because a close look at the map makes it possible to think about space from different contexts and approaches. But for this to be possible, it is necessary to make the map become a language that effectively dialogues with Geography, that overcomes the idea of being taken only as an illustration.

Well, by highlighting these ideas we hope to highlight how spatial thinking is an important element and concept that contributes to the construction of another thought, in this case, broader and more complex, known as geographic thinking. And, in this path, it is also observed the significant contribution of the map to assist and collaborate in the development of more critical readings about the geographic space.

THE SCHOOL CARTOGRAPHY IN GEOGRAPHY TEACHING: PATHS FOR SPATIAL READING AND ANALYSIS

Before going deeper on the mental map as an important product of cartographic language and its possibility to contribute with spatial reading and analysis in Geography teaching, it is considered necessary to recognize how the visual issue has been taking prominence in the debates of geographic science.

Some studies of Geography in recent years have demonstrated the strong relationship that this science has with the visual perspective (GOMES, 2013 and 2017b; BESSE, 2014). For Fonseca (2019, p. 44, author's translation), "images have historically been part of the repertoire of Geography", evidencing its tributary role in the construction of the look of this science towards the facts and phenomena that occurred and occur in a given reality. These notes have contributed to strengthening the idea that, in order to think geographically, it is pertinent to make use of resources or materials that enable the visualization of space, for its subsequent reading and analysis. Hollmann and Lois (2015, p. 29, author's translation) highlight this articulation very well, for the referred authors

The geographer's profession has been built and legitimized around practices that have a visual framework: geographers analyze, produce, search and compare images; we use images in the production, communication and teaching of geographic knowledge.

In this sense, to meet these demands, it is recognized that visual accuracy requires from this professional, the geographer, the use of specific tools that can contribute to the development and construction of a spatial analysis of different geographic situations. It is in this context that one can call attention to the map as an important graphic and visual resource that enables its reader to have an understanding of the different processes that interfere in the production of space. For, in this way, the map gains relevance for Geography not only from the point of view of illustrating or presenting a certain area but for allowing that in its reading and analysis different elements are taken into consideration and that its set enables the construction of an effectively geographic interpretation of space. That is, it is taken as a methodological reference that the map itself is not an eminently geographic product, since its nature runs through different fields of knowledge, among which Cartography itself. But it is the interpretation that is made of the map that makes it geographic since this task requires the adoption of concepts and reasoning that are proper of Geography.

Besides recognizing the importance of spatial and geographic thinking for the teaching of Geography, it is also highlighted how the work with cartographic language can contribute to this process. However, it would not be possible to indicate the numerous possible paths to accomplish this task, since Cartography is a source

Figure 1. Theoretical and methodological path of school cartography in geography teaching
Org.: Author, 2022

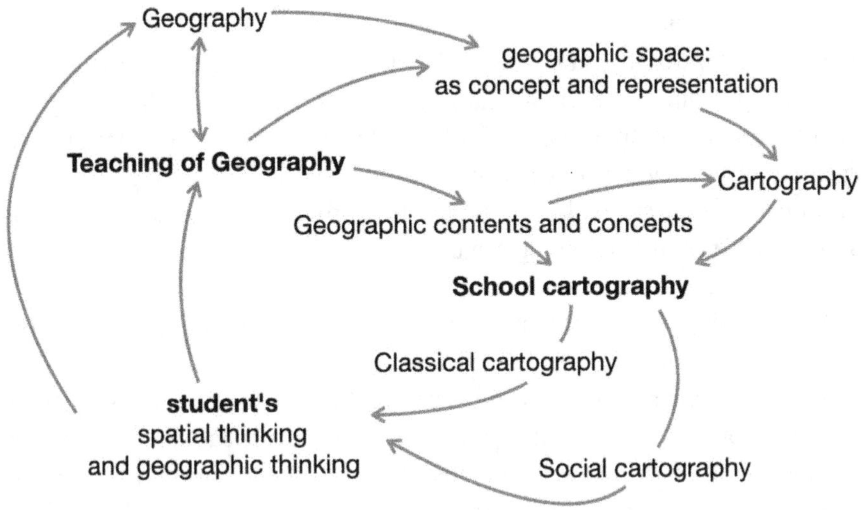

of distinctive possibilities for schoolwork. For this, it is necessary to present a theoretical-methodological structure that has subsidized the understanding of the place of school Cartography in the teaching of Geography, as well as elements that interfere and that result from this movement. As can be seen in Figure 01, below.

Regarding figure 01, it is important to consider that this schematic representation is presented cyclically, i.e., it is understood that the educational process occurs in a feedback system, which is, at the same time, contributory and tributary of their cognitive paths. This conception is based on the ideas of Vygotsky (2000) and the so-called Critical Historical Pedagogy, which is structured on the recognition that the learning of individuals is the result of an articulation between everyday and scientific knowledge, which are interdependent and inseparable. In this way, the idea is not to establish an end in itself but a movement that expands, widens, and becomes more complex as this cycle is traveled.

It can be observed in this scheme that three words are highlighted in bold for having a centrality, namely: Geography teaching, school Cartography, and student. This is justified by the fact of understanding that, firstly, all this movement stems from the presence of Geography as a school discipline. Therefore, the teaching of Geography would be the main vector in constituting itself, at the same time, as a science and as disciplinary knowledge, and establishing all the relations present in this system. In this turn, cartography is highlighted by its articulation as a powerful language of Geography, by graphically representing the space and allowing its geographic reading and analysis by the school subjects. And, last but not least, the student is established in this scheme as the target audience of this whole process, that is, it indicates that this movement exists so that the learning of this individual can effectively occur.

In this sense, we highlight how school cartography is integrated into the teaching of Geography, from the teaching-learning process and its unfolding as proposals for spatial representation (Classical Cartography and Social Cartography), with the perspective of developing students' spatial and geographic thinking.

After the appropriate explanations have been given, the scheme in figure 01 can be understood from the following route:

1. The teaching of Geography is constituted as a propeller element for the development of all the actions present in this system, but its strong link with the geographic science itself must be highlighted, for we understand that the schoolwork of this science is done from its theoretical and methodological bases and that its impact is directly related to the perspective of developing the student's spatial and geographic thinking.
2. The work of the Geography teacher, in the context of the classroom, is organized from the recognition that this disciplinary knowledge has its own concepts, such as: geographic space, place, landscape, region, territory, scale, and nature. Among these, the geographic space is taken as a reference, which can be analyzed both as a concept and as a representation. The concept is established by understanding the space as a key concept for Geography (MOREIRA, 2007) and the representation from the perspective of materializing/presenting the space from a given product, as is the case of the map. Moreover, it is noteworthy that in the didactic work the Geography teacher uses the school contents to develop the knowledge related to this science, so it is this knowledge that is effectively constituted in the pedagogical practices.
3. Following the scheme is Cartography, in which case its presence is mainly conditioned to the highlight of space as representation. The illustrative, visual, and graphic character of cartographic language makes it possible to organize spatially the elements, concepts, and analyses that are constituted in and by social practices. And from its materiality, it is possible to build different readings about space. Thus, Cartography integrates this system in the perspective of problematizing reality, by understanding that the map is, at the same time, product and producer of spatial readings and analyses. However, it is important to point out that the school context does not work with cartographic science directly. For this, it is necessary to think about the adequacy of this language

to work with students, i.e., it is essential to think about a specific approach for students, a function established by the proposal of school Cartography.

4. From the work with school Cartography it is possible to carry out different ways and methodological proposals that enable the use of the most distinctive products of cartographic language in the teaching of Geography. Of these paths, two that are considered important and quite representative are highlighted in this system, which refers to the Classical Cartography and Social Cartography. The Classical proposal concerns the use of more conventional maps, generally associated with the work with thematic cartography and even cartographic content. They are those types of maps that are organized respecting some normative established by cartographic science, such as title, scale, legend, symbology, etc., and also by the Euclidean basis in relation to the so-called background of the map, a direct result of the combination between projection, scale and metric space (FONSECA, 2019). The other proposal, the Social, is constituted in the work of construction of maps that seeks to value more the phenomenon to be represented than to respect the bases of the flat geometry of Cartography. For Gomes (2017a, p. 101, author's translation),

While conventional [Classical] Cartography privileges Euclidean space, territory as a nation-state, precision, and pretended neutrality, Social Cartography prioritizes the lived, perceived, and conceived space, territory, and the territoriality issues of the communities and social groups involved in mapping. The content of symbols and political positioning are taken as a principle.

5. Even recognizing the specificities and characteristics of each proposal, it is known that there is not a best or most appropriate route to work in Geography schools. This choice is linked to the focus that the teacher wants to highlight or give more visibility in relation to the use of the map with a particular geographic content. Regardless of the option, it is understood that both proposals are tributary to the process of developing students' spatial and geographic thinking, by making use of the integration of cartographic language with school content. With this, the system concludes its cycle but does not end in itself, because it is understood that the construction of an evidently geographical reading about a given situation or fact enables students to develop new reflections on different contexts. In other words, new geographical readings.

All this detailed path in figure 01 helps to identify the place where the mental map is in the Geography teaching-learning process. Which, in this case, refers to the proposal of Social Cartography. Its inclusion in this approach is directly linked to the contributions of Jacob (2016, p. 226) when he says that "maps make the invisible visible", starting from a sociological conception of Cartography and recognizing its potential as a cognitive tool. This same author, based on the ideas of Harley (1991), highlights that,

The power of maps lies in the way they communicate knowledge and implicitly corroborate political and social order through their efficiency as symbols. Maps are didactic devices for the socialization of individuals, for the indoctrination of soldiers, officers, colonists, and schoolchildren. (JACOB, 2016, p. 229, author's translation)

Therefore, when having clarity of the ideological function of the map and that its construction is strongly associated with a given culture, a social and political discourse, and loaded with meanings, it is considered the need to think of proposals that make this more evident and that allow exploring also this nature of Cartography in the school context, as is the case of mental maps. In this sense, we try to make clear in the scheme of figure 01 that this proposed work with cartographic language is not an alternative that distances itself from the theoretical and methodological basis of Geography or that can be read as a proposal contrary to this science. In fact, the work with mental maps is understood as a model that enhances geographic readings and analysis that are often neglected by this disciplinary field. It is based on this that are presented below some reflections on the construction of mental maps in Geography teaching, regarding some research already done that allowed highlighting its contribution to the development of students' spatial reading and analysis.

MENTAL MAP: THEORETICAL AND METHODOLOGICAL PERSPECTIVES FOR THE WORK IN SCHOOL GEOGRAPHY

The first point that can be highlighted in relation to the mind map refers to its representative potentiality, by bringing to light and making visible elements, contexts, and perceptions that would not be apparent in a first observation of a given landscape or space. This demonstrates its proximity to the act of representing, to highlight that which is the result of interpretations from social practices and that, therefore, cannot have only one version or way of being presented. For, the human gaze is multiple, diverse, and built from different approaches and conceptions. Therefore, it is observed in this relationship its strong articulation with Geography, because for Kozel (2018, p. 33, author's translation) "the most fruitful perspective of analysis for geographic studies is the one that takes representation as 'image or drawing that represents an object or a fact', or [...] that refers us to the idea of mental maps".

These ideas help to understand how the mental map is constituted as a cultural product and full of meanings coming from its builder, in which the perception about a certain place stands out concerning the norms and rules established by cartographic communication. In this sense,

> *[...] the mind map is analyzed as a product of the individual's cognition in reference to the reading and interpretation of the different landscapes that are present in the space; in other words, it is the construction of a more geographical " view" on the contexts that occur in society. (RICHTER, 2011, p. 135, author's translation)*

Features like these demonstrate the specificity of the mind map and its working possibilities for the teaching of Geography, which in this case is inserted in the context of valuing the most subjective elements that students have about a given space. It is worth mentioning that this link of the mind map with the intrinsic aspects of those who build it actually demarcates its strong relationship with the psychological field, which is at the origin of this proposal. Since the first researches that made use of the mind map were more linked to Psychology, with the so-called "Personal Geography" (GOULD; WHITE, 1974). These proposals had as focus the recognition of the "[...] personal vision of the world mixed with the fantasy that each man possesses and the egocentric of the experience of this personal vision" (CAPEL, 1974, apud NOGUEIRA, 1994, p. 68, author's translation). Of course, this approach has been altered over the years and has made it possible to bring the construction of mental maps closer to Geography studies, by being structured in the sense of giving greater emphasis to the spatial and cognitive elements and their geographic contents present in this representation.

Besides these perspectives, one can also highlight the contribution of Buzan (2002), with the publication of the book "How to mind map" (published in 1974), when he presented this concept of mind map linked to the conceptions of information, memorization, and planning, in which this resource could help in the organization of individuals' ideas by enhancing creativity and the structuring of different data and elements. In this case, the studies of this British researcher were closer to the fields of administration, entrepreneurship, strategic management, business planning, and others. Nevertheless, one cannot disregard that these ideas helped consolidate the recognition of the mind map as a product that systematizes information and enables an understanding of the subjects' reading and interpretation of particular issues or themes.

Além desse pesquisador, outro nome que merece destaque na valorização de uma mudança na metodologia de construção do mapa refere-se a Lynch (1960). Seu estudo representa um importante marco para a análise do espaço sob uma perspectiva mais ampla, a partir da interpretação dos percursos urbanos dos moradores de algumas cidades norte-americanas. Mesmo este autor não tendo feito menção direta ao termo mapa mental em seu livro "A imagem da cidade", é visível que há uma forte relação desse tipo de representação espacial com as imagens produzidas pelos citadinos sobre suas respectivas áreas urbanas.

Years later, Gould & White (1974) published the book "Mental Maps" in which they present some possible paths to the production of this spatial representation. These authors point out that the construction of this type of map does not come only from the field of individuals' experiences about the places they live; they also result from the historical-social knowledge of society itself, which is passed on from generation to generation. Therefore, besides mental maps representing individual experiences, they transcend this scale of relationship and are records of a representation of human knowledge over time, which expresses, through cartographic language, their interpretations of the environment in which they live.

However, the term mental map is far from consensus among researches that focus on the theme of more free spatial representation. It is possible to find different nomenclatures for this type of cartographic product, each one with its own specificities and theoretical and methodological bases, such as cognitive maps, which are structured, according to Tolman (1948), on the understanding that individuals acquire signs of the environment by constructing a particular mental image, or for Downs and Stea (1973) who value perception as a central element in the immediate apprehension of the senses in the production of these representations; spatial schema, having as reference the studies of Lee (1968) that highlights what has become essential for the development of our body image, to orient ourselves, to move around, and to relate with other people; and another example refers to the duality of the terms mental maps and cognitive mapping. For Bertrand (1984), this issue can be resolved by seeking specific definitions for each. Thus, he points out:

Mental maps are graphical impressions made without any aid, i.e., they do not use a reference map or any other precise cognitive mapping document. Cognitive mapping is a process composed of a series of psychological transformations by which an individual habituates, encodes, keeps, recalls, and decodes information about relative locations and attributes of phenomena in their everyday spatial environment. (p. 68, translated by the author)

Based on this, it can be established that there is an order to be respected between these two terms; in this case, the cognitive mapping is developed before the production of the mind map because it is necessary that the observer first performs an internalized interpretation so that then he can transpose to the graphic language his impressions about the environment. In the same study, Bertrand (1984) comments that the mind map should not be interpreted as a truth, an unquestionable knowledge, but it is necessary to analyze it as a singular reading of the space.

Bertrand (1984) also proposes an approach to using mental maps in school practices of Geography. However, this researcher recognizes barriers preventing the inclusion of numerous teaching methodologies in didactic and pedagogical actions, either by infrastructure conditions or by the issue of class time dynamics. The result of these limitations is that the teacher always tries to use the lesson, the task activities, as "a teaching method apparently more effective" (id., p. 02). Thus, Bertrand stresses the importance of providing students with a Geography teaching closer to map reading and construction. He justifies that this pedagogical stance contributes to the individual's understanding of the various theoretical contexts of the Geography discipline itself, mainly because the map becomes a graphic language capable of expressing spatial processes and phenomena.

These investigations contributed to identifying the singular and plural character of the mental map, as well as the recognition that this product is loaded with elements, and contexts from the social and cultural practices of individuals, which are in vogue at the time of its construction. This means that when reading a mind map one has a dimension of both the subject's point of view, his or her particular vision of the world, as well as the elements, values and standards present in society that interfered in its materiality. Therefore, the mind map has the potential to reveal the interpretations of the subjects and bring to light the spatial perceptions of those who construct it.

Having these notes as a reference, we can observe how this proposal of the mind map was taken as an investigation focus for several studies, among which we can highlight in Brazil the researches of Nogueira (1994) and Teixeira (2001), and in Spain the study of Castro Aguirre (2004). The two Brazilian works represent, effectively, the pioneer researches on this theme in the country and have contributed by systematizing the use of the mind map in the context of Geography teaching. The dissertation of Nogueira (1994) aimed to value the spatial knowledge present in the mental map and its possibility of articulation with the geographic contents learned in school. While the doctoral research of Teixeira (2001) structured a methodology of interpretation of mental maps, having as a theoretical perspective the conception of the use of the sign as social representation, based on Bakhtin. It can be recognized that these two studies opened the way in Brazil for other researches to better understand the potentiality of the mind map for the didactic-pedagogical work of Geography in Basic Education. And regarding the publication of the Spanish researcher Castro Aguirre (2004), it is worth highlighting his theoretical and methodological effort to recognize the influence of cognitive sciences in the production of the mind map, identifying the importance of the subject's lived experiences as a contribution in the construction of this cartographic product.

Even though these three investigations have the mind map as a common element, it is known that their theoretical itineraries followed different paths. However, there is a point of connection between these researches regarding the valuation of the spatial category in the construction of the mind map, by revealing how the social and spatial practices of the subjects are tributary in the development of a reading of the space and in the cognitions linked to this concept. It can be said that these and other works, such as Galvão's (2007) and Kozel's (2007), were significant for the unfolding of some studies carried out in recent years focusing on the production of mental maps and their relation with Geography teaching, among which we highlight: Richter (2010), Lopes and Richter (2013) and Campos and Richter (2020).

These researches are considered representative for this investigative course by contributing to the theme of the mind map, its articulation with spatial reading and analysis, and the development of geographic thinking. Thus, it is noteworthy that,

> *[...] the mind map is analyzed as a resource that allows the construction of a freer graphic expression, having the perspective that the student can transpose to this spatial representation the geographic contents learned throughout his school education. Thus, besides using speech, writing, image or the conventional/traditional map itself, the student will have the opportunity to present in a mind map his interpretations about a certain place, being these ones coming from more scientific readings about the reality (RICHTER, 2010, p. 27, author's translation)*

With these works as references, some reflections and proposals related to the use of mental maps in the context of geographic education are presented. These methodological approaches are based on the concept that the school knowledge needs different instruments and resources that enhance the teaching-learning process. And, in the case of Geography, there is a strong link to the cartographic language as a material that enables an understanding of what is specific to this science - spatial reading and analysis. For Duarte (2017, p. 200),

> *[...] what we mean by thinking spatially to understand socio-spatial phenomena with strong geographic content, we want to state that not always when we organize an activity in Geography class we are contributing to teach the student to think spatially or, in other words, to develop their spatial thinking.*

Therefore, the following is a proposal of how to work with mental maps in Geography classes in Basic Education, having the perspective of promoting students' spatial reading and analysis (see table 01).

Methodological Proposal of the Mind Map for Geography Classes

The indications suggested here are structured in the conception that learning about geographic knowledge is more effective and powerful when integrated with cartographic products (RICHTER and GARCIA DE LA VEGA, 2019; CASTELLAR and DE PAULA, 2020), which in this case highlights the mind map. For this, this process was organized into three main stages, namely: construction, reading, and analysis, which will unfold in a series of actions and specific activities that aim to develop students' spatial and geographic thinking.

Table 1. Mind map methodological proposal for geography classes

STEPS	ACTIONS AND ACTIVITIES
1. Construction	Select a theme/objective for the construction of the mind map
	Articulate the theme/objective of the mind-map with a school content - a geographical problem
	Establish a cartographic scale/level of analysis of a geographic situation for the representation of the mind map
	Give freedom to the author of the mind-map at the moment of its construction and be careful not to give directions or indicate which elements, objects or spaces should be considered in this representation
2. Reading	Read the mind map with reference to the theme/objective selected for its construction and identify the space represented from the symbols, texts, objects and elements present on the map
	Dialog with the constructor of the mind map about the elements and objects present in the representation, trying to understand the particular and collective contexts that justified his choices
	Present the produced mind-map to the other students of the class, as a way of socializing this representation and encouraging different readings of this cartographic product
3. Analysis	Understand the central idea of the mental map from the objects, spaces and symbols present in this representation and its articulation with the geographic knowledge
	Recognize and identify the spaces present in the map, as well as its voids (the non representation of certain areas), in order to understand the spatial reading of its constructor
	Analyze the spatial elements present in the mental map, such as distances, volumes of represented objects, dimensions, routes, flows, etc.
	Understand the geographic reading of the mental map builder on the represented problem and its articulation with the geographic contents and concepts, seeking to analyze the development of the student's spatial and geographic thinking.

Org.: Author, 2022.

The ideas presented in table 01, referring to the methodological proposal of the mind map for Geography classes, were organized taking into account some teaching and investigative experiences carried out on this theme in recent years (RICHTER, 2010; LOPES and RICHTER, 2013; CAMPOS and RICHTER, 2020). From these studies, it was observed the importance of organizing a protocol for the use of the mind map articulated to the didactic and pedagogical activities of Geography in the school environment, trying to overcome the conception that the work with a mind map does not require a systematization. On the other hand, it is recognized that even though some actions considered relevant for the development of this work are raised, the teacher can make adaptations to this proposal, including other activities that he/she considers appropriate for his/her school practice or even excluding the ones presented in table 01. Even being clear that this proposal may have adjustments or changes, it is emphasized that the application of the three steps (construction, reading, and analysis) are of fundamental importance to conceive the mind map as a material that favors the development of students' spatial and geographic thinking.

Having said this, it is considered relevant to present below (figure 02) an example of a mind map that makes it possible to highlight the potential of this cartographic product with spatial reading and analysis, as well as its articulation with Geography teaching, regarding the guidelines indicated in table 01.

The mind map present in figure 02 refers to applied research conducted between the years 2012 and 2013 with high school students in a public school in the city of Goiânia/Brazil (LOPES and RICHTER, 2013). This research aimed to identify and analyze the link between the content taught in Geography classes and the production of mental maps. Thus, according to the themes that were developed by the teacher, based on her teaching plan, it was suggested to the students the construction of a mind map that could express the knowledge learned in class with the daily practices. One of the contents referred to the transformations of the landscape in the city over the years and, based on this theme, the students were asked to produce the mind map. These notes are close to the first stage presented in table 01, the construction. In this case, the relevance of establishing criteria and methodological guidelines for the development of this activity was verified. That is, it was directed that students should build a mental map of the city, which could be of different scales (the whole city or a part of it), and that the transformations of the urban landscape should be presented in this cartographic representation, based on their experiences and life experiences. This contributes greatly to helping conduct the work and allows students to have some references in the realization of this activity, giving, at the same time, guidelines for its development but without interfering or pointing paths in the construction of the mental map. For, one of the strongest characteristics related to the mind map refers to the freedom of its author at the moment of construction.

Figure 2. Mind map produced by a high school student of the city of Goiânia/Brazil, aiming to present the landscape transformations in urban space.
Source: LOPES and RICHTER, 2013, p. 09

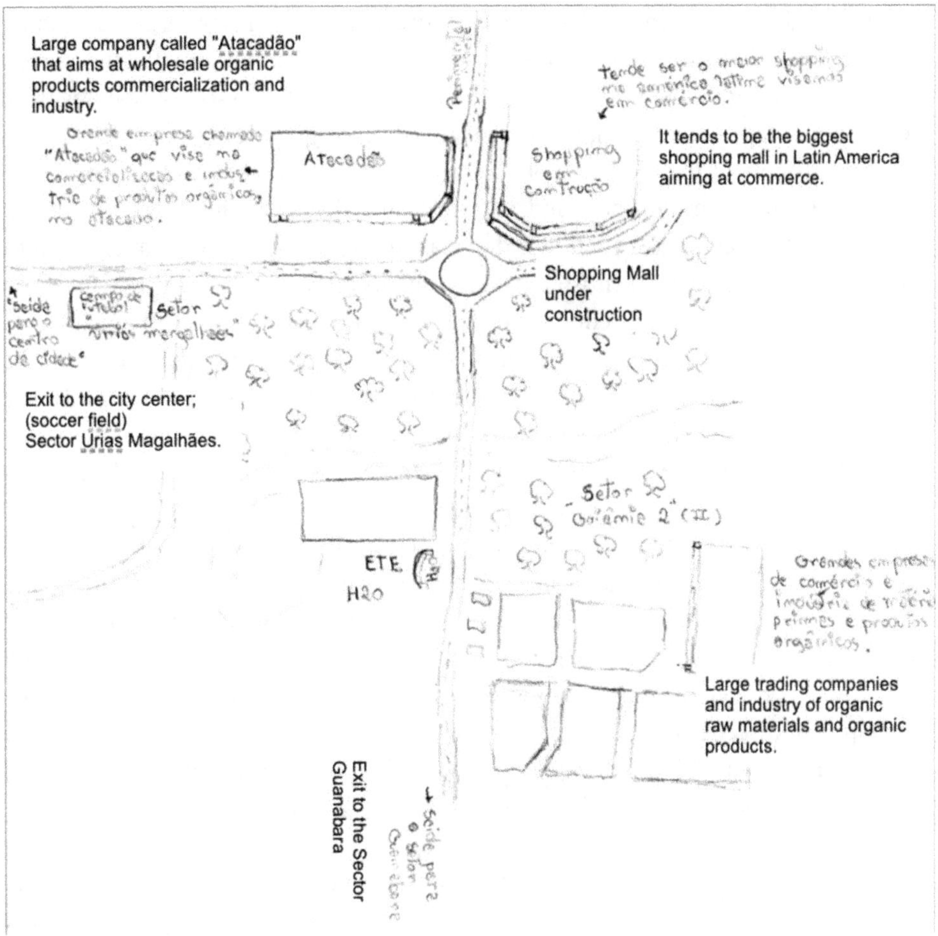

With these materials we started the reading process, referring to the second stage highlighted in table 01, trying to identify the areas of the city that were represented, the objects and symbolic elements, as well as the texts present in each mind map. This stage involved a moment of isolated reading, by the researchers and the class teacher, and also a collective reading, dialoguing with the students about their representations, questioning them about their choices and the themes highlighted in each material. After that, the students were asked to pass their produced maps to their classmates, allowing everyone to have access, and read the mental maps built. It is understood that these guidelines for reading, both content and spatial, are essential to enhance this type of cartographic representation because they allow students to have a broader dimension of how the same content can have different interpretations from its apprehension and how these depend on the social relations established by the subject that produced the map.

Finally, the last stage, focused on spatial analysis, was concerned with understanding the geographic understanding of the content worked in the classroom that the student was able to present in the production of his mental map. To do so, some notes from the previous stages were required to analyze the cartographic representation from the point of view of the elements of spatiality and its articulation with geographic knowledge. For example, the mind map in figure 02 highlights how the student appropriated the school knowledge and managed to connect it to his spatial readings and practices since the area chosen to compose this representation refers to a region very close to the student's school and that was undergoing numerous changes at the time it was produced. Of these changes, one of the most significant was the construction of a shopping center that considerably altered the urban landscape and that was evidenced by the student in his mind map. In this representation, it is possible to observe the emphasis that the student makes when indicating "shopping mall under construction". For, where it used to be a valley bottom, with a lot of bushy vegetation, it turned into a large construction site, with an intensely modified landscape that started to change the other surrounding areas of this development. Therefore, the mind map constructed by the student becomes a representation that makes it possible to dialogue the geographic contents learned in the classroom with everyday experiences, fostering a more critical spatial analysis and enhancing the development of geographic thinking.

These ideas are in line with what Gomes (2017a, p. 106) highlights when valuing the activities related to Social Cartography. For the author,

The mapping process also demystifies the map, and by denaturalizing it, it is possible to question its neutrality, objectivity, and the exclusive domain of specialists, and thus problematize the narratives about the represented space. In this way, mastering the cartographic language is to appropriate one more instrument for the critical reading of space. That is, by producing their own maps, children and young people can understand the silences and the spatial ordering prioritized in the official maps. And, in this aspect, Social Cartography is consolidated in a great potential to School Geography, although [...] its link is stricter to social movements.

However, it is important to highlight that even in the positive contexts concerning the work with mental maps in Geography classes, it is essential that the teacher is aware of the limits of this work to have a better dimension of when to require this proposal in the school environment. In this case, it is recommended that the mental map be used when one wants to value the subjective elements of space, which are often not perceptible in a direct reading of the landscape. This means that this type of cartographic representation may not always serve to help understand the production of spatial arrangements. In some cases, the mental map can become a problem, for example, when one seeks to have this representation read by many individuals and to enable them to have a clearer and more direct interpretation of a given place. For this, the so-called conventional maps contribute more effectively to the perspective of a unidirectional spatial reading.

In addition, the more technological products integrated into the process of mapping or linked to Geographic Information Systems (GIS) should also be recognized as alternatives to the Geography schoolwork, especially in a future perspective. In recent years, there has been a search to bring these resources closer to different proposals for map construction, among which one can also find the mental map itself. The research of Oliveira [et al] (2020) highlights this possibility when using area images in the process of mental map production, as well as its validation and evaluation about the represented spaces and its potentiality linked to the work of cartographic content teaching.

That being said, it is emphasized that all these theoretical-methodological approaches, favoring the use of mind map in Geography classes, seeks to consolidate one of the most powerful characteristics of this science, its ability to enable individuals to understand the transformations or permanences that occur in space over the years. In addition, the ability to integrate all this knowledge within students' social practices, broadening their view of reality and their critical reading of the world.

FINAL CONSIDERATIONS

It has been possible to observe that the teaching of Geography in Brazil has undergone numerous changes in recent years, among which one of them is strongly related to the recognition and appreciation of this science in the development of spatial and geographic thinking. This theoretical perspective has contributed to giving greater emphasis to one of the main goals of Geography: thinking about space. This scenario is considered a breakthrough to make this school subject in Basic Education more meaningful. However, for this growth to reach the classroom, it is essential to structure methodological paths that help the teaching work, focusing on the construction of a more critical spatial reading and analysis by the students.

It is in this sense that the important link between the cartographic language and the school practices of Geography is presented. For, the map has been a resource widely used by this disciplinary field over the years but is not always required in order to enhance the reasoning relevant to Geography. It is understood that this context needs to be changed to effectively contribute to the work of Geography at school. Thus, from the researches indicated in this text, it is observed that the mind map has been constituted as a possibility to collaborate with this demand, mainly by its characteristic of being a freer product of Cartography and being able to be adapted to different geographic contents. At the same time, this whole investigative journey allows us to recognize the need to organize some guidelines for the realization of this work in Geography classes. The use of the mind map without pedagogical objective or poorly articulated with the geographic themes and concepts can make this resource underutilized in school practices. Therefore, the methodological indications presented in this article seek to give more quality to the work with mental maps in Geography classes in Basic Education, as well as to provide an opportunity that the construction, reading, and spatial analysis of this language contribute significantly to the development of students' geographic thinking.

REFERENCES

Bertrand, R. (1984). *Mind mapping experiments at ITC (International Institute for Geoinformation and Earth Observation)*. Journal ITC.

Besse, J. M. (2014). *O Gosto do Mundo: Exercícios de paisagem*. EdUERJ.

Buzan, T. (2002). *How to mind map*. Harper Collins.

Campos, L. R., & Richter, D. (2020). Ensino de Geografia, escola Quilombola e mapas mentais: diálogos possíveis na construção de uma identidade territorial. In N. Rego & S. Kozel (Eds.), *Narrativas, Geografias e Cartografias: para viver é preciso espaço e tempo* (Vol. 1, pp. 211–230). Compasso Lugar-Cultura. doi:10.29327/519558.1-10

Castellar, M. V. (2020). O papel do pensamento espacial na construção do raciocínio geográfico. *Revista Brasileira de Educação em Geografia*, *10*(19), 294–322. doi:10.46789/edugeo.v10i19.922

Castro Aguirre, C. (2014). *Mapas mentales*. Universidad Pública de Navarra.

Cavalcanti, L. S. (2019). *Pensar pela Geografia: ensino e relevância Social*. C&A Alfa Comunicação.

Downs, R. M., & Stea, D. (1973). Cognitive maps and spatial behavior: process and products. In R. M. Downs & D. Stea (Eds.), *Image and environment* (pp. 1–20). Aldine.

Duarte, R. G. (2017). A linguagem cartográfica como suporte ao desenvolvimento do pensamento espacial dos alunos na educação básica. *Revista Brasileira de Educação Em Geografia*, *7*(13), 187–206. doi:10.46789/edugeo.v7i13.493

Fonseca, F. P. (2019). O espaço como representação. In A. F. A. Carlos & R. C. A. Cruz (Eds.), *A necessidade da Geografia* (pp. 42–53). Contexto.

Freitas, R., & Almeida, M. (2014). O pensamento geográfico nos alunos do ensino básico, com base na taxonomia do pensamento espacial. *GOT - Geography and Spatial Planning Journal, 6*, 135–152. doi:10.17127/got/2014.6.009

Galvão, W. (2007) *Que Geografia se ensina?: um estudo sobre representações de Geografia segundo alunos da 6a série do ensino fundamental* [Dissertação de Mestrado]. Universidade Federal do Paraná.

Gersmehl, P. (2008). *Teaching Geography* (2nd ed.). The Guilford Press.

Golledge, R. G. (1995). Primitives of spatial knowledge. In T. L. Nyerges, D. M. Mark, R. Laurini, & M. J. Egenhofer (Eds.), *Cognitive Aspects of Human-Computer Interaction for Geographic Information Systems* (pp. 29–44). Kluwer Academic Publishers. doi:10.1007/978-94-011-0103-5_4

Gomes, M. D. F. V. B. (2017a). Cartografia social e geografia escolar: Aproximações e possibilidades. *Revista Brasileira de Educação Em Geografia, 7*(13), 97–110. doi:10.46789/edugeo.v7i13.488

Gomes, P. C. C. (2013). *O lugar do olhar*. Bertrand Brasil.

Gomes, P. C. C. (2017b). *Quadros Geográficos*. Bertrand Brasil.

Gould, P., & White, R. (1974). *Mental Maps*. Penguin Books. doi:10.4324/9780203163757

Harley, J. B. (1991). A nova história da cartografia. O Correio da Unesco, 19(8), 4-9.

Hollman, V., & Lois, C. (2015). *Geo-grafías*. Editorial Paidós.

Jacob, C. (2016). Por uma história cultural da cartografia. *Espaço e Cultura, 39*(39), 221–236. doi:10.12957/espacoecultura.2016.31762

Jo, I., & Bednarz, S. W. (2009). Evaluating geography textbook questions from a spatial perspective: Using concepts of space, tools of representation, and cognitive processes to evaluate spatiality. *The Journal of Geography, 108*(1), 4–13. doi:10.1080/00221340902758401

Kozel, S. (2007). Mapas mentais - uma forma de linguagem: perspectivas metodológicas. In S. Kozel, J. C. Silva, & S. F. Gil Filho (Eds.), *Da percepção e cognição a representação: reconstrução teóricas da Geografia Cultural e Humanista* (pp. 114–138). Terceira Margem.

Kozel, S. (Org.). (2018). *Mapas Mentais. Dialogismo e Representações*. Appris.

Lee, T. (1968). Urban neighbourhood as a socio-spatial schema. *Human Relations*, *21*(3), 241–267. doi:10.1177/001872676802100303

Lopes, A. R. C., & Richter, D. (2013). A construção de mapas mentais e o ensino de Geografia: articulações entre o cotidiano e os conteúdos escolares. *Revista Territorium Terram*, *2*(3), 2-12. http://seer.ufsj.edu.br/index.php/territorium_terram/article/view/606

Lynch, K. (1960). *The Image of the City*. The MIT Press.

Moreira, R. (2007). *Pensar e ser em geografia*. Contexto.

National Research Council. (2006). *Learning to Think Spatially: GIS as a Support System in the K-12 Curriculum* (1st ed.). National Academies Press.

Nogueira, A. R. B. (1994). *Mapa mental: recurso didático no ensino de geografia no 1o grau* [Dissertação de Mestrado]. Universidade São Paulo.

Oliveira, A. F., Carvalho, A. J. E., & Fernandes, C. C. P. (2020). Mapas mentais como suporte para interpretação de áreas imageadas por veículo aéreo não tripulado no ensino de cartografia. *Geomae*, *11*(2), 45–56.

Richter, D. (2010). *Raciocínio geográfico e mapas mentais: a leitura espacial do cotidiano por alunos do Ensino Médio* [Tese de Doutorado]. Universidade Estadual Paulista.

Richter, D. (2011). *O mapa mental no ensino de Geografia: concepções e propostas para o trabalho docente*. Cultura Acadêmica.

Richter, D., & de la Vega, A. G. (2019). O mapa no ensino de geografia: Uma análise do trabalho docente em Madrid (Espanha) e em Goiânia (Brasil). *Ateliê Geográfico*, *13*(3), 46–65. doi:10.5216/ag.v13i3.60860

Richter, D., & Moraes, L. B. (2020). A Cartografia escolar na BNCC de Geografia do Ensino Fundamental: uma análise do pensamento espacial e do raciocínio geográfico. In E. M. B. Morais & D. Richter (Eds.), *Formação de professores e ensino de Geografia: contextos e perspectivas* (pp. 141–168). Editora Alfa.

Silva, P., & Roque Ascenção, V. O. (2018). Por uma construção do raciocínio geográfico para além do pensamento espacial (spatial thinking). In *Anais do 5º Colóquio Internacional da Rede Latino-Americana de Investigadores de Didática de Geografia* (pp. 73–94). UFG.

Soares, P. A. P. (2017) *O pensamento espacial: um desafio no Ensino da Geografia Escolar* [Dissertação de Mestrado]. Universidade de Coimbra.

Teixeira, S. K. (2001). *Das imagens às linguagens do geográfico: Curitiba, a "capital ecológica"* [Tese de Doutorado]. Universidade de São Paulo.

Tolman, E. C. (1948). Cognitive maps in rats and men. *Psychological Review*, *55*(4), 189–208. doi:10.1037/h0061626 PMID:18870876

Vygotsky, L. S. (2000). *A construção do pensamento e da linguagem*. Martins Fontes.

Chapter 10
River Landscape:
A Didactic Proposal for Primary Education

Elena María Muñoz Espinosa
Universidad Autónoma de Madrid, Spain

Celia Laguna Mora
University of Castilla-La Mancha, Spain

ABSTRACT

This educational proposal is built up from a varied set of disciplines with different viewpoints which manage different approaches to rivers: hydrology, ecology, land uses, cultural heritage, or fluvial landscape. Nowadays, traditional approaches are misleading, and they promote conflicts between the required knowledge and what society demands. Thus, didactic innovation on an integrated base is required from early educational ages. Here are presented some didactic strategies that could be implemented from elementary schools, with students from 6 to 12 years old. The aim is to establish an approach and procedures for river teaching and help to develop proposals for teachers, which eventually could be used in class. The development of these guidelines is implemented by the identification of fluvial functions along specific items suggested.

INTRODUCTION

Rivers are very changing systems. Their dynamic and values do not finish in the waterlogged limits like something rigid and steady. Most rivers have quite special regimes and functioning due to many factors, i.e., irregular climate, seasonal process

DOI: 10.4018/978-1-7998-9598-5.ch010

(flooding and droughts) or regulatory structures like dams, channels, or conduits, which intervene and diverse the river flow. These elements interfere in the state of the river in a particular place and fixed moment. Therefore, people/teachers are no aware because they do not completely know about the complex and the interesting fluvial processes.

This proposal presents an alternative look at rivers to understand and know them and show their values for a Primary Education audience. The authors advocate a quality and scientific education, starting from childhood. To do this, fluvial knowledge must be translated into teaching activity with real classroom approaches and an integrated perspective, which tries to leave nothing out. Each river has its own natural and cultural history and to explain it, this proposal inserts contents from Natural and Social Sciences which are included in this educational stage. But it also includes contents from other subjects like Spanish Literature and Language, Maths, Arts and Physical Education.

This job is born on the bases of Geography and the bioclimatic factors which define a watershed (scale, latitude, longitude, or altitude), as well as the characteristics of each river, such as temperature, rainfall, geomorphology, and ecosystems. It also integrates the people who have lived there and their demands from the river (water, fertile land, protection, etc.), considering that they determine cultural factors which are specifics to the river, i. e., infrastructures, ways of communication or uses of floodplains. Thorough this chapter the main function of the fluvial corridors will be studied: morphological dynamic, water flows, water quality, ecological processes, uses of the river, historical heritage, and landscape approach.

Finally, specific items are presented here as background material for teachers in accordance with Primary Spanish Curriculum.

The main objectives proposed are:

- To increase the knowledge about rivers from early educational levels.
- To show a deeper social perception of rivers to dealing with them.
- To recognize river systems as important natural and cultural heritage.
- To show a wide view of unknown fluvial landscape processes and functions.
- To establish general rules and procedures for river teaching.
- To outline proposals for teachers which could be used in class.
- To provide resources for innovative Geography teaching.

BACKGROUND

Social, Technical, and Climatic Issues for River Didactics

There are many reasons why teaching river ecosystems in the classroom is justified. From social to economic issues and technical decisions or those involving the good health of the planet. That is, climate change or sustainability, among others.

Throughout history human, human settlements have been occurred near rivers. There are a lot of benefits that human obtain from rivers: drinking water, irrigation, power, navigation, etc. Therefore, and given the proximity of the population to the rivers and the increase in river management, it is necessary to know how rivers work. Issues such as safety of people and property are becoming increasingly important in a territory (the river) that needs its own space. In this regard, some recent episodes can be cited: floods in France and Germany[1] (July and August 2021), floods in Spain[2] (December 2021) or the "cut-of-low"[3] in the West of the Iberian Peninsula.

On the other hand, what used to occur naturally (floods and torrential rains) is now being increased by other effects, such as global change, the disregard of river space, and the proliferation of bad practices in wetland and floodplain management (Florín et al, 2018). Although river intervention has always been vital for human populations, the sustainability of such actions is now even more pressing. Therefore, in order to avoid the vulnerability of the population (e.g., floods), education becomes part of the solution because *only a portion of the public responds appropriately to flood warnings, and this lack of response can have grave results* (USACE, 1994).

In addition, urban planning in inhabited areas must consider the presence of watercourses in public decision-making and provide nature-based solutions (Cohen-Shacham et al., 2016; Florín et al., 2021). The recognition of ecosystem functions that improve the quality of ecosystems, and the quality of people´s life has been technically demonstrated on numerous occasions[4]. In this sense, trends such as climate urbanism seek to mitigate the effects of climate change, both in terms of its impact on cities and its accentuation of urban inequality[5]. Practitioners and scholars may be interested in the transformation of urban governance that follows global advances in climate change and urban policy (Castán, 2017).

Furthermore, and in accordance with the European Landscape Convention (Council of Europe, 2000) (art. 1), landscape is defined as "*an area, as perceived by people, whose features are the result of the action and interaction of natural and/or human factors*".

As we can see in Figure 1, the concept of landscape must not be applied only to small restricted special areas, nor items but to each part of the territory, significances, and values. Council of Europe (2009) proposed tree paths for reading landscape with children:

Figure 1. A conceptual model for landscape
Source: Modified from Castiglioni, 2002

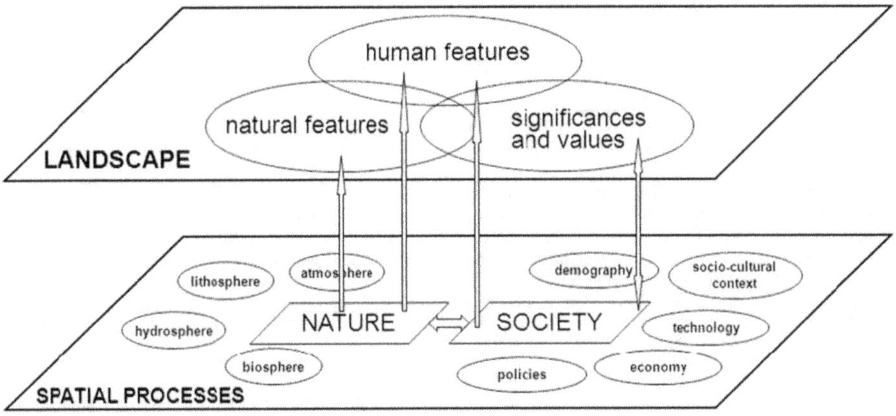

- The first path moves in a horizontal dimension, in the top plain of landscape elements and features.
- The second path goes deep "inside" the landscape, in order to explore – in a vertical dimension – the causal relationships between landscape factors and landscape features.
- The third path moves in a temporal dimension, trying to understand the landscape change, a result of different driving forces acting today as well as in the past.

The Educational Curriculum in Spain: From the First to the Sixth Year of Primary Education.

The ideas exposed above by subjects are easy to break it down by academic levels. As a Social Science, Geography is linked to History at all levels of Primary Education. In the geographical part, different specific contents related to rivers can be extracted according to the level (See Table 1). This content ranges from a general approach in the first year (hydrosphere) to more specific content in the rest of the years (names of rivers, river landscapes, differentiation of river sections).

According to classical authors such as Hannoun (1972) or Piaget and Inhelder (1947), the child's conception of space goes through different evolutionary stages. These stages, according to the different theories, would be related to the lived, perceived, or conceived space (Hannoun, 1977) or to the development of simple, concrete, or formal operations (Piaget and Inhelder (1947). Thus, the age intervals would be distributed in three stages: up to 6 years old, between 7 and 11 years old, and from 12 to 15 years old (age that already corresponds to Secondary Education).

Table 1. Extract from the curricular contents separated by levels

1st year of Primary Education
• The hydrosphere. o Knowing the existence and distribution of water on Earth: oceans, seas, rivers and lakes. o Explaining the problems of water pollution, the characteristics of drinking water and the need for responsible consumption. o Identifying some rocks and materials in the immediate environment and in photographs or videos. o Observing different types of landscape in photographs or videos. o Identifying different components of relief: mountain, mountain range, plain, plateau and valley.
2nd Year of Primary Education
• Rivers in our province and Autonomous Community. • Identifying the main relief units using the Iberian Peninsula as an example: island, archipelago, peninsula, mountain, mountain range, plain, valley, river, lake, coast, beach, gulf, cape and harbour. • Understanding the need to adopt a responsible attitude in the use of water, care for the environment and the use of natural resources, proposing measures and behaviours that lead to the improvement of the environmental conditions of our planet.
3rd Year of Primary Education
• The water cycle. • The landscape. • Map of Europe. Other continents. o Locating on the political map of Europe the different countries and know their capitals. o Locating on the physical map of Europe the most important seas, rivers, and mountain ranges. o Identifying some countries, cities, rivers and mountain ranges of non-European countries. • Physical map of Spain. Relief and main rivers. o Locating on a map of Spain the main mountain ranges and mountains. o Locating on a map of Spain the most important rivers. Identify their source and springs, as well as their main tributaries. • Plans and maps. o Interpreting the conventional signs that appear on a map. o Identifying and uses maps of different scales. o Drawing simple plans with different scales.
4th Year of Primary Education
• Inland waters: ice masses, groundwater, lakes, lagoons, and rivers. • The waters of Castilla La Mancha and the great rivers of Spain and their location on maps. • Rivers, reservoirs, and lagoons in Castilla la Mancha. • The hydrosphere. o Describing in orderly fashion the phases in which the water cycle takes place. o Knowing the different water formations and their characteristics (oceans, seas, rivers, groundwater, ...). o Understanding how groundwater is formed, how it emerges and how it is accessed. o Identifying the sections of a river and the characteristics of each section. o Differentiating between a tributary, river basin, riverbed, and watershed. • The lithosphere. Characteristics and types of rocks. o Observe, identify, and explain the composition of rocks, naming some of their types. o Understanding the difference between minerals and rocks. o Identifying some minerals according to their properties.
6th Year of Primary Education
• Knowing and locating on the map of Spain the main rivers, their source, springs, and most important tributaries. • The landscapes of Europe and Spain: elements of the landscape (relief, coasts, climate, vegetation, hydrography). • The main rivers and lakes of Europe.

Source: Author's own elaboration based on Decree 54/2014, of 10/07/2014, which establishes the Primary Education Curriculum in the Autonomous Community of Castilla-La Mancha and Decree 89/2014, of 24 July, of the Governing Council, establishing the Primary Education Curriculum for the Community of Madrid

So, the knowledge of the space will be influenced by owns child development aspects, such as visual perception, grasping, locomotion and child-like thinking to be able to represent the space mentally, orally, and graphically. The ability to read, to interpret and to learn through maps and plans is not acquired spontaneously (Rivero, 2011), but it needs training. According to this author, it is possible to work it from childhood. The complexity of the subject has already been analysed by several authors (Sebastiá and Tonda, 2015), as well as the use of the teaching of scale in traditional cartography from the first cycle of primary education (Marrón, 2012). Most authors agree on the real possibility of working in the classroom with cartography at different scales and the possibilities of real understanding of selected concepts, but always based on an adequate didactic adaptation for the child's level of development and an elaborated line of work. On the other hand, according to Piñeiro and Melón (2002*) space learning requires the development of a particular form of reasoning and to achieve this purpose they advocate, at primary education levels (10-13 years), the use of traditional maps with a problem-solving methodology.* Similarly, *Souto (2005) suggests a didactic innovation by means of problem approach (...).*

In any case, we can say that no clear line can be identified in terms of the spatial or scalar order of river-related content from the curriculum (Figure 2). In the curricular analysis by levels, a lack of continuity in the concept was detected, for example, changing the scales without a clear order or even an apparently related magnitude between the different years of the curriculum from first to sixth level (here magnitude is understood as the presence and relevance of the contents related to rivers in different courses).

Reinforcing the idea of multidisciplinary, other contents of the curriculum were analysed, in this case more related to Human Geography (See Table 2) and History (See Table 3). In both cases, it seems that all the contents that can be related to rivers are disintegrated and presented in a generic way (e.g. population and spatial distribution, productive activities, goods and services that people extract from the environment, heritage, ...). In other words, not only they are not always applied to concrete examples, but there is also an abuse of learning terms to be understood out of context.

In addition, the Objectives of Primary Education will contribute to the development of children's capacities that will enable them to achieve the key competences of this stage. Table 4 shows those of fourteen stage objectives. This proposal contributes (or could contribute) to reach 10 of them, which are highlighted in bold. In addition, the competence orientation of the educational curricula seeks to ensure that students learn to cope with the maturational stage in which they find them. But it will also prepare them for the gradual acquisition of skills for adult life (OECD, 2015).

Figure 2. River-related and scale contents in each year of primary education
Source: Author´s own elaboration

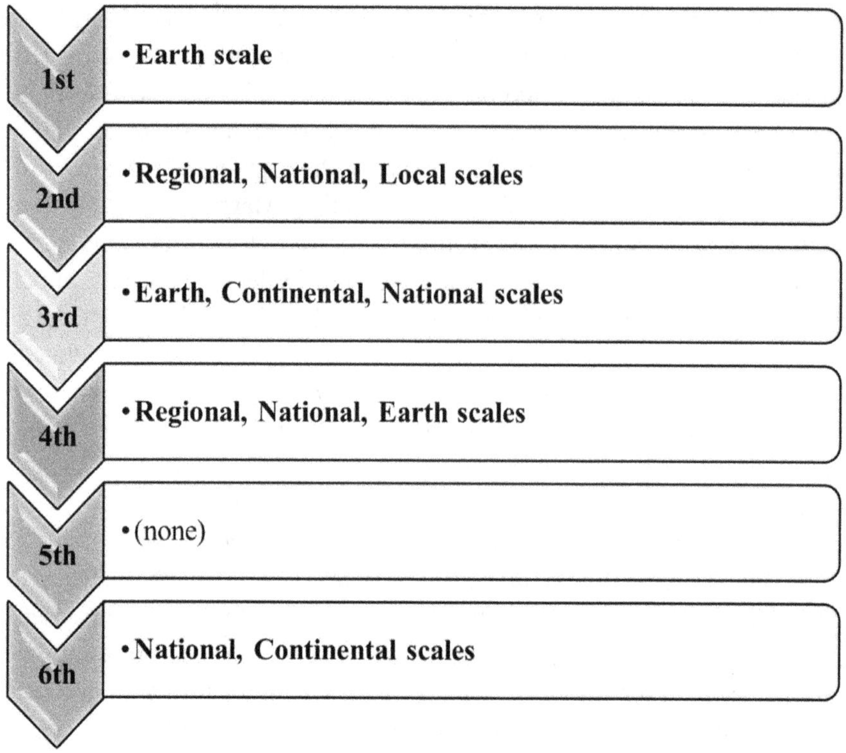

In 2003, the DeSeCo report (OECD, 2003) defined competences as the ability to respond to complex demands and to carry out diverse tasks appropriately. In other words, the concept of competence involves a combination of practical skills, knowledge, motivation, ethical values, attitudes, emotions, and other social and behavioural components that are mobilised together to achieve effective action. In the context of education, it would be appropriate to say that they are defined as the set of knowledge, skills and abilities that can be learned, enabling individuals to perform an activity or a task properly and systematically, and can be developed through learning. In this sense, this proposal would contribute to achieve some of the competences set out in the official educational curriculum for Primary Education (See Table 5).

Thus, LCC *is the result of communicative action within given social practices, in which the individual acts with other interlocutors and through texts in multiple modalities, formats and supports*, this proposal deals with and places special emphasis on the appropriate use of scientific language and the learning of a specific lexicon

Table 2. Extract of other curricular contents (by levels) in the primary related to rivers and included in the context "Living in Society"

3rd Year of Primary Education
• The production sectors. Natural products and their origin.
4th Year of Primary Education
• The population of a territory. Population by age. • Factors that modify the population of a territory: birth rate, mortality, emigration, and immigration.
5th Year of Primary Education
• Spatial distribution and population growth. Population pyramids, bar charts and pie charts. • Distribution and evolution. Migratory movements. Rural exodus. • Knowing the terms: demography, absolute population, and population density. • Locate on a map the largest centers of population in Spain and Europe, the most densely populated areas and the main migratory movements.
6th Year of Primary Education
• Productive activities (natural resources, raw materials, processed products, forms of production) *. • The production of goods and services to satisfy human needs*.

Source: Compilation based on Decree 54/2014, of 10/07/2014, which establishes the Primary Education Curriculum in the Autonomous Community of Castilla-La Mancha and Decree 89/2014, of 24 July, of the Governing Council, which establishes the Primary Education Curriculum for the Community of Madrid. (*) may appear and/or be dealt with at different levels depending on the autonomous community.

related to rivers, considering that this approach gives rigour, specificity and value to learning. See tables 6 to 10.

MCSTC could be divided into several sections. On the one hand, *the ability to apply mathematical reasoning and its tools to describe, interpret and predict different phenomena in context. (...) This requires knowledge of numbers, measures, and structures, as well as of mathematical operations and representations, and understanding of mathematical terms and concepts (operations, numbers, measures, quantity, space, shapes, forms, data, etc.). (...).* On the other hand, the attainment of these competences *provides an approach to the physical world and a responsible interaction with it through actions, both individual and collective, aimed at the conservation and improvement of the natural environment, which are decisive for the protection and maintenance of the quality of life and the progress of people. (...) they contribute to the development of scientific thought, as they include the application of the methods of scientific rationality and technological skills, leading to the acquisition of knowledge, the testing of ideas and the application of discoveries to social welfare. Thus, they train responsible and respectful citizens who develop critical judgements on the scientific and technological facts that occur throughout the ages, past and present.* In other words, the use of the scientific method, the training of thoughtful and environmentally respectful citizens or the use of appropriate technology for this purpose would be within the given approach. Thus, these aspects would be worked on in the classroom with contents related to measurement (e.g., granulometry), geometry (e.g., valley type or sinuosity) (Table 6), altitude or population ranking (Table 10), among others. In addition, if technologies are used as tools to help in the classroom, students would work on aspects related to ICTs.

DC as *involving the creative, critical, and confident use of information and communication technologies to achieve goals related to work, employability, learning, use of leisure time, inclusion and participation in society*. This competence would not be specifically addressed in the proposal. However, if teachers consider it, it could be included as possible classroom activities which include different technological resources treated in an appropriate manner from a didactic point of view and adapted to the different ages of development. This would be ideal for the achievement of the competence.

Table 3. Extract of other curricular contents (by levels) in the primary related to rivers and included in the context "The traces of time"

1st and 2nd Year of Primary Education
• Use and measurement of time. Basic notions: before, after, past, present, future, duration. • Past and present events. • Means of reconstructing the past. • Remains of the past: care and preservation
3rd year of Primary Education
• Changes in a locality throughout history (housing, clothing, food, family and social organisation, forms of work). • Knowing different techniques to locate, in time and space, events of the past. Recognising the working techniques of the archaeologist. • Recognising cultural heritage as a wealth to be known, preserved, and cared for.
4th year of Primary Education
• Includes the contents of previous courses. • Our Heritage. • Prehistory*, Ancient Age*, Romanisation*.
5th curso de Educación Primaria
• Middle Age*. • Modern Age*.
6th year of Primary Education
• Contemporary Age*. • SS. XX and XXI*. • The timeline. Place the ages of Spanish history on a timeline and indicate the different stages.

Source: Author's own elaboration based on Decree 54/2014, of 10/07/2014, which establishes the Primary Education curriculum in the Autonomous Community of Castilla-La Mancha and Decree 89/2014, of 24 July, of the Governing Council, which establishes the Primary Education Curriculum for the Community of Madrid. (*) may appear and/or be dealt with at different levels depending on the autonomous community.

Table 4. Objectives of primary education

1. To know and appreciate the values and rules of coexistence, learning to act in accordance with them, preparing for the active exercise of citizenship and respect human rights, as well as the pluralism of a democratic society. 2. To develop individual and teamwork habits, effort, and responsibility in study, as well as attitudes of self-confidence, critical sense, personal initiative, curiosity, interest, and creativity in learning, and entrepreneurial spirit. 3. Acquire skills for the prevention and peaceful resolution of conflicts, which will to enable them to develop autonomy in the family and domestic environment, as well as in the social groups with which they interact. 4. To know, understand and respect different cultures and the differences between people, equal rights and opportunities for men and women and non-discrimination of people with disabilities. 5. To know and use proper Spanish language and/or co-official language of the Autonomous Community and developing reading habits. 6. To acquire basic communicative competence in at least one foreign language that will enable them to express and understand simple messages and to cope in everyday situations. 7. To develop basic mathematical competences and to begin to solve problems that require the performance of elementary calculation operations, geometric knowledge, and estimation, as well as to be able to apply them to everyday life situations. 8. To know the fundamental aspects of Natural Sciences, Social Sciences, Geography, History and Culture. 9. To initiate in the use of Information and Communication Technologies for learning purposes, developing a critical spirit towards the messages that they receive and elaborate. 10. To use different artistic representations and expressions, and to initiate in the construction of visual and audio-visual proposals. 11. To value hygiene and health, accepting one's own body and other´s respecting differences and using physical education and sport as means to develop personal and social skills. 12. To know and value the closest animals to human beings and adopting ways of behaviour that favour their care. 13. To develop their affective capacities in all areas of their personality and in their relations with others, as well as an attitude against violence, prejudices of any kind and sexist stereotypes. 14. To promote road safety education and attitudes of respect that have an impact on the prevention of traffic accidents.

Source: Author's own elaboration based on Royal Decree 126/2014, of 28 February, which establishes the basic curriculum for Primary Education

Table 5. Curricular competences of primary education that the proposal contributes to achieve

1. Linguistic Communicative Competence (LCC) 2. Mathematical Competence and basic Competences in Science and Technology (MCSTC). 3. Digital Competence (DC). 4. Learning to Learn Competence (LLC) 5. Social and Civic Competence (SCC). 6. Sense of Initiative and Entrepreneurship Competence (SIEC). 7. Cultural Expressions and Social Competences (CESC).

Source: Author's own elaboration based on Royal Decree 126/2014, of 28 February, which establishes the basic curriculum for Primary Education

LLC *requires to know and to control the one's own learning processes to adjust them to the times and demands of the tasks and activities which lead to learning. The competence of learning to learn leads into increasingly effective and autonomous learning*. Although this is not the subject of the proposal, it is dealt with specifically insofar as the steps marked out for its design include activities that motivate self-learning and lead to the carrying out of autonomous tasks.

SIEC: *transforming ideas into actions*. The identification of specific fluvial spaces through the proposals that students can make according to their interests or the transformation into activities of personal initiative would fall within this competence. Again, this will depend on the orientation of the teacher, the interests of the pupils and the entrepreneurial initiative of the classroom. For example, Table 10 shows local people's knowledge of the river (e.g. indicator 4). Furthermore, linking students with local people through surveys and conversations can easily lead to joint activities outside the classroom.

CESC *involves knowing, understanding, appreciating, and critically valuing, with an open and respectful attitude, the different cultural and artistic manifestations, using them as a source of personal enrichment and enjoyment and considering them as part of the wealth and heritage of peoples.* This is one of the skills most worked on here. Both natural and cultural heritage and the knowledge, appreciation, and respect for both are part of the approach, including the design and development of the proposal. On the other hand, in its social part, *relates to personal and collective well-being. It requires an understanding of how people can achieve optimal physical and mental health for themselves, for their families and for their immediate social environment, and how a healthy lifestyle can contribute to this. (...). Civic competence is based on critical knowledge of the concepts of democracy, justice, equality, citizenship, and human and civil rights, as well as their formulation in the Spanish Constitution, the Charter of Fundamental Rights of the European Union and international declarations, and their implementation by various institutions at local, regional, national, European and international level. This includes knowledge of contemporary events, as well as of the most outstanding events and main trends in national, European and world history, and an understanding of the social and cultural processes of a migratory nature that imply the existence of multicultural societies in the globalised world.* Likewise the previous competence, this one works from a social and civic perspective that requires reflection, responsibility, and assumption of social and cultural processes for the respect and enhancement of the fluvial heritage. In Figure 6, it can be seen different cultural activities (agriculture, water supply, territorial communication, settlements, ...) which represent a theoretical pattern of a given territory. But from the classroom, first identifying and then getting to know this cultural and territorial pattern of activities is something that must be worked on *in situ*.

RIVER UNDERSTANDING BASED ON NATURAL AND CULTURAL ASPECTS

The proposal was born from the principal premise that rivers are not just channels, but a unique and very valuable ecosystems. There are many points of view on them. So, their maintenance and their values do not finish in the flooding borders. Besides, most rivers have their own development conditions: irregular climates, stational droughts, dam constructions, channels, permeability of borders, etc. That conditions sometimes made forget about rivers and their floodplain. As a result, rivers have been modified in artificial channels which have given the misconception of being easily manageable.

There are many points of view and interests about rivers due to the resources and services they offer. So, this proposal pretends to submit a new view about rivers based on the knowledge of the function that they provide and the values that they hold.

A typical geomorphologic dynamic exists in each riverbed. Also, hydric flows (quality and quantity) are particular too. So, each river maintains their own ecosystem and ecological processes. Furthermore, the economic productivity which is supported by their historical heritage show us how these systems have been managed, their uses, special landscapes, and the recreation areas they bring us. So, patrimonial value of rivers is considered from lithological and morphological base, their discharge regime, their water quality, the ecological processes that they maintain, their uses, their history or how people look at them. It is assumed that rivers have certain values, thus river values can be defined homogeneously by five conceptual groups: geology, geomorphology, hydrology, ecology, culture, and landscape (Muñoz, 2010).

Defining Conceptual Groups

Geology and Geomorphology

All those elements that are related with the structures and the geological and geomorphologic dynamics because they are the solid base of the fluvial system. Figure 3 represents the forms of relieve, how the water flows in the surface and subsoil, geological materials, flooded areas, etc, are included in this group.

Thus, the river fulfils certain functions structured on three main scales:

- Valley-Basin: this scale starts from a given water flow and a hydrological regime and defines a profile, a structure and a longitudinal continuity of the riverbed and the liquid-solid flow (Figure 3, yellow colour).

Figure 3. River geomorphology
Source: Modified from Muñoz, 2010

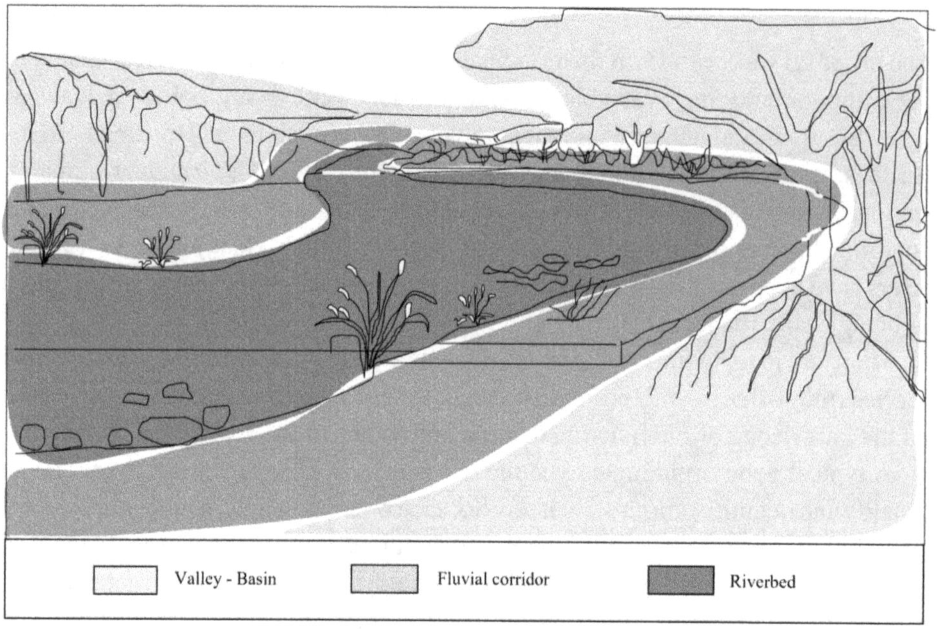

- Fluvial corridor: this scale would inform us of the morphology, cross-section, and structure, as well as the underground connections of the river (Figure 3, orange colour).
- Riverbed: the third scale would be defined by the lateral and vertical mobility of the riverbed. Solid flows and sediment mobility will explain the structure and dimensions of the riparian corridor (Figure 3, red colour).

Hydrology

Figure 4 represents concepts such as flow, water regime and water quality are included. The concepts of surface flow regime (flow rates; in Figure 4, blue-dark colour) and groundwater (aquifers; in Figure 4, blue-light dotted area), as well as water quality conditions with all their parameters and variables would not be the subject of in-depth study at these educational levels. Other concepts have been included as the appreciation of the quantitative and qualitative differentiation of how much water falls into the river, how much circulates on its surface or infiltrates into the ground. The approach to the concepts of surface and underground circulation, river drainage basin and temporal distribution of flows (annual, seasonal, hydrological cycle) are also included.

Figure 4. River hydrology
Source: Modified from Muñoz, 2010.

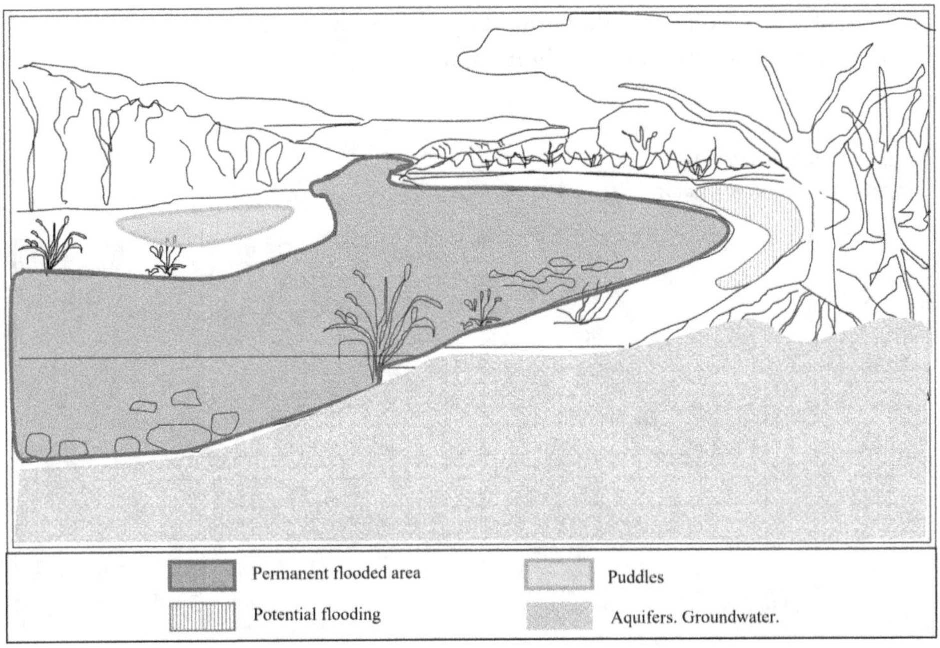

The tools and teaching materials available today are many and varied: from small weather stations and gauging stations to satellite photos and images, geographic information software for schoolchildren and others. On the other hand, it is also possible to introduce in the classroom the variation of the hydrological river levels, processes of erosion and sedimentation, removal (natural clearing of watercourses) of substances or vegetation or puddles (flooded areas; in Figure 4, light and striped areas). In other words, those characteristics that keep the river alive and dynamic. River has fluctuating structure that will change depending on each season and each hydrological moment. In this way, children aged 6 to 12 can begin to relate to the scientific variables that characterise the river system and may even begin to identify when these parameters are key to the formation of the river.

Ecology

Living species present in rivers are characterised by specific water requirements that keep them associated with that watercourse. This dependence is spatially and temporally variable. And it can be direct or indirect. In other words, there are species that need to be permanently waterlogged (Figure 5 yellow central area)[6] and others that only need certain degrees of soil moisture (Figure 5, red and green oval-shaped areas). In other cases, species may depend on the river only at certain times of the day, season and/or time of the life cycle (Figure 5, purple striped areas). Still others will depend on a certain periodicity of waterlogging or wetting, which will be specific to the communities (Figure 5, blue dotted areas).

Figure 5. River ecology
Source: Modified from Muñoz, 2010

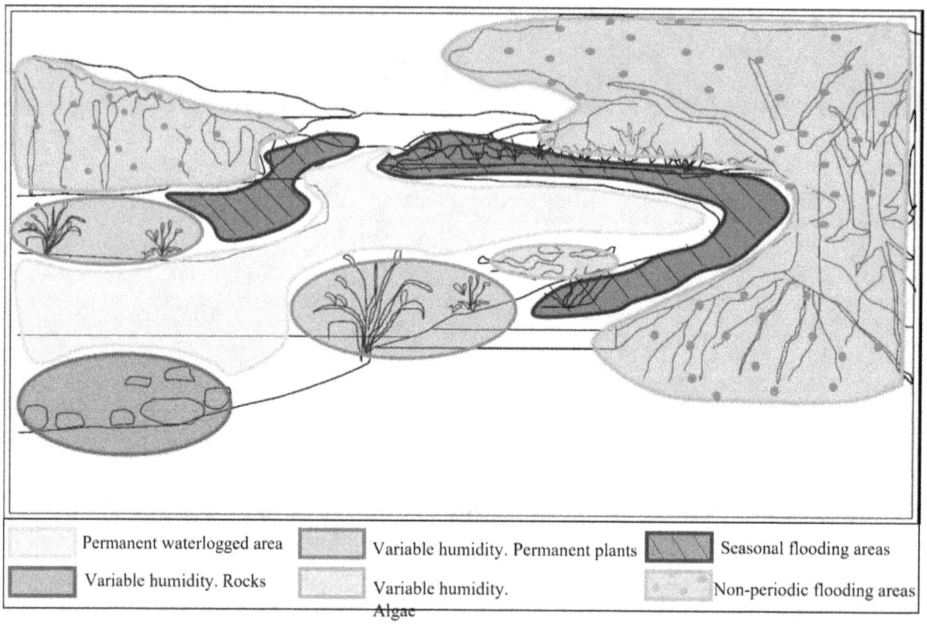

There are many sources of information at our disposal (bibliography, botanical and faunal inventories, photographs and satellite images, field guides, ...) for collecting vegetation and fauna data from the ecosystems related to the river territory. It will be interesting to know those species that are most emblematic of the river in question, and it will be necessary to recognise that there is a territorial arrangement of the species according to the river's own regime, which, again, is unique to each river.

Culture

This group of characters incorporates the actions that different civilisations have carried out on the river system, as well as their repercussions on its characteristics. This vision has varied throughout history. At each moment in time, each civilisation has had its own "needs and demands" of/from the rivers. Therefore, the contemporary vision and recent changes are included as the product of a long historical and diachronic process (Cruz and Español, 2008). It includes aspects related to the historical uses of the rivers, the interpretations that each society and each community shaped with their collective attitudes, the current and past uses of the riverbanks: for agriculture (Figure 6, crops), for water extraction and discharges (Figure 6, grey broken line), as human habitat (Figure 6, town), for energy generation (Figure 6, smokestack) or communication systems associated with the river (Figure 6, path and bridge), the structure of the settlement around the river and its dynamics and the superposition of all the above and its current functionality.

Figure 6. River culture
Source: Modified from Muñoz, 2010.

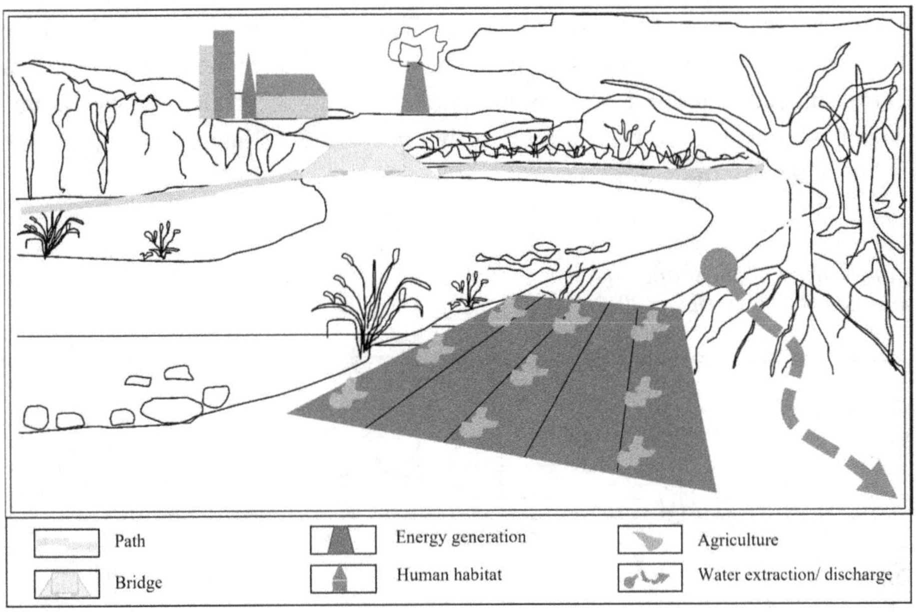

In this way, children can identify these historical clues from the past or their current state of conservation as elements, features, relationships, and even territorial patterns around the river.

Landscape

This set of characteristics is referred to the perception that human communities have of the river system. That is understood in terms of landscape meanings, character, and scenography conditions.

These disciplines include those that provide us with a reading of the river and its environment from the point of view of perception (geomorphology, ecology, history, and its reading) as set out in the European Landscape Convention (known as Florence Convention; Council of Europe, 2000). River functions 'supply and their appreciation and perception by people does not always have a direct relationship. This relationship is more difficult to establish when sciences and disciplines from very different fields are involved.

On the other hand, the appreciation of landscape will also be strongly influenced by territorial accessibility (relief; Figure 7, brown curve line). From where it is seen, how much space this image covers, how many people have a direct view of the river in its most complete sense or the frequency of accessibility to it (access to and crossing to the riverbed; Figure 7, red straight line), will mark each person's and society's own perception (Cruz y Español, 2008). Thus, elements to be included in this section are also the continuity or discontinuity of the riverbed (Figure 7, blue dotted line), the different points of view, the orography of the terrain, the state of the atmosphere, the lighting, the possibility of observing the landscape (viewpoints or visual itineraries), the vegetation covering or leafiness (Figure 7, green broken line), the extent of the views, etc.

Figure 7. River landscape
Source: Modified from Muñoz, 2010

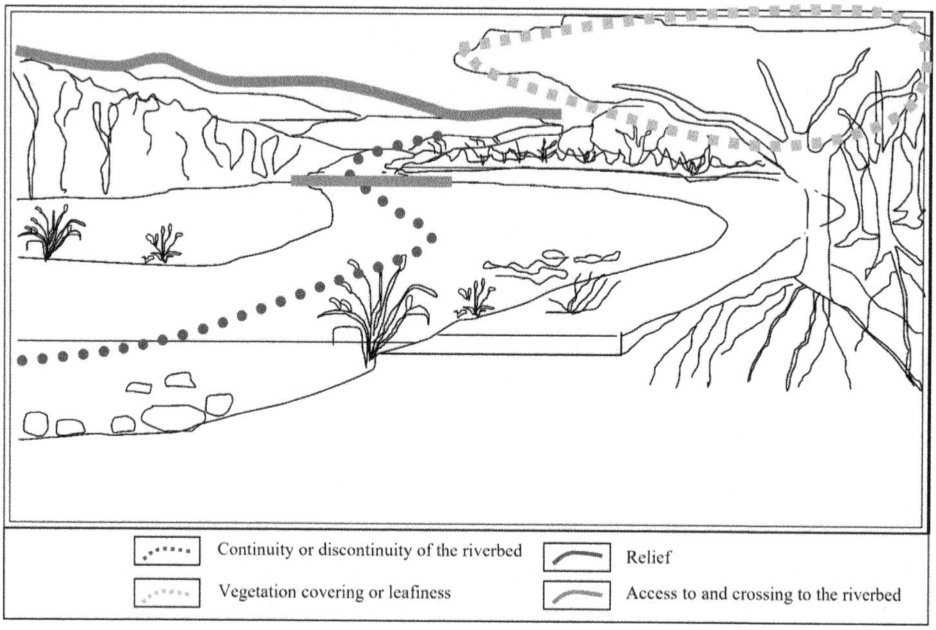

Didactic Transposition to Work on the Topic "Rivers" in the Classroom

In this proposal, there are only some items selected. All of them have been transposed and simplified. So, Tables 6 to 10 include the main river attributes to be understood in Primary Education. All of them are associated to the previous images. Furthermore, all of them are complementary to the understanding of the river.

In addition, taking into account the reflection on scales made on the basis of Figure 2, it is considered that the contents should be presented not only organised by academic levels, but on related scales and in progression, from the closest to the children's environment to the furthest away at more advanced ages. That is to say, it would start with the riverbank, continuing in progression by cross-channel, river section, tributary stream or river, main riverbed and basin. In this way, the six academic levels of Primary Education would be covered.

Table 6. Summary of geology-geomorphology indicators.

# Indicator	Explanation
1	Material(s) which characterise the river.
2	Granulometry: differentiated by size (small, medium, large depending on the materials we find).
3	Valley type: U, V, ᴗ, ‿, – (front view).
4	Altitude of the riverbed / catchment: <600m, 600m-800m, >800m
5	Width and Depth of the riverbed: can be crossed on foot, infrastructure for pedestrian travel exists, vehicle and associated road transport infrastructure is required.
6	Sinuosity: –‿ᴗ, Ƨ, ɿ (in flat representation).
7	Aquifers: presence, absence (upwellings or springs, if any; cartographic representation in the last levels.
8	Characterisation of the erosion-sedimentation process: review of river margins, presence or absence of bars or islands and their location.

Source: Author's own elaboration from Muñoz, 2010.

All the previous indicators can be worked at classroom. Notice that teacher can select any of them in the subjects of Social Sciences, Natural Sciences, Mathematics (by identifying geometric concepts, planning the problem-solving process, numbers, graphic representation, measurement of time, measures of length, equivalencies, geometry,...), Spanish Language and Literature (mainly, reading and writing information), Art Education (by identifying the immediate environment, explaining its characteristics with an appropriate plastic language, elaborating three-dimensional works with different materials, ...), Physical Education (outings outside the classroom coordinated with motor actions in situations of adaptation to the physical environment, values related to the conservation of the natural environment, marches and excursions on foot or by bicycle, and so on) and Foreign language (with specific rivers vocabulary).

In the end, what you get are integrative activities that involve more than one subject from what have been called here *conceptual groups* (Fig. 8): geology and geomorphology, hydrology, ecology, culture, and landscape.

Table 7. Summary of fluvial hydrology indicators

# Indicator	Explanation
1	Water quality monitoring parameters: date and time of sampling. Characteristics such as: colour, transparency/clarity, presence of solids, room temperature, water temperature, volume, etc. They can be visual observed and with very little resources.
2	Flows. Directly in the riverbed: water level, with vegetation as reference, topography or elements built around the watercourse (bridge, building, ...).
3	Hydrological cycle and fluvial regime. Correspondence of flows with the season and/or from the hydrological year (the latter at higher levels).
4	Climate: precipitation, Temperature.
5	Isolated elements included in the hydrological cycle: springs and sinks.

Source: Author's own elaboration from Muñoz, 2010.

Table 8. Summary of fluvial ecology indicators

# Indicator	Explanation
1	Habitat identification. Spatial distribution.
2	Description of the territorial grid of habitats: core (main habitats), ecotones (contact between two different habitats) and corridors (territorial connection axes).
3	Identification of the river-dependent habitats and description of the local grid (small scale).
4	Age and size of riparian vegetation (herbaceous, shrub, tree; small, medium, large) and arrangement and distance from the axis (flooded, waterlogged, wet, dry).
5	Soil: with/without vegetation, tillage/agriculture, use/utilisation).
6	Biological quality elements (identification of specific and recognisable species, if any).

Source: Author's own elaboration from Muñoz, 2010.

Table 9. Summary of river culture indicators

# Indicator	Explanation
1	Urban, industrial and agricultural uses (supply and sanitation of the different uses).
2	Energy uses (mills, fulling mills, mini-power stations, large hydroelectric power stations, ...).
3	Leisure (fishing, bathing, recreation, leisure, water sports, ...).
4	Intervention, rectifications and modifications of the riverbed (identification of intervened river stretches).
5	River as a current or historical barrier between two territories (physical boundary wadeable or not wadeable).
6	Communications system in relation to the river (transversal: bridges, fords, ...; longitudinal: tracks, walkways, roads, ...).
7	Network of settlements. Hierarchisation (isolated construction, grouping, small population, large population). It is also possible to identify the distance/approach to the river of these population center and their temporal evolution from the use/abandonment of the constructions.
8	Specialisation and functionalities of the activities (housing, tertiary, industrial, ...).
10	Growth-stoppage trends of river areas.

Source: Author's own elaboration from Muñoz, 2010.

Table 10. Summary of fluvial landscape indicators

# Indicator	Explanation
1	Compilation of works that different artistic or cultural movements have made of the river.
2	Existence of tourist activity.
3	Appearance of news in the media.
4	Opinions of local people (surveys of the population on their knowledge about the river).
5	Typical relief (predominant).
6	Water body type (predominant).
7	Habitat type (predominant).
8	Typical uses and exploitations (predominant).
9	Construction typology: buildings and settlement, type of structures and characteristic materials.
10	Scenography and/or types of scenography (graphical representation by hand or by taking images).
11	Visual organisation. Accessibility and points of visit to the watercourse. Itineraries and viewpoints.

Source: Author's own elaboration from Muñoz, 2010.

Figure 8. Integrative learning activities
Source: Author's own elaboration.

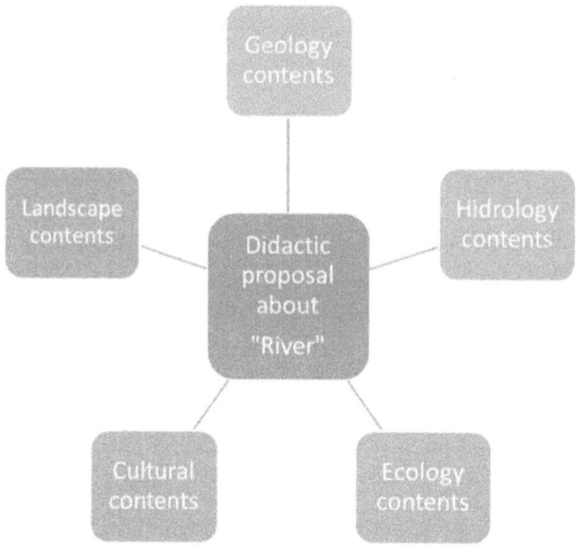

Outline for the Teacher's Didactic Proposal on Rivers

Here is present a scheme adjusted to the type of traditional didactic proposals that include the curricular elements from the educational laws (See Table 11). The proposed model scheme would include a total of fifteen curricular elements.

SOLUTIONS AND RECOMMENDATIONS

Curricular adaptation in each school will always be necessary. However, some steps are proposed to follow in order to implement this proposal (Table 12). The following are the main steps to be taken in the classroom:

FUTURE RESEARCH DIRECTIONS

This proposal has viability with national and regional curricula adjustment for Primary Education in Spain but also in other countries. So, it would be very feasible to put in practice and achieve feed-back to check the proposal. The adaptation to school curricula in any other country should consider that the definition and adaptability of the proposal presented is very open and flexible. All the items included in the five conceptual groups are applicable to any river in the world, since adaptation to any river or stretch of river and educational level is implicit in the work.

Table 11. Summary for the development of the didactic proposal about rivers for different grades of primary education.

Level	1st to 6th de EP
Title	River Landscape
Rationale	The presence of fluvial system in a disintegrated manner in the school curriculum does not allow us to know the rivers well nor to study them adequately. Given the territorial importance of these river corridors, it is necessary to include a didactic proposal in the classrooms and it would fit within the curricular sections present in each one of the educational levels of Primary Education (See tables 6 to 10).
Overall Stage Objectives (OSO)	The main OSO worked on are a, b, c, d, e, g, h, i y j.
Didactics Objectives	To be developed for each classroom activity, considering the context and the particular teaching objectives.
Competences	Mainly CCL, CMCT, SIE, CEC y CSC (See table 5).
Contents	Select contents of tables 1 to 3 and tables 6 to 10 in a complementary way
Activities	See example in Appendix 1. They are recommended: introductory activities, development or settlement activities, synthesis activities and evaluation activities.
Timing	It will be adapted to the needs of the classroom, the number and type of activities selected.
Methodology	*Mixed*: it should be flexible in terms of planning and development because we are working with variable and changing natural systems. *Innovative*: a continuous work outside the classroom and the implementation of integrative activities involving different subjects is not common in classrooms. *Active*: it involves movement.
Evaluation criteria	Those set by the education authority [7].
Evaluation tools	Those set by the education authority [8].
Transversality	It will be derived from the number of subjects involved (see Appendix 1).
Attention to diversity	It must be adapted to the needs of the students.
Resources and materials	They will be adapted to the possibilities of the classroom. The importance of the natural, cultural, and social environment is emphasised as a first-rate teaching resource through excursions outside the classroom: museums, interpretation centres, the lived experience of the landscape that is the object of study, etc. See some examples in Appendix 2.
Bibliography	See bibliography at the end of the chapter.

Source: Author's own elaboration.

Table 12. Main steps

Sequence	Tasks Description
Step 1: Select de river.	Select de river in which you want to work in the classroom. It is recommended to choose a stretch of river close to the school environment.
Step 2: Assess the possibilities.	It is advisable to visit the stretch of river. Therefore, it is necessary to know beforehand several variables in order to plan the visit in advance, such as: distance, available means of transport (if applicable), existence of educational infrastructures (nature classrooms, observatories, trails, information panels, ...), responsible administration, accessibility, contents to be worked on, time you want to spend at class, ...
Step 3: Reconnaissance visit.	It is recommended that teacher bis familiar with the study area in advance.
Step 4: Classroom activities contents.	1. Get to know your river (name, location, main sites, traditions, …) 2. We know the main attributes of the river (one activity per indicator group or one activity per indicator): • Geology and Geomorphology (see table 6). • Hydrology (see table 7). • Ecology (see table 8). • Culture (see table 9). • Landscape (see table 10). 3. Curricular contents (see tables 1 and 2).
Step 5: Field trip.	Data collection.
Step 6: Review activities.	Testing of the attributes worked on in the classroom.

Source: Author's own elaboration.

Finally, to improve this work, a new much more extensive work should include many sheets for different attributes and type of river: scale, type of regime, cultural values, … (see an example sheet in Appendix 1). Moreover, training programs and new educational proposals could be designed for any River Basin.

On the other hand, the importance of innovative learning methods is understood and geographic information systems (GIS), satellite photographs or images can be worked in the classroom. In addition, if there is a technological approach to learning methods, other suggestions such as those proposed in Appendix 2 can also be considered.

CONCLUSION

The focus of the conclusions is to reflect on the different ways of understanding the river reality. How we interpret the natural and cultural heritage of rivers is something that will determine the evolution of the system and its long-term sustainability.

To this end, work from early ages is a fundamental issue for future generations to understand, value and coexist with their landscape in a conscious way. In conclusion, these lines of reflection can be summarised as follows:

1. Each river system is unique.
2. There are inseparable natural and cultural values in the interpretation of river heritage.
3. There are numerous examples of good natural and cultural river conservation with an associated heritage that is also very valuable.
4. The intention and characteristics of each approach to the river is unique.
5. As very dynamic systems, rivers offer new interpretative possibilities that can also be very dynamic, providing a lot of motivation for children to learn.
6. Based on the natural and cultural values of rivers, they become a didactic resource that can be used in all the subjects included in Primary Education.
7. Didactic adaptation to the individual classroom will always be necessary.
8. The work sessions with a river theme can be developed from a single activity (working on any of the indicators included in the chapter), throughout a didactic (made up of several activities) or during a complete term as a project (deepening the integration of all the river values from the different subjects).

Finally, the approach to river understanding presented here does not depend on the different legislative updates, but only on the age of the learners and the depth of understanding that teacher intends to give in their classroom. In this sense, the recommendation would be to start from the immediate river environment and move towards larger territorial scales based on individual learning and the students' own experiences.

REFERENCES

Castán, V. (2017). Urban Governance and the Politics of Climate change. *World Development, 93*(May), 1–15. doi:10.1016/j.worlddev.2016.12.031

Castiglioni, B. (2009). Education on landscape for children. In *5th Council of Europe Conference on The European Landscape Convention*. Council of Europe.

Cohen-Shacham, E., Walters, G., Janzen, C., & Maginnis, S. (Eds.). (2016). *Nature-based solutions to address global societal challen-ges*. IUCN. doi:10.2305/IUCN.CH.2016.13.en

Council of Europe. (2000). *European Landscape Convention* (ETS No. 176). Author.

Cruz, L., & Español, I. M. (2008). *El paisaje: de la percepción a la gestión*. Ed. Liteam.

Ferreira, R. N. (2021). Cut-Off Lows and Extreme Precipitation in Eastern Spain: Current and Future Climate. *Atmosphere, 2021*(12), 835. doi:10.3390/atmos12070835

Florín, M., Gosálvez, R., Laguna, C., Sánchez, D., Falomir, J., González, A., Muñoz, E. M., Chicote, A., Sánchez, G., Velasco, A., & Becerra, R. (2021). *Efecto de superioridad en la mejora de servicios ecosistémicos de regulación por humedales y llanuras de inundación* (Vol. 127). Revista Ambienta. https://dialnet.unirioja.es/servlet/articulo?codigo=8070009

Florín, M., Gosálvez, R. U., Velasco, A., Laguna, C., Falomir, J. P., Chicote, A., & Muñoz, E. M. (2018). Estudio sobre Efectos Constatados y Percepción del Cambio Climático en el Medio Rural de Castilla-La Mancha. Propuestas de Medidas de Adaptación. Junta de Comunidades de Catilla-La Mancha.

Hannoun, H. (1977). *El niño conquista el medio*. Ed. Kapelusz.

Maltby, E. (Ed.). (2009). *Functional Assessment of Wetlands. Towards Evaluation of Ecosystem Services*. CRC Press. doi:10.1201/9781439829141

Marrón, M. J. (2012). Aproximación al concepto de escala en el primer ciclo de educación primaria. Presentación de un juego para su tratamiento desde la enseñanza active. *Didáctica Geográfica, 13*, 93–112.

Muñoz, E. M. (2010). *Propuesta metodológica para el reconocimiento de los sistemas fluviales. Una base para su aplicación en la toma de decisions* [Tesis de Doctorado]. Universidad de Castilla La Mancha. https://www.educacion.gob.es/teseo/imprimirFicheroTesis.do?idFichero=n%2BY%2FKH0JwYU%3D

Muñoz, E. M. (2016). Los cauces fluviales como ejes didácticos. In La investigación e innovación en la enseñanza de la Geografía (pp. 291-304). Universitat d'Alacant.

Muñoz, E. M., Español, I. M., & Florín, M. (2007). Fundamental Boundaries of the Fluvial Space. Application to environmental planning. Application to environmental planning. In *III International Symposium on Transboundary Water Management* (pp. 335-348). UCLM. UNESCO.

Muñoz, E. M., & Jerez, O. (2021). El itinerario didáctico como recurso para conocer los proyectos del programa Nosotros Proponemos. In Una forma diferente de educar a través de la ciudad. El Proyecto ¡Nosotros Proponemos! Ediciones de la Universidad de Castilla-La Mancha. doi:10.18239/atenea_2021.30.00

Olcina, J. (2013). *Gota fría*. https://www.levante-emv.com/opinion/2013/10/27/gota-fria-12856313.html

Piaget, J., & Inhelder, B. (1947). *La représentation de l'espace chez l'enfant* (Ed1972). Ed. PUF.

Rivero, M. P., & Gil, J. (2011). Didáctica de las ciencias sociales para educación infantil. Mira.

Sebastiá, R., & Tonda, E. M. (2015). El concepto y representación del espacio geográfico en la enseñanza de la Geografía en los niveles educativos no universitarios. Análisis bibliométrico. In J. de la Riva, P. Ibarra, R. Montorio, & M. Rodrigues (Eds.), *Análisis espacial y representación geográfica: innovación y aplicación* (pp. 1505–1514). Universidad de Zaragoza-AGE.

USACE. (1994). *Sharing the challenge: floodplain management into the 21st century*. Report of the Interagency Floodplain Management Review Committee to the Administration Floodplain Management Task Force. https://usace.contentdm.oclc.org/digital/collection/p266001coll1/id/2471/

ADDITIONAL READING

Costanza, R., d'Arge, R., de Groot, R., Farber, S., Grasso, M., Hannon, B., Limburg, K., Naeem, S., O'Neill, R. V., Paruelo, J., Raskin, R. G., Sutton, P., & van den Belt, M. (1997). The value of the world's ecosystem services and natural capital. *Nature*, *387*(6630), 253–260. doi:10.1038/387253a0

K. S. Hiemstra, S. van Vuren, F. S. R. Vinke, R. E. Jorissen & M. Kok (2020). Assessment of the functional performance of lowland river systems subjected to climate change and large-scale morphological trends, *International Journal of River Basin Management*. doi:10.1080/15715124.2020.1790580

OECD. (2003). Definición y selección de competencias clave: Proyecto DeSeCo. Resumen ejecutivo. https://www.deseco.ch/bfs/deseco/en/index/03/02.parsys.78532.downloadList.94248.DownloadFile.tmp/2005.dscexecutivesummary.sp.pdf

OECD. (2015). *Education at a Glance 2015: OECD Indicators*. OECD Publishing., doi:10.1787/eag-2015-

OECD. (2015), OECD Skills Strategy Diagnostic Report Spain 2015. https://www.oecd.org/skills/nationalskillsstrategies/Diagnostic-report-Spain.pdf

Rychen, D. S., & Salganik, L. H. (Eds.). (2003). *Key Competencies for a Successful Life and a Well-Functioning Society External website. Content opens in new window. DeSeCo's final report 2003*. Hogrefe & Huber Publishers.

Ward, J. V., & Standford, J. A. (1995). Ecological connectivity in alluvial river ecosystems and its disruption by flow regulation. *Regulated Rivers*, *11*(1), 105–119. doi:10.1002/rrr.3450110109

Ward, J. V., Tockner, K., & Schiemer, F. (1999). Biodiversity of floodplain river ecosystems: Ecotones and connectivity. *Regulated Rivers*, *15*(1-3), 125–139. doi:10.1002/(SICI)1099-1646(199901/06)15:1/3<125::AID-RRR523>3.0.CO;2-E

KEY TERMS AND DEFINITIONS

Educational Innovative Strategies for Rivers: Those didactic strategies that present the river as an integrated and integrating system, beyond the legal definition and always in relation to the students' immediate environment.

Fluvial Ecology: Support of trophic networks; biomass consumption, export and circulation associated with the river's capacity to transport, recycle and assimilate nutrients and maintain biological diversity.

Fluvial Function: River's capacity to provide goods and services in an intrinsically and solely way by its essence and existence, regardless of whether these goods or services are used by people, other living beings and/or the surrounding environment.

Fluvial Geology and Geomorphology: Profile and structure of the materials and their arrangement. They provide the solid basis of the river.

Fluvial Heritage: Heritage that, at some point in history, has been generated by the river or by people from river resources and is reflected in today's landscape.

Fluvial Hydrology: Surface and underground water flows, as well as their quality, quantity and composition and the river's own temporal cycles.

Fluvial Landscape: Perception of the river by both river inhabitants and river users, including all the values and services associated with the river territory.

Functional River Delineation: Physical and conceptual identification of the river territory along its entire length and beyond the flooded channel.

River: Natural origin system whose main attributes are defined by a specific geology and geomorphology, the circulation and conveyance of water through a surface and underground channel (or several) and the longitudinal and transversal connectivity which establishes with other natural and human ecosystems. It is also dynamic in space and time and a source of abiotic, biotic and cultural resources (based on Muñoz, 2010).

River Corridor: Territorial structure whose presence, discontinuity and appearance may vary along its course, connecting two or more centres of natural and/or cultural importance linked to river functions.

ENDNOTES

1 https://edition.cnn.com/2021/08/23/europe/germany-floods-belgium-climate-change-intl/index.html

2 https://www.bbc.com/news/world-europe-59608785

3 The origin of the term "cut-of-low" appears in 1886 in the German school, which introduced the idea of "kaltlufttropfen", roughly translated as "cold air drop" (https://aemetblog.es/2018/10/17/las-gotas-frias-danas-ideas-y-conceptos-basicos/). Term defined by the German meteorologist Scherhag in 1937 (Ferreira, 2021).

4 Millennium Ecosystem Assessment. http://millenniumassessment.org/en/index.html

5 Prof. Vanesa Castán Broto: https://youtu.be/Ny6XOMrpn64

6 In these cases, in addition to the degree of flooding of the species, other factors also play a role in selecting the presence or absence of the species, such as river bed configuration, water velocity or depth, etc.

7 In Spain, set from Royal Decree 126/2014, of 28 February, which establishes the basic curriculum for Primary Education.

8 Íbidem.

APPENDIX 1

Table 13. Example sheet: "The landscape of the upper course of the Segura River, Spain.

Level: 1st to 6th	**GSO:** b, d, e, h, i y j.
Competences: CCL, CMCT, CD, CEC	**Timing:** Quarterly.
Justification	
Knowledge of a section of the Segura River allows us to study the territorial and historical key role that this fluvial system has played. To do so, the work will be done on the hydrological, geological, ecosystemic and landscape functions. All these functions that shaped the landscape as it is perceived today.	
Didactic objectives	
1. To identify the historical key stages of the Segura River. 2. To learn the territorial organisation provided by rivers as penetration routes from protohistoric times, i.e., the Segura River as a border between Aragon and Castilla. 3. To learn how the exploitation of natural river resources was carried out, i.e., water and forests for Segura River. 4. To learn about the protection figures: natural and cultural values of the rivers, i.e., ZEC/ZEPA ES4210008 - Sierras de Alcaraz y de Segura y Cañones Del Segura y Del Mundo. 5. To reflect on the hydrological regulation of the riverbed and its main tributaries and how they condition the functioning of the system today. 6. To identify a territory with low demographic occupation and its relationship with the process of rural depopulation. 7. Reflect on how populations far from the point of water abstraction benefit from the resource. Sustainability of human actions (water transfers).	
Specific curricular content (per year) (from tables 1 to 3)	
1st Primary Education • Know the existence and distribution of water on Earth: oceans, seas, rivers, and lakes. • Explain the problems of water pollution, the characteristics of drinking water and the need for responsible consumption. *2nd Primary Education* • Rivers in our province and Autonomous Community. *3rd Primary Education* • The water cycle. • The landscape. • Physical map of Spain. Relief and the main rivers. *4th Primary Education* • Inland waters: ice masses, groundwater, lakes, lagoons, and rivers. • The waters of Castilla-La Mancha and the major rivers of Spain and their location on maps. • Rivers, reservoirs, and lagoons in Castilla la Mancha. • The hydrosphere. o Describes in orderly fashion the phases in which the water cycle takes place. o Knows the different water formations and their characteristics (oceans, seas, rivers, groundwater...). o Understanding how groundwater is formed, how it emerges and how it is accessed. o Identifies the sections of a river and the characteristics of each section. o Establishes the differences between tributary, river basin, riverbed, and watershed. *5th Primary Education* • Spatial distribution and population growth. Population pyramids, bar graphs and pie charts. • Distribution and evolution. Migratory movements. Rural exodus. • Knows the terms: demography, absolute population, and population density. • Locate on a map the largest population centres in Spain and Europe, the most densely populated areas and the main migratory movements. • The Middle Ages in the Segura River Basin. • The Modern Age in the Segura River Basin. *6th Primary School* • Know and locate on the map of Spain the main rivers, their source, spring, and most important tributaries. - The landscapes of Europe and Spain: Elements of the landscape (relief, coasts, climate, vegetation, hydrography).	

continued on following page

Table 13. Continued

<table>
<tr><th colspan="2">Specific indicators contents (from tables 6 to 10) (*)</th></tr>
<tr><td colspan="2">1st Primary Education: 7.1.
2nd Primary Education: 8.1, 8.2
3rd Primary Education: 10.1, 10.6, 10.7, 10.10, 10.11
4th Primary Education: 6.3, 6.5, 6.7, 7.3,
5th Primary Education: 9.1, 9.5, 9.6,
6th Primary School: 6.1, 6.8, 8.3, 8.6, 9.7, 10.4, 10.10
(*) The first number corresponds to the table and the second number corresponds to the indicator</td></tr>
<tr><th colspan="2">Methodology</th></tr>
<tr><td colspan="2">Mixed. It should be flexible in terms of planning and development because we are working with variable and changing natural systems. Moreover, it is innovative in that work outside the classroom is not a common practice in schools. Therefore, it will be an active methodology. In addition, introductory, developmental, synthesis and evaluation activities are recommended to achieve more meaningful learning.</td></tr>
<tr><th colspan="2">Development</th></tr>
<tr><td colspan="2">The Segura River has lost some of its functionality in terms of flow maintenance. The massive hydrological regulation of the riverbed and nowadays its main tributaries greatly conditions the functioning of the system. Due to the hydrological exploitation, it maintains permanent extreme conditions which, in addition to the climatological conditions, increase to the extreme the fragility of the river ecosystem and the processes which maintains.
For this reason, in the face of this fragility, the river maintains very high values of natural quality in terms of its ecological state. This is partly due to the scarce population settled in the area, the scarce access to the main riverbed and its immediate surroundings and, all of this, linked to the special orography of the area, which makes it inaccessible.
The existing uses in the section are also subject to the hydrographic exploitation of the basin: high flow fluctuations, regulated inflows and outflows and export of the resource to points far away from the area. This means that the perception of the inhabitants and visitors differs in terms of the resources themselves: water, nature, and landscape, largely due to the different relationship between them and the river and their direct knowledge (or not) of river management.
The so marked inaccessibility and orography generates an impressive scenography of gorges, canyons, meanders and very active processes of erosion and sedimentation, water circulation, etc, that remain alive, which, together with the limestone materials, also give rise to important exokarstic formations, impressive from the point of view of the observer's perception.
• Activity 1: Classroom research (small groups). In the case of the first years of Primary Education (1st and 2nd), the research could be extended to families, each of them contributing their perception of the river.
• Activity 2: Sharing in the classroom.
• Activity 3: Allocation of areas of study. Each group will be responsible for one aspect of the river landscape, understanding as such the conceptual groups marked: hydrology, geology-geomorphology, ecology, culture, and landscape.
• Activity 4: Visit to the villages and the associated river sections.
Activity 5: Classroom work (small groups) on the representation of the Segura River landscape. Oral presentation. Recommended for5th and 6th levels.
• Activity 6: Elaboration of common conclusions among all the groups. Depending on the level (age), the elaboration may have different formats and/or depth. From a poster in which the conclusions are graphically represented, to a paginated document produced with ICT, and the elaboration of a common manifesto for the recognition of the functions of the river, a travelling photographic exhibition, etc.</td></tr>
<tr><th>Evaluation criteria</th><td>Teacher should select those criteria set by the education authority that are most appropriate to the activities to be carried out.</td></tr>
<tr><th>Evaluation tools</th><td>Teacher should select those evaluations tools set by the education authority that are most appropriate to the activities to be carried out.</td></tr>
<tr><th colspan="2">Transversality</th></tr>
<tr><td colspan="2">• Social Sciences: Specifically Presented (See Table 3).
• Natural Sciences: Specifically presented (See Table 2).
• Mathematics: Identifying geometric concepts in the reality around the student with their graphic application. Through different blocks, numerous contents can be introduced: processes (e.g. planning the problem-solving process), numbers (e.g. graphic representation), measurement (e.g. measurement of time, measures of length, equivalencies, ...), geometry (e.g. the situation in the plane and in space), statistics and probability (e.g. collection and classification of qualitative and quantitative data, making and interpreting simple graphs, construction of tables of absolute and relative frequencies, ...).
• Spanish Language and Literature: among other contents, reading and writing are the instruments through which the cognitive processes that elaborate knowledge of the world are set in motion (...).
• Art Education: Identifying the immediate environment (...), explaining its characteristics with an appropriate plastic language, elaborating three-dimensional works with different materials, (...).
• Physical Education: Outings outside the classroom coordinated with motor actions in situations of adaptation to the physical environment (movement with or without materials, carried out in the natural environment, regulation of the intensity of effort, values related to the conservation of the natural environment, marches and excursions on foot or by bicycle, camping, orienteering activities, nature games, track games, approach games and others, etc.).
• Foreign language: comprehension, narration of remote and recent past events, ...</td></tr>
</table>

continued on following page

Resources and materials	Habitats - Natura 2000 - Nature - Environment - European Commission (europa.eu) http://en.www.turismocastillalamancha.es/naturaleza/parque-natural-los-calares-del-mundo-y-de-la-sima-sierra-del-segura-53372/descripcion/ https://areasprotegidas.castillalamancha.es/rap/espacios-naturales-protegidos/enp-parque-natural/parque-natural-calares-del-mundo-y-de-la-sima https://www.castillalamancha.es/gobierno/desarrollosostenible/estructura/dgapfyen/actuaciones/plan-director-de-la-red-natura-2000-en-castilla-%E2%80%93-la-mancha https://cultura.castillalamancha.es/arte-rupestre-levantino-en-CLM
Bibliography	Verdiell-Cubedo, D., Parrondo-Celdrán, P. & Murcia-Requena, J. (eds.). 2019. Guía sobre las reservas naturales fluviales de la cuenca del río Segura. Murcia. Asociación Columbares. file:///C:/Users/EM.5058159/Downloads/GUIA_RESERVAS_FLUVIALES_RIO_SEGURA.pdf

Source: Author's own elaboration.

APPENDIX 2

Some materials for working on rivers in the Primary Education classroom in other geographical contexts:

BBC Teach. Geography KS1/KS2: Rivers. https://www.bbc.co.uk/teach/class-clips-video/geography-ks1ks2-rivers/z6qsf4j

Create or open a map. My Maps. https://support.google.com/mymaps/answer/3024454?hl=en&amp%3Bref_topic=3188329

Geographical UK Association. https://www.geography.org.uk/Teaching-Resources

Guía didáctica. Viñuales A. & Jiménez M. (2003). Adopta un río. AEMS-Ríos con Vida. Junta de Castilla y León. https://es.calameo.com/read/004047217393e38182478

Instituto Geográfico Nacional. Teaching and learning material. https://www.ign.es/web/ign/portal/recursos-educativos

Proyecto geoLAB. La acción de los ríos. Junta de Extremadura. https://emtic.educarex.es/crea/ccss/formas_relieve/la_accin_de_los_ros.html

Chapter 11
The Spatial Experience for Teaching Geography in the Anthropocene

Sylvie Joublot Ferré
https://orcid.org/0000-0003-0122-8982
Haute École Pédagogique de Lausanne, Switzerland

ABSTRACT

The contemporary period is disrupted by multiple and interdependent crises. The Anthropocene is the concept designating the fact that, for the first time, a geological epoch would be defined by human action. It would be a matter of rethinking the relationship to knowledge and the methods of its transmission, particularly relevant to geography and its teaching in schools, both at primary and secondary levels. What strategies should be developed for teaching geography in a way that is civic-minded, responsible, participatory, and forward-looking, but also embodied, sensitive, and humanistic? The question that now arises is: Why and how can scholars take into account juvenile spatiality when thinking about learning? What avenues could be favoured to bring the relational process and the environmental paradigm into the geography classroom? The idea is to revive a traditional geography practice of field investigation, while allowing participants to have spatial experiences.

INTRODUCTION

The contemporary period is disrupted by multiple and interdependent crises. As the Covid-19 pandemic shows, the scale of diffusion of perils and threats is now rapidly global. Major ecological challenges, in particular climate change, the depletion of certain resources, threats to biodiversity, soil degradation, multiple forms of pollution, and attacks on living organisms in general are no longer disputed.

DOI: 10.4018/978-1-7998-9598-5.ch011

Besides, the term Anthropocene seems increasingly consensual for thinking about the social sciences, indicating "the entry of the Earth system into a new geological period affecting ecosystems" (Wallenhorst, 2020, p. 9). This indicates that, for the first time, a geological epoch would be defined by human action, with unpredictable effects in terms of climate and sea level, and even the possibility of a fateful planetary threshold (Steffen et al., 2015).

As far as education systems are concerned, under the international impetus of the United Nations and UNESCO, from the 1980s onwards, education for sustainable development (ESD) constituted a major step towards taking account of environmental and societal issues. However, the rather prescriptive and programmatic nature of ESD could slow down the objectives to be achieved among students, while the challenges are global and rapid. The Anthropocene is indeed characterized by an acceleration of social and biophysical changes on a planetary scale, which constitutes an unprecedented threat to humanity itself. It is the meaning of human life and human attachments to the Earth that are directly questioned (Albrecht, 2019). It is also the awareness that the place of humans on the planet is established within a community of all living beings (Louv, 2011). Two decades ago, some people already believed that the questions posed by this uncertain world could not be reserved for experts, elucidated by authority speeches or decided by political decision-makers alone (Barthe et al., 2001). On the contrary, ordinary citizens and inhabitants should be involved.

The Anthropocene constitutes a real paradigm shift, which also has consequences for education. More and more people are thinking that one of the levers for action and for preparing another future would indeed be education (for example Hétier, 2021). The Anthropocene therefore calls for a change in educational responses, particularly didactic ones. It could be a matter of becoming more aware of human responsibility, not just for the benefit of sustainable development, but rather from the point of view of a new relationship with the Earth, as the environmental philosopher Glenn Albrecht (2019) proposed. The conditions of the Earth's habitability could thus be at the center of the approaches, in order to invite the development of a new relationship with the world and with living beings, according to both anthropological and relational perspectives. Educational proposals must therefore be invented, beyond ESD, and more like *education in the Anthropocene*.

Several works recognize precisely this new field and support it from an epistemological and didactic point of view, opening up avenues such as education in politics (Barthes, 2022). Other proposals are situated at the interface between the limits of ESD and the new field of education in the Anthropocene, valuing the democratic participation of students and the systemic understanding of problems, and involving their affective relationships in their daily lives (Kowasch & Lippe, 2019). Tracing a path to a brighter future, this certainly means taking the paths of connection with the world to foster contact and creativity: The daily walk, urban gardens, and exposure to light would be ways of this connection (Louv, 2011). How to do this with students? With the pandemic, there has been a craze for outdoor education. Forest schools of Scandinavian origin, outdoor education in Scotland and Switzerland, and indigenous land-based education in Canada are all possible options. However, it is not enough to do school outside, nor to limit the construction of links to emblematic places, to establish an education in the Anthropocene. Devices for connecting with others and with the world can be deployed everywhere, and in particular in the city, where a large part of humanity lives, using all possible teaching methods, including digital and virtual ones. Children learn about the nature through digital means. They inhabit virtual countries and even places in digital games. It has been shown that their relationship with nature could be enhanced by technology, but not replace direct contact (Kahn Jr., 2011).

It is a question of rethinking the relationship to knowledge and the methods of its transmission. All these concerns are relevant to geography and its teaching in schools, both at primary and secondary level. Indeed, on the one hand, geography is concerned with the space of societies, and inhabited space is thus at the center of its questioning. On the other hand, its teaching must give students certain keys to understanding the world. It would seem that the geography taught could thus lead pupils, first of all, to love the world and build a sufficiently positive relationship with the place in which they live, while being aware of the current issues and risks, in order to better prepare for the future. What strategies should be developed for teaching geography in the Anthropocene, to encourage students to take care of the Earth and the community of living beings? How can people contribute to giving meaning to their presence on Earth?

The author's hypothesis is that the teaching of geography could be enriched to meet current challenges by more systematically mobilizing pupils' imaginations and sensibilities. The question that now arises is: Why and how can scholars take into account juvenile spatiality when thinking about learning? What avenues could be favored, to bring the relational process and the environmental paradigm into the geography classroom? The idea is to revive a traditional geography practice of field investigation, while allowing participants to have spatial experiences. In order to shed light on the potential of spatial experience for the teaching of geography, with regard to current challenges, and in particular from the perspective of education in the Anthropocene, the author mobilizes different empirical data: On the one hand, the results of a participatory research she conducted with adolescents, and, on the other hand, the cartographic and narrative realization of a student within the framework of her training at the Haute École Pédagogique of Lausanne.

THE SPATIAL EXPERIENCE FOR A RETURN TO REALITY AND SENSITIVITY

Several authors denounce the loss of physical contact of today's youth with nature. As early as 1990, the American author and journalist Richard Louv denounced the significant collapse of children's engagement with nature, particularly because their parents themselves no longer foster or transmit this relationship. Yet, he wrote: "The natural world is the ultimate web, and essential to the emotional health of children" (p. 173). In a more recent book, this author (Louv, 2011) pointed to ways forward for a better future. The American psychologist Peter Kahn (2002) defines the idea of "generational environmental amnesia" and in *Children and Nature: Psychological, Sociocultural, and Evolutionary Investigations*, Kahn and Kellert (2002) distinguished cognitive, affective, and evaluative modes of learning, related to three types of experience with nature, namely direct (e.g., going to the fields), indirect, and symbolic.

The challenge of the following proposal, articulating investigation and spatial experience, would be to trigger a contact, and a conscious contact, of the pupils with the world. From then on, it is obviously not only a question of nature and experiences in the forest or in the mountains, but of multiple experiences of the world. Today, according to World Bank's (2020) figures,

> *55% of the world's population, or 4.2 billion people, live in cities. This trend is expected to continue: By 2050, the current number of city dwellers is expected to double, and almost 7 out of 10 people in the world will live in urban areas. (para. 1)*

The issues at stake are therefore not at all to restrict the relationship to the world to a relationship to nature, but rather to reflect on the ways of living everywhere on the planet, starting with the daily living environment of students, the majority of whom are urban. Spatial experience is the proposed method of doing geography by experiencing, surveying, and *feeling the world*, before learning it in a more theoretical way.

It can include the landscape experience (Joublot Ferré, 2021), and, above all, it can be integrated into the survey approach, traditional to the social sciences, and in particular to geography, or even school geography. It is a question of revisiting the field survey and enriching it by putting spatial experience into perspective.

The Heuristic Dimension of Spatial Experience

The notion of spatiality has made it possible, from a relational point of view, to focus on the individual themselves, and precisely on their relationship to space.

The notion gives a reality to the spatial actor and not only to the social actor. The notion tools a new field for geography, and a new way of doing geography from the geography of the individual. The approach is documented, for example, by the study of the geography of the famous musician Franz Schubert (Lazzarotti, 2017). Some geographers give spatiality a primary role and space a secondary role (Lussault, 2019). Spatiality would be inseparable from the concept of "*habiter*" (the author can suggest the English word "inhabiting," designating the idea of experiencing the world in which people live); due to the "geographical condition" of the individual (Lazzarotti, 2006). "To study a man or woman geographically is to approach him or her as an inhabitant" (Lazzarotti, 2017, p. 145), which refers to "the human experience deployed in its geographical dimension" (Lazzarotti, 2017, p. 145). Now, inhabited places are in fact, for individuals, bearers of fundamental issues, namely political and existential. This approach through the concepts of spatiality and inhabitation would therefore be quite eloquent in geography classes in order to respond to the challenges of the Anthropocene, which require to question both the political (Barthes, 2022) and the existential (Albrecht, 2019). Individual spatiality would include spatial elements, constitutive of "the geographical part of identities" (Lazzarotti, 2017, p. 146), both from a political and existential point of view. These are, for example, the places where people live, of course, but also the places they dream about. Thus, investigating students' dream places should give access to this constitutive part of being, linked to spatiality, and should therefore be considered as a choice material for doing geography with them, precisely with the idea of touching them, engaging and involving them.

"There is no body without location" (Lazzarotti, 2017, p. 149), which leads to the reminder that the body is permanently confronted with space, it is always in a place, whatever it may be, including on a microscale, such as the bed or the bedroom for example. The place is nested in a plurality of locations up to the Earth itself. By the same token, every spatial experience is determined by the context of place, but also by the immediate or wider social context, the temporal context, and finally by the context of the individual and their own beliefs and interpretations (i.e., discourses, imaginings, and convictions). Spatial experience could thus be defined by this dynamic of an experience that has a spatial, social, and temporal context; it is therefore localized, socialized, and temporalized. It is also influenced by the discourses internalized by the inhabitant, such as imaginary, spatial representations, political, aesthetic, and cultural convictions (Figure 1). It involves the body, the senses, the emotions, and relationships with others. It instructs the construction and deconstruction of an animated and mobile relationship to the world, both political and existential. It could contribute conceptually to the understanding of the individuation process in geography. Finally, spatial experience, like all experience, is reflexive, testing oneself in space and oneself and others. It is therefore a physical, sensitive, and social experience.

Figure 1. The components of the spatial experience

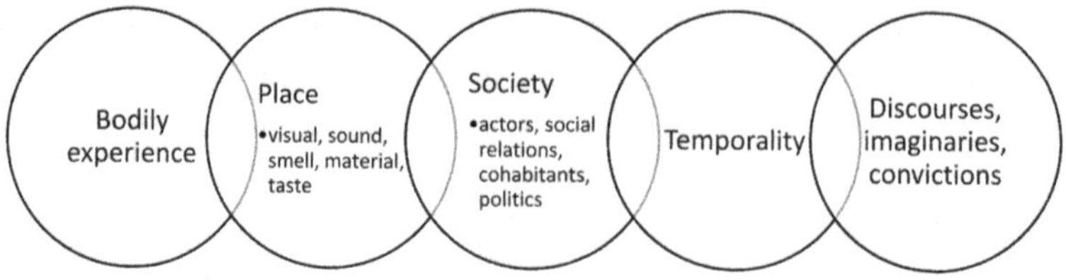

The Field Survey

The survey "has the value of a marker of scientificity for the social sciences" (Demanze, 2019, p. 16) and "has, once again, become a major paradigm in the contemporary era, for tying disciplines together" (p. 17). The survey would thus find its legitimacy within a double movement that is both epistemological and also "postdisciplinary" according to Annick Louis's (2013,on line) expression. In other words, in the future, the enquiry could be presented as a revisited device, conducive to decompartmentalization, in a school system too strongly rigidified by the traditional disciplinary organization, even though the major societal challenges of today bring into play multiple and interdependent processes, actors, and scales. Inspired by the procedures of microstoria, a current of Italian origin, which favours taking into account the experience lived by individuals, particularly through traces, discourses and clues (this current is notably founded by two emblematic works from Ginzburg (1980) and Levi (1989)); geography enquiry could contribute to making geography "down to earth", reference to the preface by the historian Jacques Revel: "*L'histoire au ras du sol*", preface to the French edition of *Pouvoir au village* by Giovanni Levi. It would be a matter of reducing the scale of observation by favoring the lived space, the inhabited spaces in the pupil's proximity. The assembly of traces can reveal the social, cultural, and even ideological practices of societies with regard to space and the world, as long as scholars endeavor to identify in the data the indices of values or societal narratives. People are also in the wake of the culture of enquiry in literature in the nineteenth century (Kalifa, 2010), following the writers who "question the blind spots of the common space" (Demanze, 2019, p. 24). With the survey, it is thus a matter of rediscovering, literally and figuratively, "the material thickness of the world" (Demanze, 2019, p. 19).

Fieldwork could restore a dialogical relationship with today's world and its contemporaries. It is the act of recognizing "the right of each person to participate in the search for truth, to propose his or her interpretation in a collective and dialectical fabrication of reality. It captures the fragmentation of perspectives and the negotiated search for consensus" (Demanze, 2019, p. 26).

In francophone Switzerland, the "*moyens d'enseignements romands* (i.e., the official teaching resources distributed to schools) specify in concrete terms the investigative approach to be developed in the human and social sciences. In the spirit of the curriculum, it is a question of "inviting the student to conceive of natural or social phenomena in the manner of a geographer, historian or scientist" (Conférence intercantonale de l'instruction publique de la Suisse romande et du Tessin, 2014, p. 28). It should be noted, however, that the enquiry process is still defined by its main purpose to provide problem-based teaching. It is thus included among the stages of the teaching-learning sequence inspired by the work of geography didactician Philippe Hertig (2012). Framed by a primer that serves as the basis for the problematization and a synthesis that serves as a summary, the investigation consists of looking for clues to answer a question. The process involves consulting resources, analyzing and verifying hypotheses, possibly in the field, or even with stakeholders, as illustrated, for example, in the Cycle 1 teaching guide for years 3 and 4 (Figure 2).

In geography, fieldwork usually involves three operations: Walking, observing, and describing.

In principle, going to a site generally involves walking, and there is always, more or less, a walking time. However, walking is valued by many works in the social and artistic sciences (De Baeque, 2019; Demailly et al. 2021; Le Breton, 2000; Thomas, 2007). It would constitute a privileged tool for identifying what happens in places; also, it is an effective mode of investigation and exploration, for example with regard to landscapes (MacFarlane, 2013). According to the anthropologist Michel de Certeau (quoted by de Baeque, 2019), the act of walking leads to:

The appropriation of the topographic system by the pedestrian.
The spatial realization of the place.
Relations between differentiated positions, in other words, spatial movements.

Going for a walk is thus already *inhabiting* the world (Ingold, 2018).

Observation could also play an important role in the investigation and be part of the cognitive process, in different dimensions from spontaneous observation to analytical observation, perceptions, and emotions.

Finally, in geography, description has traditionally been attached to landscape in geography. Echoing the "return to description," this practice is again legitimized in the social sciences as a mode of data production. Indeed, before explaining the processes at work, it seems relevant to identify and describe them. To describe would be to have recourse to an enunciation whose quality, precision, and expressiveness bear witness to a cognitive process. In other words, in the age of the Anthropocene, description could be a means of making people's modes of inhabitation visible and audible, since they are generally immersed in blindness and inattention. On the contrary, it might seem essential to reconnect with their living spaces, and description could be one of the possible acts among others.

Figure 2. Stages of the enquiry process in rem cycle 1, years 3 and 4 (Conférence intercantonale de l'instruction publique de la Suisse romande et du Tessin , 2014)

In the survey, a series of knowledge and skills would thus be coconstructed by the participants (Table 1). With reference to taxonomies, and in particular Bloom's revised taxonomy (Krathwohl, 2002), the whole range of tasks, from the simplest to the most complex, is covered.

Table 1. Cognitive virtues of the survey

Tasks	Attitudes, Skills, and Knowledge
Walk	Locating, finding your way, navigating, orienting, surveying, recognizing, understanding, and crossing. Sensitive, open and curious attitude, toponymic knowledge, navigation skills, and sociospatial knowledge
Observe	Look, listen, smell, touch, and taste. Perceptual, sensory and attention skills, and spatial knowledge.
Describe	Specify, verbalize, name, categorize, inventory, measure, and express. Documentary skills and geographical knowledge.
Collect	Informing, documenting, photographing, filming, interviewing, collecting, selecting, and drawing. Research and collection attitude, and index competence.
Telling	Collate, confront, assemble, arrange, organize, link, articulate, argue, discuss, interpret, apply, analyze, and evaluate. Analytical, compositional and narrative skills, and geographical and spatial knowledge.
Create	Modelling, conceptualizing, spatializing and making visible in different formats of textual, cartographic, iconographic, documentary, photographic, and sound writing. Creative and forward-looking attitudes and skills, and geographical and spatial knowledge.
Share	Debate, confrontation, institutionalization, mutualization, emancipation, and commitment. Building a relationship with the world and others. Societal and citizen attitude, multidisciplinary and indigenous knowledge, coconstructed on the world.

There would thus be a sensitive force on the ground, particularly crucial, to awaken benevolent attitudes towards the world.

ADOLESCENTS AND THE WORLD: REFLECTIVE DRAWINGS

At the initiative of the cultural and scientific association F93 (http://www.f93.fr) and within the framework of a partnership between this structure and the Solveig Anspach secondary school in Montreuil (France, Seine-Saint-Denis, East of Paris), a collaborative research project was conducted. The team consisted of two researchers (Colin Roche, researcher and composer, and the author of the chapter), a cultural mediator, the principal of the school, three teachers (two history-geography teachers, a mathematics teacher, and the class's head teacher) and the 23 students of a fourth-grade class, aged about 14. The project, entitled *Grandir la ville (Growing Up in the City)*, is promoted by the Conseil Général de Seine-Saint-Denis as part of a culture and art initiative in secondary schools, which relies heavily on the presence of an artist and a scientist in the school to support a research-creation process with the students. The problematic that the project had to answer was to question the construction by the pupils of their relationship to the city of Montreuil and to what extent the pandemic had had detrimental effects on this process. The methodology is multifaceted (Table 2):

Table 2. Different methodological steps

Part 1: Collection of adolescents' representations and discourse

- Individual questionnaires.
- Mind maps.
- Reflective drawings.
- Reflexivity on the exercises.
- Recordings.

Part 2: Using students' spatial experience

- Urban tour in Montreuil around historical and cultural markers, outside the home-school travel.
- Sharing of the experience.
- Creation of a representation of Montreuil in duos-trios: sensitive map, psycho-geographic map, drawing board.

Part 3: Community feedback

- Presentation by the students of their cartographic creations.
- Presentation of the final sound composition by Colin Roche.
- Restitution of the main results.

In this chapter, the aim is to link the author's proposal for a spatial experience in the geography classroom to the results obtained in the framework of this research. Indeed, these results allow the researchers to better understand the universe of adolescents in the context of a pandemic. They reveal a certain distance from the world, increased by the period of total confinement imposed in France between 17 March and 11 May 2020. A series of data is analyzed, consisting of reflective drawings about the dream place. These productions operate as a survey on the way these young inhabitants of the Parisian outskirts feel and even be in the world, a survey about their spatiality.

Graphic representations are recognized by various disciplines (e.g., geography, psychology, and urban planning) as eloquent in understanding the relationship that individuals have with the world. Following the seminal work of the American urban planner Kevin Lynch (1960) on the mental map, this tool is credited with the ability to reproduce a mental and internalized representation of space. Thomas Saarinen's survey conducted in 49 countries (André et al., 1989) confirmed the interest of mental maps as a means of accessing individuals' spatial representations. However, beyond the mere representations of spatial structures, it is indeed the relational dimension of individuals to space that it would be possible to collect with this method, drawing inspiration from the proposals of psychic geography, for example, or from work on the geographical imaginary, as a set of mental references, which influences the individual in their way of representing the world (Debarbieux, 2013). The geographical imaginary, as a "way of relating to space" (Debarbieux, 2013, p. 534) could thus be involved in the construction of the individual's relationship to the world, particularly on an ideal level. It could partly inform the understanding of the links that individuals form with the world. It also constitutes a method for better understanding individual spatialities, in the sense of the material and ideal relations established with space. Imagination can be aroused by real or virtual travel, literature, cinema, series, and video games. In other words, having children or adolescents draw spaces could make it possible to document the spatial representations of the lived or imagined space (Gumuchian,1989) by incorporating affects.

Recent innovative initiatives continue this orientation: The cartography of emotions (Griffin & McQuoid, 2012), to represent a territory (Muis, 2016) or the contemporary world (Lascaux, 2020); sensitive cartography to evoke the space experienced during confinement in France (Jolivet et al, 2021), participatory cartography using information on the territory produced by the inhabitants themselves, informing the meaning of places for the contributors (Palsky, 2013), cartography of migratory routes (Mekdjian & Amilhat Szary, 2015), and the subjective cartography projects of the psychoanalyst and documentary artist Catherine Jourdan (www. geographiesubjective.or). At present, the investigations carried out with students by means of mental maps or reflexive drawings tend to be carried out within didactic or geographical research protocols: spatial representations of gentrified neighborhoods (Authier & Lehman- Frisch, 2012), of practised spaces (Leroux & Verherve, 2014) or even political spaces on a global scale (Ott, 2020) or of high school pupils' vision of the world (Rigollet, 2017). However, pedagogical scenarios in school geography could much more regularly involve pupils in the production of images to understand their relationship to space, in support of different learning approaches and themes. This would lead to taking better account of pupils' experiences and their way of representing the world, and therefore to giving more meaning to learning objectives. However, it is worth noting, without any concern for completeness, the proposal for pupils to make drawings, which can be used to relay their spatial representations in the *moyens d'enseignement officiels romands*1, for example. In fact, in class 5[e] (children aged 7-8) in geography, the program focuses on *Inhabiting* and the activity that opens the school year consists of answering the question "Where would you like to live? How?" with a drawing.

In this creative process, the pupils are authors of an image, and they call upon one or more geographical imaginations. It is these two registers (i.e., showing and imagining) that enrich the exercise and reflect the pupils' spatiality. First, the drawing produced by the pupil is presented as a representation of the world, a "spatial device" (Besse, 2018, p. 93) made by the pupil, which has to be deciphered by a procedure of denotation. The image obtained is interpreted as a "system of signs that mediates the individual's relationship to the world" (Lussault, 2013, p. 533). Then, drawing is also a projection "into the unreal, the virtual, the imaginary" (Castellotti & Moore, 2009, p. 45) and therefore gives access to the reference universes, the geographical imaginary of the student

The conditions of possibility of this reflexive image are therefore cognitive (e.g., knowledge of the territory and local knowledge), technical (i.e., ability to draw), semiological (i.e., symbolic codification), scalar (i.e., choice of a scale), spatial (i.e., perspective and structure), social (i.e., reality of links and universes of reference), and political (i.e., values and civic commitment).

Below, Figures 3 to 7 are collected from pupils in class of the Montreuil middle school the author mentioned above. The series responds to the instruction: "Draw and write words that represent a dream place for you. If this place exists, indicate its name and location." Five drawings from a series of 23 are presented.

These five drawings and the texts that accompany them provide excellent information about the place of imaginary and spatial representations in the creation of reference spaces for pupils (Table 3).

First of all, there is the change of scenery, the imagined Elsewhere or the known Elsewhere. The imagined Elsewhere is located in Japan, the Bahamas, and in space; the known Elsewhere is represented by the United States. It should be noted, however, that the dream location for each of these five students would be a place of expatriation from their current home.

Figure 3. Dreaming of Japan

Figure 4. A house in the United States

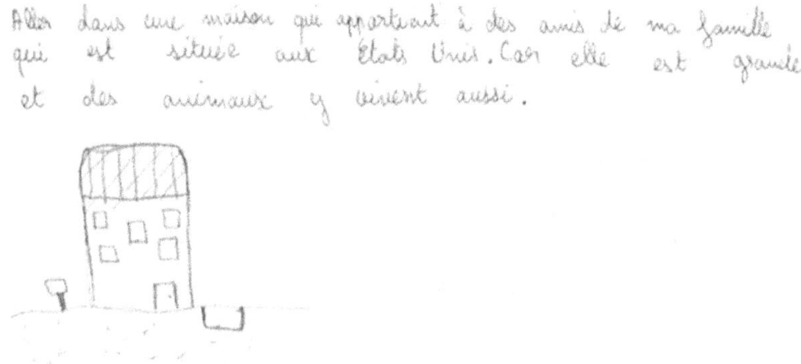

In Figures 3 and 7, knowledge of Japan is not based on direct experience, but on virtual experience, instructed by Japanese manga and animation series, or even through a recently discovered legendary figure such as Yasuke, the first black samurai. The cultural, gastronomic, and even landscape appeal is formulated without direct knowledge of the country itself, but through symbolic and imported images. In Figures 3 and 7, the focus is on figures or objects from Japanese *anime*, specifically the Japanese animation series *One Piece*, adapted from the manga of the same name. One could recognize Monkey D. Luffy, a pirate with a straw hat and a character from

One Piece (Figure 3) or the Thousand Sunny, inspired by the world of pirates and a ship of the *One Piece* crew (Figure 7). These two drawings testify to a juvenile universe shared with the *One Piece* series, a marker for imaginations and dreams, a playful universe linked to adventure, leisure, and fictional places.

Figure 5. The Bahamas Archipelago

Figure 6. Another planet

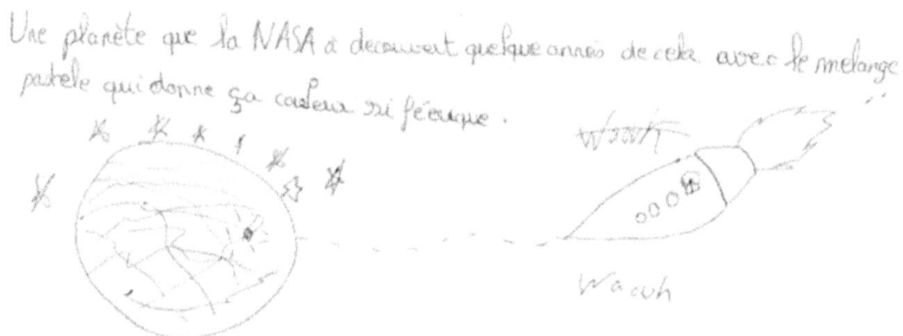

Figure 7. A pirate ship

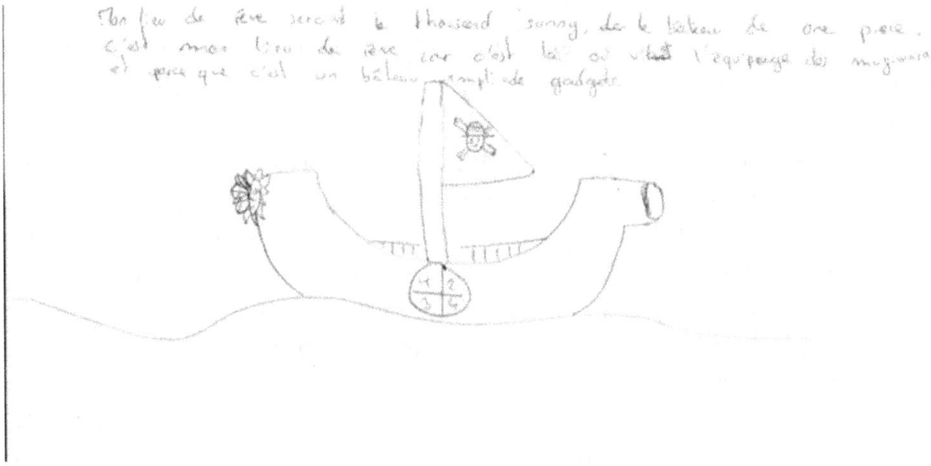

Table 3. Analysis

Figures	Iconography	Denotation
Figure 3. Dreaming of Japan	The flag of Japan. A character bigger than the building, people, a restaurant?	Japan is explicitly identified through the flag. The character seems to resemble the straw- hatted pirate Monkey D. Luffy, hero of the manga and anime *One Piece*, unless it is the hero Yasuke, the first black samurai, mentioned by the student. What appears to be a building is perhaps a sushi bar?
Figure 4. A house in the USA	A multistorey house, a mailbox, a garden or a meadow.	The American house is identifiable by the specific-looking mailbox placed at the edge of the road.
Figure 5. The Bahamas Archipelago	Two surfaces colored yellow and blue are juxtaposed.	To lignes drawn: Sand by the sea to symbolize a beach of the Bahamas archipelago (Caribbean, Atlantic Ocean).
Figure 6. Another planet	A rocket with an astronaut or passenger, a planet, and stars.	The pink planet discovered in 2013 by Nasa, with pink and magenta colors.
Figure 7. A pirate ship	A Norman drakkar pirate ship, with a flag at the top of the mast with a Jolly Roger skull and crossbones, and ornaments at the stern and bow of the ship.	The Thousand Sunny is the crew ship in the Japanese animation series *One Piece*, adapted from the manga of the same name.

Figure 5 shows the imaginary of the tropical island, which may be associated with representations of a pleasant way of life and surroundings, beaches and good weather; the social or anthropological conditions of island life would thus be rather idealized. The symbolism mobilizes the natural elements, they speak of the body's relationship with the world and even the seaside landscape, certainly seen more as resources, an assurance of well-being and amenities.

Figure 6 is perhaps influenced by the high-profile private spaceflight projects funded by billionaires. It also reflects the appeal of planets as possible extensions of the ecumene.

The known Elsewhere is present in Figure 4, an American family friend's house, presumably in the country, spacious and with animals.

These five drawings have in common that they situate the dream place outside the perimeter of life. For these five pupils, the dream place is a place that takes them outside their usual living area. Only one of the 23 drawings in this series locates the dream place in Montreuil, and that is in the house, the bed.

These drawings confirm the importance of the imaginary for geography, both in terms of the object of study and the methodology (Schoepfer, 2021). Indeed, these productions testify to the fact that the individual possesses a set of mental images of space, derived from direct experience, or instructed through a variety of media (e.g., cinema, song, literature, travel reports, social networks, and video games). As an object of study, these drawings are fruitful in documenting the references at work in the making of spaces by adolescents. The author recognizes exoticism, the role of which has already been identified as a determining factor in the construction of the Western imaginary of Elsewhere (Staszak, 2012), but also heliotropism and the seaside landscape. Video games or animated series seem to provide several students with a source of symbolic inspiration for imagining their dream places, which, by definition, do not exist, are virtual, unreal, but could constitute means of escape from reality, from everyday life. These drawings show the interest of knowing more about the spatiality of the pupils, in order to propose a geography in touch with their worlds and their spatial knowledge or representations. This consideration is part of the concern to connect pupils to their ways of being in the world. In particular, it is a question of legitimizing the effect of direct or indirect, and therefore virtual, experiences of space, as contributing to people's subjectivities towards the world.

THE POTENTIAL OF SPATIAL EXPERIENCE IN THE ANTHROPOCENE

As the author highlighted above, the relationship to the world is constructed within a referential bundle. This bundle develops through spatial experiences in real or imagined places, internalized through real or virtual and/or mediated spatial experiences.

The didactic interest of a singularization of spatial experience, particularly within fieldwork, is to recognize its thickness, texture, and cognitive implication. This hypothesis is mainly inspired by the work of the American philosopher John

Dewey, one of the pillars of pragmatism and author of a theory of enquiry. In his essay *Experience and Education,* Dewey (2018) contrasts "the teaching that proceeds from textbooks and books, with that of experience; .../... with the static goals and manner of the curriculum, the commerce with a world in perpetual change" (p. 461). Further on, Dewey refers to "the idea that there is an intimate and necessary relationship between the processes of experience and education" (p. 462). The transposition of Dewey's principles to questions of space and spatiality seems very inspiring, and could now be approached through experimentation, which would seem to be promising for the teaching of geography. Despite the spatial turn, no didactic theory has really considered geographical education in the format of spatial experimentation. One can also refer to the transpositions offered by anthropologist Tim Ingold (2018) of Dewey's work for education. According to Dewey's (2018) idea, the experience retained by the educator must be fruitful and creative. Spatial experiments should be conducted in order to assess their richness and potential for inventiveness.

The exploration in initial training allows to evaluate the spatial experience through the prism of education in the Anthropocene. The aim is not to give an exhaustive account. Rather, it is a question of highlighting a transversal device in the training of students and pupils: The use of enquiry and the conceptualization of spatial experience. The challenge is to strengthen and maintain the desire of students to enter into direct contact with the outside world.

Figure 8 is produced in the context of an interdisciplinary module based on experiences of urban wandering in Lausanne, institutionalized by a narrative of the city, documented by the participants' spatial experiences.

Following the discoveries and encounters made during this module, I decided to work around creativity and experience, which are both transversal aspects but which are systematically found in Visual Arts.

For this purpose, I mixed wandering in the city and observation drawings with the aim of making art. Here, the wandering is meant to be a starting point and a source of inspiration.

First, I chose five objects that I find most of the time in Lausanne: Tree leaves, bicycles, signs, facades, and cobblestones. I then looked for the silhouettes of these objects on a map of the city, following the street lines, a bit like looking for constellations in the sky. I went to the site and followed the path traced by these silhouettes. During each route, I looked for the object in question in order to observe and draw it from all angles. This enabled me to build up a bank of sketches. With this repertoire of drawings, I then made a collage on a map of Lausanne, gathering all these details which represent, for me, this city.

Figure 8. Constellations lausannoises, draw me a collage

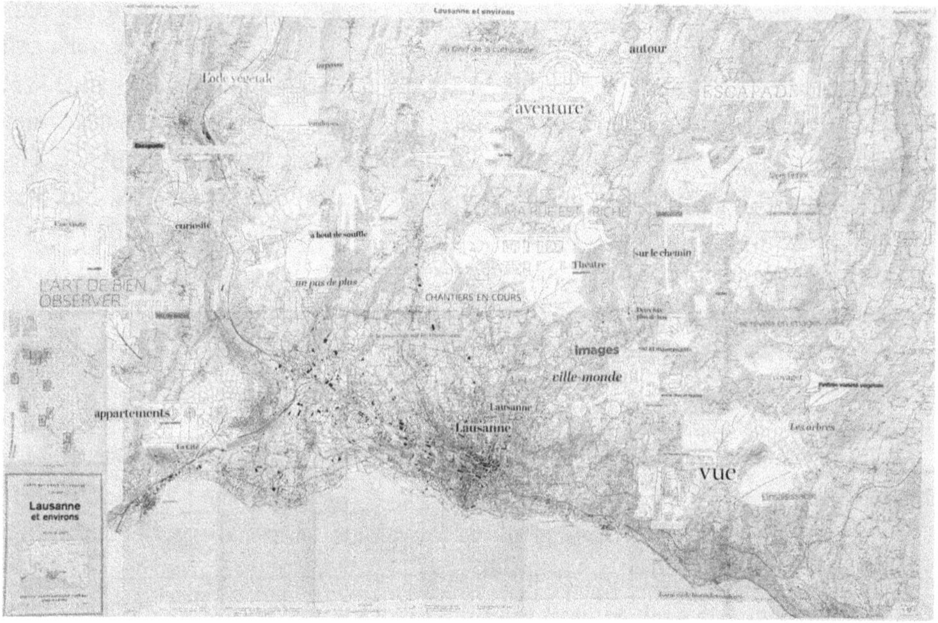

My final work is thus articulated between, on the one hand, a sketchbook listing my different routes and my observations in the field, and, on the other hand, a sort of poster, a visual interpretation of the city.

Thus, this work is spatial, as it is marked out by a walking tour of the city, scientific, as it involves observations and the creation of a sort of repertoire, and finally aesthetic, as it results in a visual creation. This approach was also a pretext for me to observe my city in a different way. The city is of course a whole, a whole, but it is also all these small details to look at, observe and spy on. From a didactic point of view, the same applies: Through the experience of wandering, the idea would be to invite pupils to discover their city in a different way, to take a closer look at a neighborhood, a street or a corner. (A student's reflective commentary on their process and achievement.)

These *constellations of Lausanne* (Figure 8) have several advantages. On the one hand, they reflect a real investigation, that is a pedestrian investigation, which engaged the student in a corporeal and sensitive manner. The cartographic implementation here is conditioned by the direct experience of space, by different movements, and consequently it testifies to the spatial appropriation. On the other hand, careful observation of places is mobilized, which is explicitly expressed by the student. Observation is obviously a perceptive process that produces new knowledge about places. This knowledge could very well be involved in a prospective operation, in order to solve problems. Finally, the realization constitutes in itself a descriptive operation, in addition to the base map itself, there are elements collected which participate in the location added by the student. Thus, data are produced and stated.

The procedure contributes directly to activating attention to the world (Hétier, 2021), to the world precisely crossed and inhabited. The conscious contact operates and allows one to experience from within, to realize, and to feel the places. The space traversed is that of bodily experience, of people's sensibilities and representations. It is possible to evoke the perceptive, creative, and therefore political possibilities of this whole device, which associates investigation and spatial experience. Referring to Glenn Albrecht's (2019) work and, particularly, to the idea of the "Symbiocene," the author's procedure could have the ambition to arouse positive emotions towards the Earth and the desire for an alternative and happier future in reaction to the stressful and negative environmental changes that characterize the Anthropocene. This procedure could also rebalance catastrophism in favor of an awareness of the social and ecological resilience of societies (Adger, 2000). Spatial appropriation, the internalization of an attachment to place, could lead to a desire for resilience. From another perspective, the meaning of the sustainable city and nature in the city, for example, between policy (Alexandre, 2013) and spatial experience, could be assessed according to the categories of geography.

CONCLUSION

The present problem is to determine the relationship *within* experience between the works of the past and the events of the present. We must be able to discriminate how knowledge of the past can become the instrument of our efficiency in the future. (Dewey, 2018, p. 463-464)

In the Anthropocene, this knowledge issue is considerable, since the emergencies of contemporary era require to better prepare the future with the young inhabitants.

In this chapter, the author supported the idea that fieldwork and spatial experience could constitute new avenues for this exploration, to be explored in the teaching of geography, without any dogmatism. It is not at all a question of asserting that there is only one way of doing geography, as this would be very impoverishing for education and the challenges it must meet.

Pragmatically, spatial experience as experimentation with space is one possible modality, but not the only one. Then, it seems eloquent, including from the point of view of students, to make geography in the Anthropocene, which contributes to conscious links between future citizens and their environment, connecting young individuals to their way of being in the world.

REFERENCES

Adger, W. N. (2000, September). Social and Ecological Resilience: Are They Related? *Progress in Human Geography*, *24*(3), 347–364. doi:10.1191/030913200701540465

Albrecht, G. (2019). *Earth Emotions: New Words for a New World.* Cornell University Press.

Alexandre, F. (2013). The Role of Vegetation in the Urban Policies of European Cities in the Age of the Sustainable City, *European Spatial. Research Policy*, *20*(2), 11–27.

André, Y., & Bailly, A. (Eds.). (1989). *Représenter l'espace : L'imaginaire spatial à l'école*. Diffusion, Economica.

Authier, J., & Lehman-Frisch, S. (2012). Il était une fois... des enfants dans des quartiers gentrifiés à Paris et à San Francisco. *Actes de la Recherche en Sciences Sociales*, *195*(5), 58–73. doi:10.3917/arss.195.0058

Barthe, Y., Callon, M., & Lascoumes, P. (2001). *Agir dans un monde incertain. Essai sur la démocratie technique*. Seuil.

Barthes, A. (2022). Quels curricula d'éducation au politique dans les questions environnementales et de développement? *Éducation et socialisation*, *63*.

Besse, J.-M. (2018). *La nécessité du paysage*. Éditions Parenthèses.

Castellotti, V., & Moore, D. (2009). Dessins d'enfants et constructions plurilingues. Territoires imagés et parcours imaginés. In *Molinié, M., Le dessin réflexif: élément d'une herméneutique du sujet plurilingue* (pp. 45–85). CRTF, Encrage, Belles Lettres.

Conférence Intercantonale de l'Instruction Publique de la Suisse romande et du Tessin (CIIP), (2014). Géographie, histoire, sciences de la nature. Guide pour l'enseignement. 3e-4e, cycle 1. Neuchâtel : CIIP.

De Baeque, A. (2019). *Une histoire de la marche*. Perrin.

Debarbieux, B. (2013). Imaginaire géographique. In Dictionnaire de la géographie et de l'espace des sociétés, (pp. 530-534). Belin.

Demailly, K.-E., Monnet, J., Scapino, J., Deraeve, S. (2021). *Dictionnaire pluriel de la marche en ville*. Paris: Éditions L'œil d'or.

Demanze, L. (2019). *Un nouvel âge de l'enquête*. Éditions Corti.

Dewey, J. (2018). Démocratie et éducation, suivi deExpérience et Éducation. Malakoff: Éditions Armand Colin.

Griffin, A., & McQuoid, J. (2012). At the intersection of maps and emotion: The challenge of spatially representing experience. *Kartographische Nachrichten*, *62*(6), 291–299.

Gumuchian, H. (1989), Les représentations en géographie. Définitions, méthodes et outils. In Représenter l'espace : L'imaginaire spatial à l'école (pp. 29-56). Diffusion, Economica.

Hétier, R. (2021). *L'humanité contre l'Anthropocène. Résister aux effondrements.* Presses universitaires de France.

Ingold, T. (2018). *L'anthropologie comme éducation.* Presses Universitaires de Rennes.

Jolivet, L., Dominguès, C., Mermet, E., & Seten, S. (2021). Lived and perceived space during lock-down in a sensitive map approach, *Proc. International Cartographic Association*, *4*, 50. doi:10.5194/ica-proc-4-50-2021

Joublot Ferré, S. (2021), L'enquête paysagère, une méthode d'intelligibilité politique du monde? Étude de cas en formation d'enseignants. Projets de paysage, 24.

Kahn, P. H. Jr. (2011). *Technological Nature. Adaptation and the Future of Human Life*. MIT Press. doi:10.7551/mitpress/7983.001.0001

Kahn, P. H., & Kellert, S. R. (Eds.). (2002). *Children and nature: Psychological, sociocultural, and evolutionary investigations*. MIT Press. doi:10.7551/mitpress/1807.001.0001

Kalifa, D. (2010). Enquête et « culture de l'enquête » au XIXe siècle. *Romantisme*, *149*(3), 3–23. doi:10.3917/rom.149.0003

Kowasch, M., & Lippe, D. F. (2019). Moral impasses in Sustainability Education? Empirical results from school geography in Austria and Germany. *Environmental Education Research*, *25*(7), 1066–1082. doi:10.1080/13504622.2018.1557112

Krathwohl, D. R. (2002). A Revision of Bloom's Taxonomy: An Overview. *Theory into Practice*, *41*(4), 212–218. doi:10.120715430421tip4104_2

Lazzarotti, O. (2006). *Habiter : La condition géographique*. Paris: Belin.

Lazzarotti, O. (2017). *Une place sur Terre ? Franz Schubert, de l'homme mort à l'habitant libre*. Auxerre: HDiffusion.

Le Breton, D. (2000). *Éloge de la marche*. Éd. Métailié.

Leroux, X., & Verherve, M. (2014). Ma petite géographie ou la fabrique de la représentation des lieux chez de jeunes élèves. *Mappemonde, 113*. http://mappemonde.mgm.fr/num41/articles/art14101.html

Louis, A. (2013). Ce que l'enquête fait aux études littéraires: à propos de l'interdisciplinarité. *Fabula / Les colloques*. https://www.fabula.org/colloques/document2097.php

Louv, R. (1990). *Childhood's future*. Houghton Mifflin.

Louv, R. (2011). *The nature principle. Human restoration and the end of nature-deficit disorder*. Algonquin Books of Chapel Hill.

Lussault, M. (2019). La spatialité d'abord. In Y. Calbérac, O. Lazzarotti, J. Lévy, & M. Lussault (Eds.), *Carte d'identités : L'espace au singulier* (pp. 7–24). Hermann. doi:10.3917/herm.povoa.2019.01.0007

Lynch, K. (1960). *The Image of the City*. MIT Press.

MacFarlane, R. (2013). *The Old Ways. A Journey on Foot*. Penguin Ed.

Mekdjian, S., & Amilhat Szary, A.-L. (2015). *Cartographies traverses, des espaces où l'on ne finit jamais d'arriver*. Visioncarto.

Muis, A.-S. (2016). Psychogéographie et carte des émotions, un apport à l'analyse du territoire? *Carnets de géographes, 9*. https://journals.openedition.org/cdg/713

Olmedo, E. (2017). L'expérimentretien comme méthode d'enquête. Cartographie sensible et terrains de recherche collaboratifs entre art et géographie. *Mappemonde, 121*.

Ott, A.-C. (2020). L'organisation de l'espace mondial chez les enfants: les contours d'une géopolitique enfantine. *L'Espace Politique*, *40*(1). https://journals.openedition.org/espacepolitique/7997

Palsky, G. (2013). Cartographie participative, cartographie indisciplinée. *L'Information Geographique*, *77*(4), 10–25. doi:10.3917/lig.774.0010

Rigollet, N., A., (2017). Adolescences "Em Sampa". Représentations urbaines de lycéens français expatriés à Sao Paulo. *Carnets de géographes, 10*.

Schoepfer, I. (2021). Éditorial. GeoAgenda, 4, 3.

Staszak, J.-F. (2012). La construcción del imaginario occidental del 'allá' y la fabricación de las 'exótica': El caso de los koi moko maoris. In *Geografia de los imaginerarios* (pp. 179–210). Anthropos/Universidad Autónoma Metropolitana Iztapalapa.

Steffen, W., Broadgate, W., Deutsch, L., Gaffney, O., & Ludwig, C. (2015). The trajectory of the Anthropocene: The great acceleration. *The Anthropocene Review, 2*(1), 81–98. doi:10.1177/2053019614564785

Szanto, C., & Bergsjö, A. (2018). Représenter l'expérience spatiale. Description et analyse d'une expérimentation graphique. *Projets de paysage, 18.* https://journals.openedition.org/paysage/1119

Thomas, R. (2007). La marche en ville. Une histoire de sens. *L'Espace Geographique, 36*(1), 15–26. doi:10.3917/eg.361.0015

Wallenhorst, N. (2020). *La vérité sur l'Anthropocène*. Éditions Le Pommier.

KEY TERMS AND DEFINITIONS

Geographical Imagination: All the spatial images and images of the world that the individual possesses, that contribute to constructing his spatial representations but also to nourishing his desires for elsewhere.

Inhabit: This verb is now a central concept in geography and the social sciences in the French-speaking world. Introduced also in school geography, to inhabit informs the way in which societies and individuals occupy space. Beyond housing, the concept thus concerns all spaces up to the Earth itself, it includes material and symbolic relations, while raising the question of habitability. The notion refers to the individual as much as to the universal.

Spatial Experience: This expression designates both the fact of experiencing space in the sense to make with space, but also the fact of expanding one's spatial knowledge and skills. It contains the idea of an actualizable spatial capital.

Spatiality: It is the spatiality of individuals. Each person has his or her own spatiality, linked to his or her geographical condition, which integrates his or her material and symbolic relations to space.

ENDNOTE

1 See: https://www.plandetudes.ch/web/mer

Compilation of References

Abrami, P. C., Bernard, R. M., Borokhovski, E., Waddington, D. I., Wade, C. A., & Persson, T. (2015). Strategies for teaching students to think critically: A meta-analysis. *Review of Educational Research*, *85*(2), 275–314. doi:10.3102/0034654314551063

Adedoyin, O. B., & Soykan, E. (2020). Covid-19 pandemic and online learning: the challenges and opportunities. In *Interactive Learning Environments*. Routledge. doi:10.1080/10494820.2020.1813180

Adger, W. N. (2000, September). Social and Ecological Resilience: Are They Related? *Progress in Human Geography*, *24*(3), 347–364. doi:10.1191/030913200701540465

Adkins, J. K. (2018). Active Learning and Formative Assessment in a User-Centered Design Course. *Information Systems Education Journal*, *16*(4), 34–40.

Agnew, J. (1987). Place and Politics: the Geographical, Mediation of State and Society, Boston: Allen and Unwin.

Agudo, S., Pascual, M. A., & Fombona, J. (2012). Uses of Digital Tools among the Elderly. *Comunicar*, *20*(39), 193–201. doi:10.3916/C39-2012-03-10

Akçayır, G., & Akçayır, M. (2018). The flipped classroom: A review of its advantages and challenges. *Computers & Education*, *126*, 334–345. doi:10.1016/j.compedu.2018.07.021

Albrecht, G. (2019). *Earth Emotions: New Words for a New World.* Cornell University Press.

Alexandre, F. (2013). The Role of Vegetation in the Urban Policies of European Cities in the Age of the Sustainable City, *European Spatial. Research Policy*, *20*(2), 11–27.

Alexson, R., & Kemnitz, C. (2001). The World Bank Scenario - A Problem-Based Learning Activity in Human Geography and Environmental Science. *Planet*, *4*(1), 25–26. doi:10.11120/plan.2001.00040025

Allen, P. (2008). From landscapes into geological history. *Nature*, *451*. doi:10.1038/nature06586

Al-Maktoumi, A., Al-Ismaily, S., & Kacimov, A. (2016). Research-based learning for undergraduate students in soil and water sciences: A case study of hydropedology in an arid-zone environment. *Journal of Geography in Higher Education*, *40*(3), 321–339. doi:10.1080/03098265.2016.1140130

Álvarez, J., & de Lázaro, M. L. (2017). Spatial Data Infrastructures and Geography learning. *European Journal of Geography*, *8*(3), 19–29. http://eurogeojournal.eu/articles/2.Spatial data infrastructures and Geography learning.pdf

Amirian, P., Basiri, A., Gales, G., Winstanley, A., & McDonald, J. (2015). The Next Generation of Navigational Services Using OpenStreetMap Data: The Integration of Augmented Reality and Graph Databases. In J. J. Arsanjani, A. Zipf, P. Mooney, & M. Helbich (Eds.), *Openstreetmap in Giscience: Experiences, Research, and Applications* (pp. 211–228). Springer-Verlag Berlin. doi:10.1007/978-3-319-14280-7_11

Amosun, P. A. (2016). Why Nigerian Geography Teachers Scarcely and Scantly Teach Map Reading and Why Students Are Scared of It. *American Educational Research Journal*, *4*(2), 42–48.

André, Y., & Bailly, A. (Eds.). (1989). *Représenter l'espace : L'imaginaire spatial à l'école*. Diffusion, Economica.

Andrew, J. T., & Sauquet, E. (2017). Climate Change Impacts and Water Management Adaptation in Two Mediterranean-Climate Watersheds: Learning from the Durance and Sacramento Rivers. *Water (Basel)*, *9*(2), 126. doi:10.3390/w9020126

Arends, R.I. (1995). *Uczymy się nauczać*. WSiP.

Arsanjani, J. J., & Bakillah, M. (2015). Understanding the potential relationship between the socio-economic variables and contributions to OpenStreetMap. *International Journal of Digital Earth*, *8*(11), 861–876. doi:10.1080/17538947.2014.951081

Arsanjani, J. J., Zipf, A., Mooney, P., & Helbich, M. (Eds.). (2015). *OpenStreetMap in GIScience: Experiences, Research, andApplications*. Springer-Verlag Berlin. doi:10.1007/978-3-319-14280-7

Ashmore, B. (2003). Arno Peters changed the world! Development education and the Peters' projection. *The Cartographic Journal*, *40*(1), 57–5. doi:10.1179/000870403235002097

Attfield, I., Tamiru, M., Parolin, B., & DeGrauwe, A. (2002). *Improving Micro Planning in Education Through a GIS: Studies on Ethiophia and Palestine*. UNESCO.

Ausubel, D. P. (1960). The use of advance organizers in the learning and retention of meaningful verbal material. *Journal of Educational Psychology*, *51*(5), 267–272. doi:10.1037/h0046669

Ausz, M., & Ausz, A. (2007). Lublin moje miasto – propozycja projektu interdyscyplinarnego. *Wiadomości Historyczne*, *5*, 55–60.

Authier, J., & Lehman-Frisch, S. (2012). Il était une fois... des enfants dans des quartiers gentrifiés à Paris et à San Francisco. *Actes de la Recherche en Sciences Sociales*, *195*(5), 58–73. doi:10.3917/arss.195.0058

Baena, J. M. (2016). Patrimonio y educación obligatoria. Deseos y realidades. *Revista UNES: Universidad, Escuela y Sociedad.*, *1*, 6–20. https://www.revistaunes.com/index.php/revistaunes/article/view/1/1

Baker, T., Kerski, J., Huynh, N., Viehrig, K., & Bednarz, S. (2012). Call for an agenda and center for GIS education research. *Review of International Geographical Education Online*, *2*(3), 254–288.

Balram, S. (2019). Teaching and learning pedagogies in higher education Geographic Information Science. In S. Balram & J. Boxall (Eds.), GIScience Teaching and Learning Perspectives (pp. 1-8). Springer. doi:10.1007/978-3-030-06058-9_1

Bandy, J. (2011). *What is service learning or community engagement?* Vanderbilt University Center for Teaching. Retrieved 4th Dec 2011 from https://cft.vanderbilt.edu/guides-sub-pages/teaching-through-community-engagement/

Barbadillo, F. (2016). *Manual para observar e interpretar paisajes*. Tundra.

Barell, J. (2006). *Problem-Based Learning: An Inquiry Approach*. Corwin Press.

Barkley, E. F., Cross, K. P., & Major, C. H. (2005). *Collaborative learning techniques: A handbook for college faculty*. Jossey Bass.

Barrows, H. S. (1986). A Taxonomy of problem-based learning methods. *Medical Education*, *20*(6), 481–486. doi:10.1111/j.1365-2923.1986.tb01386.x PMID:3796328

Barrows, H. S. (1996). Problem-Based Learning in medicine and beyond: A brief overview. In L. Wilkerson & W. H. Gijselaers (Eds.), *Bringing Problem-Based Learning to Higher Education: Theory and Practice* (pp. 3–12). Jossey-Bass Pub. doi:10.1002/tl.37219966804

Barthes, A. (2022). Quels curricula d'éducation au politique dans les questions environnementales et de développement? *Éducation et socialisation, 63*.

Barthe, Y., Callon, M., & Lascoumes, P. (2001). *Agir dans un monde incertain. Essai sur la démocratie technique*. Seuil.

Bartoschek, T., & Keßler, C. (2013). VGI in Education: From K-12 to Graduate Studies. In D. Sui, S. Elwood, & M. Goodchild (Eds.), *Crowdsourcing Geographic Knowledge: Volunteered Geographic Information (VGI) in Theory and Practice* (pp. 341–360). Springer Netherlands. doi:10.1007/978-94-007-4587-2_19

Basaraner, M., & Cetinkaya, S. (2019). New measures for analysis and comparison of shape distortion in world map projections. *Cartography and Geographic Information Science*, *46*(6), 518–531. doi:10.1080/15230406.2019.1567394

Bautista-Cerro, M. J., Murga-Menoyo, M. A., & Novo, M. (2019). La Educación Ambiental en el S. XXI (página en construcción, disculpen las molestias). *Revista de Educación Ambiental y Sostenibilidad, 1*(1), 1103.

Baytak, A., & Land, S. (2011). An investigation of the artifacts and process of constructing computers games about environmental science in a fifth grade classroom. *Educational Technology Research and Development*, *59*(6), 765–782. doi:10.100711423-010-9184-z

Beane, J. A. (1995). Curriculum Integration and the Disciplines of Knowledge. *Phi Delta Kappan*, *76*(8), 616–622.

Bear, J. B., & Collier, B. (2016). Where are the Women in Wikipedia? Understanding the Different Psychological Experiences of Men and Women in Wikipedia. *Sex Roles*, *74*(5), 254–265. doi:10.100711199-015-0573-y

Bednarz, S. W., & Van Der Schee, J. (2006). Europe and the United States: The implementation of geographic information systems in secondary education in two contexts. *Technology, Pedagogy and Education*, *15*(2), 191–205. doi:10.1080/14759390600769573

Bednarz, S., Heffron, S., & Huynh, N. (2013). *A Road Map for 21st Century Geography Education. Recommendations and Guidelines for Research in Geography Education*. Association of American Geographers, National Geographic Society.

Behrens, J., van Elzakker, C. P. J. M., & Schmidt, M. (2015). Testing the Usability of OpenStreetMap's iD Tool. *The Cartographic Journal*, *52*(2), 177–184. doi:10.1080/00087041.2015.1119457

Bell, D. M., & Pahl, K. (2018). Co-production: Towards a utopian approach. *International Journal of Social Research Methodology*, *21*(1), 105–117. doi:10.1080/13645579.2017.1348581

Bell, R. L., Smetana, L., & Binns, I. C. (2005). Simplifying inquiry instruction. *Science Teacher (Normal, Ill.)*, *72*(7), 30–33.

Benayas, J., Heras, F., De Lucio, J. V., Marcén, C., & Ruiz, J. P. (1994). *Viviendo el paisaje: guía didáctica para interpretar y actuar sobre el paisaje*. Fundación Natwest.

Bergmann, J., & Sams, A. (2012). *Flip your classroom: reach every student in every class every day*. ISET.

Bertrand, R. (1984). *Mind mapping experiments at ITC (International Institute for Geoinformation and Earth Observation)*. Journal ITC.

Besse, J. M. (2014). *O Gosto do Mundo: Exercícios de paisagem*. EdUERJ.

Besse, J.-M. (2018). *La nécessité du paysage*. Éditions Parenthèses.

Bing, Z., Hai-qiao, T., & Li-juan, W. (2007). A New Approach to Urban Rainwater Management. *J China Univ Mining & Technol*, *17*(1), 82–84. doi:10.1016/S1006-1266(07)60018-2

Bisquerra, R. (2014). *Metodología de la investigación educativa* (4ª). La Muralla.

Bittner, C. (2017). OpenStreetMap in Israel and Palestine - 'Game changer' or reproducer of contested cartographies? *Political Geography*, *57*, 34–48. doi:10.1016/j.polgeo.2016.11.010

Blair, E., Maharaj, C., & Primus, S. (2016). Performance and perception in the flipped classroom. *Education and Information Technologies*, *21*(6), 1465–1482. doi:10.100710639-015-9393-5

Blankman, M., Schoonenboom, J., Van der Schee, J., Boogaard, M., & Volman, M. (2016). Learning to teach geography for primary education: Results of an experimental programme. *Journal of Geography in Higher Education*, *40*(3), 425–441. doi:10.1080/03098265.2016.1144731

Blazques Salom, M., & Blanco-Romero, A. (2020). Fieldwork in ground zero, Mallorca. A methodological fieldwork proposal for a geographical analysis of tourist areas. *Investigaciones Geográficas*, *75*(75), 43–59. doi:10.14198/INGEO2020.BB

Bogacz, A., Woźniczka, P., Burszta-Adamiak, E., & Kolasińska, K. (2013). Methods of enhancing water retention in urban areas. *Scientific Review Engineering and Environmental Science*, *22*(1), 27–35.

Boisvert, J. (2004). *La formación del pensamiento crítico. Teoría y práctica.* Fondo de Cultura Económica.

Boix, G., & Olivella, R. (2007). Los Sistemas de Información Geográfica (SIG) aplicados a la educación. El proyecto PESIG (Portal Educativo SIG). In M. J. Marrón, J. Salom, & X. M. Souto (Eds.), *Las competencias geográficas poara la educación ciudadana* (pp. 23–32). Grupo de Didáctica de la Geografía de la Asociación de Geógrafos Españoles.

Bothwell, E. (2020). Blended research methods 'will become more prevalent' post-Covid | Times Higher Education (THE). *Times Higher Education.* https://www.timeshighereducation.com/news/blended-research-methods-will-become-more-prevalent-post-covid?utm_source=THE+Website+Users&utm_campaign=705eb14464-EMAIL_CAMPAIGN_2020_11_30_02_52&utm_medium=email&utm_term=0_daa7e51487-705eb14464-73364557

Boud, D., & Cohen, R. (2013). *Peer learning in higher education.* Routledge., doi:10.4324/9781315042565

Bovill, C. (2020). Co-creation in learning and teaching: The case for a whole-class approach in higher education. *Higher Education*, *79*(6), 1023–1037. doi:10.100710734-019-00453-w

Bradbeer, J. (1996). Problem-based learning and fieldwork: A better method of preparation. *Journal of Geography in Higher Education*, *20*(1), 11–18. doi:10.1080/03098269608709340

Bringle, R., Hatcher, J., & McIntosh, R. (n.d.). Analyzing Morton's Typology of Service Paradigms and Integrity. *Michigan Journal of Community Service Learning*, *2006*(13), 5–15.

Brinkley-Etzkorn, K. E. (2018). Learning to teach online: Measuring the influence of faculty development training on teaching effectiveness through a TPACK lens. *Internet and Higher Education*, *38*, 28–35. doi:10.1016/j.iheduc.2018.04.004

Brown, G., & Kyttä, M. (2014). Key issues and research priorities for public participation GIS (PPGIS): A synthesis based on empirical research. *Applied Geography (Sevenoaks, England)*, *46*, 122–136. doi:10.1016/j.apgeog.2013.11.004

Bruner, J. (1960). *The process of education.* Harvard University Press.

Bruns, A. (2008). *Blogs, Wikipedia, Second Life, and Beyond: From Production to Produsage*. Peter Lang.

Bryson, J. R., & Andres, L. (2020). Covid-19 and rapid adoption and improvisation of online teaching: Curating resources for extensive versus intensive online learning experiences. *Journal of Geography in Higher Education, 44*(4), 1–16. doi:10.1080/03098265.2020.1807478

Budhathoki, N. R., & Haythornthwaite, C. (2013). Motivation for Open Collaboration: Crowd and Community Models and the Case of OpenStreetMap. *The American Behavioral Scientist, 57*(5), 548–575. doi:10.1177/0002764212469364

Busquets, J., & Rubert, J. (2011). *La sensibilización en paisaje: un reto para el siglo XXI. Awareness-raising on landscape: a challenge for the 21st Century* (J. Busquets & J. Rubert, Eds.). Generalitat de Catalunya - Pays.Med.Urban. http://www.paysmed.net/pdf/paysurban_attivita_sensibilizzazione_es.pdf

Busquets, J. (2011). La importància de l'educació en paisatge. In J. Nogué, L. Puigbert, G. Bretcha, & À. Losantos (Eds.), *Paisatge i Educació. Plecs de Paisatge: reflexions 2* (pp. 69–88). Observatori del Paisatge de Catalunya. http://www.catpaisatge.net/fitxers/publicacions/paisatge_educacio/paisatge_i_educacio.pdf

Butcher, K. R., & Sumner, T. (2011). Self-directed learning and the sensemaking paradox. *Human-Computer Interaction, 26*(1–2), 123–159. doi:10.1080/07370024.2011.556552

Butt, G. (2010). Which methods are best suited to the production of high-quality research in geography education? *International Research in Geographical and Environmental Education, 19*(2), 103–107. doi:10.1080/10382046.2010.482189

Butt, G., & Lambert, D. (2014). International perspectives on the future of geography education: An analysis of national curricula and standards. *International Research in Geographical and Environmental Education, 23*(1), 1–12. doi:10.1080/10382046.2013.858402

Buzan, T. (2002). *How to mind map*. Harper Collins.

Bybee, Taylor, Gardner, Van Scotter, Powell, Westbrook, & Landes. (2006). The BSCS 5E instructional Model: Origins and Effectiveness. Academic Press.

Caballero, P. A., Sánchez, S., & Belmonte, M. L. (2019). Análisis de la creatividad de los estudiantes universitarios. Diferencias por género, edad y elección de estudios. *Educatio XX1, 22*(2), 213–234. doi:10.5944/educxx1.22552

Calaf, R. (2009). *Didáctica del patrimonio*. Trea.

Calbó, M., Juanola, R., & Vallés, J. (2011). *Visiones interdisciplinares en educación del patrimonio: Artes, Culturas, Ambiente*. Documenta Universitaria.

Campos, L. R., & Richter, D. (2020). Ensino de Geografia, escola Quilombola e mapas mentais: diálogos possíveis na construção de uma identidade territorial. In N. Rego & S. Kozel (Eds.), *Narrativas, Geografias e Cartografias: para viver é preciso espaço e tempo* (Vol. 1, pp. 211–230). Compasso Lugar-Cultura. doi:10.29327/519558.1-10

Care, E., Scoular, C., & Griffin, P. (2016). Assessment of collaborative problem solving in education environments. *Applied Measurement in Education, 29*(4), 250–264. doi:10.1080/08957347.2016.1209204

Carraro, V., & Wissink, B. (2018). Participation and marginality on the geoweb: The politics of non-mapping on OpenStreetMap Jerusalem. *Geoforum, 90*, 64–73. doi:10.1016/j.geoforum.2018.02.001

Casas, M., Erneta, L., & Puig, J. (2018). La encuesta como herramienta para conocer las ideas previas sobre paisaje del alumnado de ESO: una experiencia piloto en Navarra. *Didáctica Geográfica, 19*(2018), 47–76. https://www.age-geografia.es/didacticageografica/index.php/didacticageografica/article/view/416/389

Castán, V. (2017). Urban Governance and the Politics of Climate change. *World Development, 93*(May), 1–15. doi:10.1016/j.worlddev.2016.12.031

Castellar, M. V. (2020). O papel do pensamento espacial na construção do raciocínio geográfico. *Revista Brasileira de Educação em Geografia, 10*(19), 294–322. doi:10.46789/edugeo.v10i19.922

Castellotti, V., & Moore, D. (2009). Dessins d'enfants et constructions plurilingues. Territoires imagés et parcours imaginés. In *Molinié, M., Le dessin réflexif: élément d'une herméneutique du sujet plurilingue* (pp. 45–85). CRTF, Encrage, Belles Lettres.

Castiglioni, B. (2009). Education on landscape for children. In *5th Council of Europe Conference on The European Landscape Convention.* Council of Europe.

Castro Aguirre, C. (2014). *Mapas mentales.* Universidad Pública de Navarra.

Catling, S. (2018). To know maps: Primary school children and contextualized map learning. *BoletimPaulista de Geografia, 99*, 268–290.

Cavalcanti, L. S. (2019). *Pensar pela Geografia: ensino e relevância Social.* C&A Alfa Comunicação.

Cho, S., & Nam, S. (2015). An exploration on the significance and the effect of place-based education (PBE) in Social Studies Education. *The Journal of The Korean Association of Geographic end Environmental Education, 23*(1), 1-17.

Cho, C. (2020). *Place-Based Geography Education.* Kyongbuk University Press.

Cichoń, M. (2015). Metoda projektu edukacyjnego. In M. Beczkiewicz & M. Szyszkiewicz-Golis (Eds.), Ekologia Krajobrazu cz. II. Przewodnik metodyczny (pp. 38-47). IŚRiL PAN.

Cichoń, M., & Piotrowska, I. (2018). Level of academic and didactic competencies among students as a measure to evaluate geographical education and preparation of students for the demands of the modern labour market. *Quaestiones Geographicae, 37*(1), 73–86. doi:10.2478/quageo-2018-0006

Cobas, I. (2016). El concepto de paisaje cultural como recurso para la educación patrimonial en la educación secundaria. *Cadernos de Arqueoloxía e Patrimonio, 37*, 60. http://digital.csic.es/handle/10261/134694

Cohen, L., Manion, L., & Morrison, K. (2011). Surveys, longitudinal, cross-sectional and trend studies. In *Research Methods in Education* (6th ed., pp. 205–225). Routledge Falmer.

Cohen-Shacham, E., Walters, G., Janzen, C., & Maginnis, S. (Eds.). (2016). *Nature-based solutions to address global societal challen-ges*. IUCN. doi:10.2305/IUCN.CH.2016.13.en

Collins, L. (2018). The impact of paper versus digital map technology on students' spatial thinking skill acquisition. *The Journal of Geography, 117*(4), 137–152. doi:10.1080/00221341.2017.1374990

Conférence Intercantonale de l'Instruction Publique de la Suisse romande et du Tessin (CIIP), (2014). Géographie, histoire, sciences de la nature. Guide pour l'enseignement. 3e-4e, cycle 1. Neuchâtel : CIIP.

Conway, R., Kember, D., Sivah, A., & Wu, M. (1993). Peer assessment of an individual's contribution to a group project. *Assessment & Evaluation in Higher Education, 18*(1), 45–56. doi:10.1080/0260293930180104

Corbin, J., & Strauss, A. (1990). Grounded Theory Research: Procedures, Canons and Evaluative Criteria. *Zeitschrift für Soziologie, 19*(6), 418–427. doi:10.1515/zfsoz-1990-0602

Council of Europe. (2000). *European Landscape Convention* (ETS No. 176). Author.

Council of Europe. (2000). *European Landscape Convention.* https://rm.coe.int/1680080621

Council of Europe. (2009). *5th Council of Europe Conference on the European Landscape Convention. Education on Landscape for Children.* https://rm.coe.int/16806b081b

Council of Europe. (2014). Recommendation CM/Rec (2014)8 of the Committee of Ministers to member States on promoting landscape awareness through education. Brussels.

Council of Europe. (2015). Recommendation CM/Rec (2015) of the Committee of Ministers to member States on pedagogical material for landscape education in primary school. Strasbourg.

Crampton, J. W., & Krygier, J. (2005). An Introduction to Critical Cartography. *Acme-an International E-Journal for Critical Geographies, 4*(1), 11-33.

Crampton, J. W. (2001). Maps as social constructions: Power, communication and visualization. *Progress in Human Geography, 25*(2), 235–252. doi:10.1191/030913201678580494

Crawford, J. (2020). COVID-19: 20 countries' higher education intra-period digital pedagogy responses. *Journal of Applied Learning & Teaching, 3*(1), 09–28. doi:10.37074/jalt.2020.3.1.7

Cresswell, T. (2004). *Place: a short introduction.* John Wiley & Sons, Ltd.

Cruz, L., & Español, I. M. (2008). *El paisaje: de la percepción a la gestión.* Ed. Liteam.

Cuenca, J. M., Estepa, J., & Martín, M. (2015). La enseñanza de las sociedades actuales e históricas a través del patrimonio. *Íber. Didáctica de Las Ciencias Sociales. Geografía e Historia, 79*, 33–40.

Das, M., Hecht, B., Gergle, D., & Assoc Comp, M. (2019). The Gendered Geography of Contributions to OpenStreetMap: Complexities in Self-Focus Bias. *Proceedings of the 2019 Chi Conference on Human Factors in Computing Systems.* 10.1145/3290605.3300793

De Baeque, A. (2019). *Une histoire de la marche.* Perrin.

de Souza Cavalcanti, L. (1998). Geografia, escola e construção de conhecimentos (14th ed.). São Paolo: Papirus.

Debarbieux, B. (2013). Imaginaire géographique. In Dictionnaire de la géographie et de l'espace des sociétés, (pp. 530-534). Belin.

Demailly, K.-E., Monnet, J., Scapino, J., Deraeve, S. (2021). *Dictionnaire pluriel de la marche en ville.* Paris: Éditions L'œil d'or.

Demanze, L. (2019). *Un nouvel âge de l'enquête.* Éditions Corti.

DeSeCo. (2002). *Definition and selection of competences (DeSeCo): Theoretical and conceptual foundations. Strategy paper.* http://deseco.ch/bfs/deseco/en/index/02.parsys.34116.downloadList.87902

Deslauriers, L., McCarty, L. S., Miller, K., Callaghan, K., & Kestin, G. (2019). Measuring actual learning versus feeling of learning in response to being actively engaged in the classroom. *Proceedings of the National Academy of Sciences of the United States of America, 116*(39), 19251–19257. doi:10.1073/pnas.1821936116 PMID:31484770

Dewey, J. (2018). Démocratie et éducation, suivi deExpérience et Éducation. Malakoff: Éditions Armand Colin.

Dhawan, S. (2020). Online Learning: A Panacea in the Time of COVID-19 Crisis. *Journal of Educational Technology Systems, 49*(1), 5–22. doi:10.1177/0047239520934018

Doering, A., & Veletsianos, G. (2008). An investigation of the use of real-time, authentic geospatial data in the K-12 classroom. *The Journal of Geography, 106*(6), 217–225. doi:10.1080/00221340701845219

Dolan, E. L., Elliott, S. L., Henderson, C., Curran-Everett, D., St. John, K., & Ortiz, P. A. (2018). Evaluating Discipline-Based Education Research for Promotion and Tenure. *Innovative Higher Education, 43*(1), 31–39. doi:10.100710755-017-9406-y

Downs, R. M., & Stea, D. (1973). Cognitive maps and spatial behavior: process and products. In R. M. Downs & D. Stea (Eds.), *Image and environment* (pp. 1–20). Aldine.

Duarte, R. G. (2017). A linguagem cartográfica como suporte ao desenvolvimento do pensamento espacial dos alunos na educação básica. *Revista Brasileira de Educação Em Geografia, 7*(13), 187–206. doi:10.46789/edugeo.v7i13.493

Duggan, M. (2019). Cultures of Enthusiasm: An Ethnographic Study of Amateur Map-Maker Communities. *Cartographica, 54*(3), 217–229. doi:10.3138/cart.54.3.2018-0002

Dumford, A. D., & Miller, A. L. (2018). Online learning in higher education: Exploring advantages and disadvantages for engagement. *Journal of Computing in Higher Education, 30*(3), 452–465. doi:10.100712528-018-9179-z

Dylak, S. (2000). Konstruktywizm jako obiecująca perspektywa w kształceniu nauczycieli. In H. Kwiatkowska, T. Lewowicki, & S. Dylak (Eds.), Współczesność a kształcenie nauczycieli. WSP ZNP.

Dylak, S. (Ed.). (2013b). Strategia kształcenia wyprzedzającego. Ogólnopolska Fundacja Edukacji Komputerowej, UAM.

Dylak, S., & Duda, L. (2010). *The implementation strategy of Collegium Sniadeckich.* Ogólnopolska Fundacja Edukacji Komputerowej, UAM.

Dylak, S. (2013a). *Architektura wiedzy w szkole*. Difin.

Egan, K. (2008). Learning in depth. *Educational Leadership, 66*(3), 58–63.

Ehrlich, T. (1996). Foreword. In B. Jacoby & ... (Eds.), *Service-Learning in Higher Education: Concepts and Practices* (pp. xi–xvi). Jossey-Bass.

Environment, U. N. (2019). *Global Environment Outlook - GEO6: healthy Planet.* Healthy People., doi:10.1017/9781108627146

Ertmer, P. A., & Simons, K. D. (2006). Jumping the PBL implementation hurdle: Supporting the efforts of K–12 teachers. *The Interdisciplinary Journal of Problem-Based Learning, 1*(1), 40–54. doi:10.7771/1541-5015.1005

European Commission. (2018). *Proposal for a Council Recommendation on Key Competences for LifeLong Learning*. Author.

Exley, K., & Dennick, R. (2004). *Small group teaching: Tutorials, seminars and beyond.* Routledge Falmer.

Fargher, M. (2017). GIS and Other Geospatial Technologies. In M. Jones (Ed.), *Handbook of Secondary Geography* (pp. 244–259). Geographical Association.

Fazlagić, J. (2021). Edukacja klimatyczna. *Dyrektor Szkoły, 4*, 53–56.

Fernández, V. (2005). Finalidades del patrimonio en la educación. *Revista de Investigación En La Escuela, 5*, 7–18. http://www.investigacionenlaescuela.es/articulos/56/R56_1.pdf

Ferreira, R. N. (2021). Cut-Off Lows and Extreme Precipitation in Eastern Spain: Current and Future Climate. *Atmosphere*, *2021*(12), 835. doi:10.3390/atmos12070835

Florín, M., Gosálvez, R. U., Velasco, A., Laguna, C., Falomir, J. P., Chicote, A., & Muñoz, E. M. (2018). Estudio sobre Efectos Constatados y Percepción del Cambio Climático en el Medio Rural de Castilla-La Mancha. Propuestas de Medidas de Adaptación. Junta de Comunidades de Catilla-La Mancha.

Florín, M., Gosálvez, R., Laguna, C., Sánchez, D., Falomir, J., González, A., Muñoz, E. M., Chicote, A., Sánchez, G., Velasco, A., & Becerra, R. (2021). *Efecto de superioridad en la mejora de servicios ecosistémicos de regulación por humedales y llanuras de inundación* (Vol. 127). Revista Ambienta. https://dialnet.unirioja.es/servlet/articulo?codigo=8070009

Fonseca, F. P. (2019). O espaço como representação. In A. F. A. Carlos & R. C. A. Cruz (Eds.), *A necessidade da Geografia* (pp. 42–53). Contexto.

Fonte, C. C., Basin, L., See, L., Foody, G., & Estima, J. (2015). *Good practice guidelines for assessing VGI data quality*. Paper presented at the AGILE 2015, Lisbon, Portugal. https://agile-online.org/conference_paper/cds/agile_2015/posters/95/95_Paper_in_PDF.pdf

Fonte, C. C., Patriarca, J. A., Minghini, M., Antoniou, V., See, L., & Brovelli, M. A. (2017). Using OpenStreetMap to Create Land Use and Land Cover Maps: Development of an Application. In Volunteered Geographic Information and the Future of Geospatial Data (pp. 113-137). IGI Global.

Ford, H., & Wajcman, J. (2017). Anyone can edit', not everyone does: Wikipedia's infrastructure and the gender gap. *Social Studies of Science*, *47*(4), 511–527. doi:10.1177/0306312717692172 PMID:28791929

Forghani, M., & Delavar, M. R. (2014). A Quality Study of the OpenStreetMap Dataset for Tehran. *ISPRS International Journal of Geo-Information*, *3*(2), 750–763. doi:10.3390/ijgi3020750

Forman, R. T. T. (1990). The Beginnings of Landscape Ecology in America. In I. S. Zonneveld & R. T. T. Forman (Eds.), *Changing Landscapes: An Ecological Perspective* (pp. 35–44). Springer-Verlag. doi:10.1007/978-1-4612-3304-6_3

Fournier, E. J. (2002). World regional geography and problem-based learning: Using collaborative learning groups in an introductory-level world geography course. *The Journal of General Education*, *51*(4), 293–305. doi:10.1353/jge.2003.0011

Fraile, P., Sánchez, D., & Medina, B. (2016). El uso de la fotografía como recurso docente para la interpretación del territorio en el aula: las potencialidades de Geophotopedia. *Didáctica Geográfica*, *17*, 63–78. https://www.age-geografia.es/didacticageografica/index.php/didacticageografica/article/view/376/349

Frank, M., & Barzilai, A. (2004). Integrating alternative assessment in a project-based learning course for preservice science and technology teachers. *Assessment & Evaluation in Higher Education*, *29*(1), 41–61. doi:10.1080/0260293042000160401

Freitas, R., & Almeida, M. (2014). O pensamento geográfico nos alunos do ensino básico, com base na taxonomia do pensamento espacial. *GOT - Geography and Spatial Planning Journal, 6*, 135–152. doi:10.17127/got/2014.6.009

Fu, B. (2020). Promoting geography for sustainability. *Geography and Sustainability, 1*(1), 1–7. doi:10.1016/j.geosus.2020.02.003

Fuertes, C. (2014). ¿La historia oral en las aulas de infantil y primaria? Difusión, aportaciones y propuestas de aplicación. *CLÍO. History and Teaching, 10*. http://clio.rediris.es/n40/articulos/Fuertes2014.pdf

Galloway, Ch. (1988). *Psychologia uczenia się i nauczania*. PWN.

Galvão, W. (2007) *Que Geografia se ensina?: um estudo sobre representações de Geografia segundo alunos da 6a série do ensino fundamental* [Dissertação de Mestrado]. Universidade Federal do Paraná.

García de la Vega, A. (2004). El itinerario geográfico como recurso didáctico para la valoración del paisaje. *Didáctica Geográfica, 6*, 79–95. http://www.didacticageografica.es/index.php/didacticageografica/article/view/182/173

García de la Vega, A. (2012). El aprendizaje basado en problemas en los itinerarios vinculados al patrimonio. *Educación y Futuro, 27*(2012), 155–175. https://cesdonbosco.com/documentos/revistaeyf/EYF_27.pdf

García de la Vega. (2015). The Geographical Scenario. An Innovative Methodological Resource in Landscape Teaching. In *Book of Abstracts 100th Annual Meeting of the National Council for Geographic Education (NCGE)*. Academic Press.

García de la Vega, A. (2012). El aprendizaje basado en problemas en los itinerarios vinculados al patrimonio. *Educación y Futuro, 27*, 155–175.

García de la Vega, A. (2018). El escenario geográfico en el estudio del paisaje. Propuestas metodológicas en la enseñanza de la geografía. In A. García de la Vega (Ed.), *Reflexiones sobre educación geográfica. Revisión disciplinar e innovación didáctica* (pp. 129–152). Ediciones UAM y Edições Lagim.

García de la Vega, A. (2019b). Perspectivas de futuro en el aprendizaje del paisaje. *Didáctica Geográfica*, (20), 55–77. doi:10.21138/DG.441

García de la Vega, A. (2021). Design-Based Implementation Research for Exploring the Ocean: A Geographical Perspective. In K. C. Koutsopoulos & J. H. Stel (Eds.), *Ocean Literacy: Understanding the Ocean* (pp. 115–147). Springer. doi:10.1007/978-3-030-70155-0_6

García de la Vega, A. (2022). A Proposal for Geography Competence Assessment in Geography Fieldtrips for Sustainable Education. *Sustainability*, *2022*(14), 1429. doi:10.3390u14031429

García de la Vega, A. (2919a). Spatial thinking acquisition through geospatial technologies for lifelong learning. In R. de Miguel, K. Donert, & K. Koutsopoulos (Eds.), *Geospatial Technologies in Geography Education* (pp. 21–40). Springer.

García, A., Martín, D. A., Costafreda, J. L., Nuñez, E., & Rodríguez, J. A. (2019). Aprendizaje-Servicio (ApS) como metodología para alcanzar los Objetivos de Desarrollo Sostenible (ODS). *V Congreso Internacional Sobre Aprendizaje, Innovación y Competitividad*, 441–446.

General Navigation. (n.d.). Retrieved August 20, 2021 from https://www.nordian.net/REPOSITORY/110_easa_general_navigation_demo.pdf

Gersmehl, P. (2008). *Teaching Geography* (2nd ed.). The Guilford Press.

Gibbs, B. (2020). What are we learning about assessment? In B. Gibbs & G. C. Wood (Eds.), *Emerging Stronger: Lasting Impact from Crisis Innovation* (pp. 3–18). Engineering Professors' Council.

Gibbs, B., & Wood, G. C. (2020). How do we partner with students in learning design at a distance? In B. Gibbs & G. C. Wood (Eds.), *Emerging Stronger: Lasting Impact from Crisis Innovation*. Engineering Professors' Council.

Gibbs, B., & Woods, G. (2020). Emerging Stronger: Lasting impact from crisis innovation. In B. Gibbs & G. C. Woods (Eds.), *Engineering Professors Council*. Engineering Professors' Council.

Gillies Robyn, M. (2020). *Inquiry-based Science Education*. CRC Press Taylor & Francis Group. doi:10.1201/9780429299179

Gilmartin, P. P. (2017). Showing the Shortest Routes-Great Circles. In Handbook of Choosing a Map Projection. Lecture Notes in Geoinformation and Cartography. Subseries: Publications of the International Cartographic Association (ICA) (pp. 1-48). Springer.

Glasze, G., & Perkins, C. (2015). Social and Political Dimensions of the OpenStreetMap Project: Towards a Critical Geographical Research Agenda. In J. J. Arsanjani, A. Zipf, P. Mooney, & M. Helbich (Eds.), *Openstreetmap in Giscience: Experiences, Research, and Applications* (pp. 143–166). Springer-Verlag Berlin. doi:10.1007/978-3-319-14280-7_8

Goldstein, D., & Alibrandi, M. (2013). Integrating GIS in the middle school curriculum: Impacts on diverse students' standardized test scores. *The Journal of Geography*, *112*(2), 68–74. doi:10.1080/00221341.2012.692703

Golledge, R. G. (1995). Primitives of spatial knowledge. In T. L. Nyerges, D. M. Mark, R. Laurini, & M. J. Egenhofer (Eds.), *Cognitive Aspects of Human-Computer Interaction for Geographic Information Systems* (pp. 29–44). Kluwer Academic Publishers. doi:10.1007/978-94-011-0103-5_4

Gomes, M. D. F. V. B. (2017a). Cartografia social e geografia escolar: Aproximações e possibilidades. *Revista Brasileira de Educação Em Geografia*, *7*(13), 97–110. doi:10.46789/edugeo.v7i13.488

Gomes, P. C. C. (2013). *O lugar do olhar*. Bertrand Brasil.

Gomes, P. C. C. (2017b). *Quadros Geográficos*. Bertrand Brasil.

González, M. E. (2012). *Las Infraestructuras de Datos Espaciales (IDE) como un recurso educativo TIC*. https://oa.upm.es/11015/1/MARIA_ESTER_GONZALEZ.pdf

Goodchild, M. F. (2008). Assertion and authority: the science of user-generated geographic content. In G. Navratil (Ed.), *Proceedings of the Colloquium for Andrew U. Frank's 60th Birthday* (pp. 1-18). Vienna: Department of Geoinformation and Cartography, Vienna University of Technology.

Goodchild, M. F. (2007). Citizens as sensors: The world of volunteered geography. *GeoJournal*, *69*(4), 211–221. doi:10.100710708-007-9111-y

Górka, A. (2020). Landscape perception and the teaching of it in Poland. *World Transactions on Engineering and Technology Education, 18*(2), 123–128. http://www.wiete.com.au/journals/WTE&TE/Pages/Vol.18,No.2 (2020)/07-Gorka-A.pdf%0A

Gould, P., & White, R. (1974). *Mental Maps*. Penguin Books. doi:10.4324/9780203163757

Griffin, A., & McQuoid, J. (2012). At the intersection of maps and emotion: The challenge of spatially representing experience. *Kartographische Nachrichten*, *62*(6), 291–299.

Gruenewald, D. A., & Smith, G. A. (Eds.). (2008). *Place-Based Education in the Global Age: Local Diversity*. Routledge.

Gryl, I., & Jekel, T. (2012). Re-centring Geoinformation in Secondary Education: Toward a Spatial Citizenship Approach. *Cartographica: The International Journal for Geographic Information and Geovisualization*, *47*(1), 18–28. doi:10.3138/carto.47.1.18

Gumuchian, H. (1989), Les représentations en géographie. Définitions, méthodes et outils. In Représenter l'espace : L'imaginaire spatial à l'école (pp. 29-56). Diffusion, Economica.

Gutstein, E. (2013). Math, Maps, and Misrepresentation. In E. Gutstein & B. Peterson (Eds.), *Rethinking Mathematics Teaching Social Justice by the Numbers* (2nd ed., pp. 187–200). Rethinking Schools.

Haklay, M. (2010). How Good is Volunteered Geographical Information? A Comparative Study of OpenStreetMap and Ordnance Survey Datasets. *Environment and Planning. B, Planning & Design*, *37*(4), 682–703. doi:10.1068/b35097

Haklay, M., Singleton, A., & Parker, C. (2008). Web Mapping 2.0: The Neogeography of the GeoWeb. *Geography Compass*, *2*(6), 2011–2039. doi:10.1111/j.1749-8198.2008.00167.x

Halvorson, S. J., & Wescoat, J. L. Jr. (2002). Problem-based inquiry on world water problems in large undergraduate classes. *The Journal of Geography*, *101*(3), 91–102. doi:10.1080/00221340208978480

Hannoun, H. (1977). *El niño conquista el medio*. Ed. Kapelusz.

Hargittai, E., & Shaw, A. (2015). Mind the skills gap: The role of Internet know-how and gender in differentiated contributions to Wikipedia. *Information Communication and Society*, *18*(4), 424–442. doi:10.1080/1369118X.2014.957711

Harley, J. B. (1991). A nova história da cartografia. O Correio da Unesco, 19(8), 4-9.

Harris, T., & Tweed, F. (2010). A research-led, inquiry-based learning experiment: Classic landforms of deglaciation, Glen Etive, Scottish Highlands. *Journal of Geography in Higher Education*, *34*(4), 511–528. doi:10.1080/03098265.2010.486851

Healey, M., Flint, A., & Harrington, K. (2014). *Engagement through partnership: students as partners in learning and teaching in higher education*. https://www.advance-he.ac.uk/knowledge-hub/engagement-through-partnership-students-partners-learning-and-teaching-higher

Healey, M., Flint, A., & Harrington, K. (2016). View of Students as partners: Reflections on a conceptual model. *Teaching & Learning Inquiry*, *4*(2), 8–20. doi:10.20343/teachlearninqu.4.2.3

Henderson, C., Connolly, M., Dolan, E. L., Finkelstein, N., Franklin, S., Malcom, S., Rasmussen, C., Redd, K., & St. John, K. (2017). Towards the STEM DBER Alliance: Why we Need a Discipline-Based STEM Education Research Community. *International Journal of Research in Undergraduate Mathematics Education*, *3*(2), 247–254. doi:10.100740753-017-0056-3 PMID:30631670

Hennig, S. (2017). OpenStreetMap used in protected area management. The example of the recreational infrastructure in Berchtesgaden National Park. *Eco Mont-Journal on Protected Mountain Areas Research*, *9*(2), 30–41. doi:10.1553/eco.mont-9-2s30

Hennig, S. (2020). Motivation and Its Consideration in Participatory Spatial Data Contribution. *The Professional Geographer*, *72*(2), 238–252. doi:10.1080/00330124.2019.1676799

Henry, H., Tawfik, A., Jonassen, D., Winholtz, R., & Khanna, S. (2012). "I know this is supposed to be more like the real world, but…": Student perceptions of a PBL implementation in an undergraduate materials science course. *The Interdisciplinary Journal of Problem-Based Learning*, *6*(1), 43–81. doi:10.7771/1541-5015.1312

Herfort, B., Eckle, M., Albuquerque, J. P. d., & Zipf, A. (2015). *Towards assessing the quality of volunteered geographic information from OpenStreetMap for identifying critical infrastructures*. Paper presented at the ISCRAM.

Hernandez-Encuentra, E., Pousada, M., & Gomez-Zuniga, B. (2009). ICT and older people: Beyond usability. *Educational Gerontology*, *35*(3), 226–245. doi:10.1080/03601270802466934

Hétier, R. (2021). *L'humanité contre l'Anthropocène. Résister aux effondrements*. Presses universitaires de France.

Hicks, D., Lee, J., Berson, M., Bolick, C., & Diem, R. (2014). Guidelines for using technology to prepare social studies teachers. *Contemporary Issues in Technology & Teacher Education*, *14*(4), 433–450.

Hmelo-Silver, C. E., & Barrows, H. S. (2006). Goals and Strategies of a Problem-based Learning Facilitator. *The Interdisciplinary Journal of Problem-Based Learning*, *1*(1), 21–39. doi:10.7771/1541-5015.1004

Hmelo-Silver, C. E., & Barrows, H. S. (2008). Facilitating Collaborative Knowledge Building. *Cognition and Instruction*, *26*(1), 48–94. doi:10.1080/07370000701798495

Hollman, V., & Lois, C. (2015). *Geo-grafías*. Editorial Paidós.

Holmes, J. H. (2002). Geography's Emerging Cross-Disciplinary Links: Process, Causes, Outcomes and Challenges. *Australian Geographical Studies*, *40*(1), 2–20. doi:10.1111/1467-8470.00157

Holyoak, K. J. (1995). Problem solving. In E. E. Smith & D. N. Osherson (Eds.), *An invitation on cognitive science* (2nd ed., Vol. 3, pp. 267–296). MIT Press.

Holyoak, K. J., & Thagard, P. (1997). The analogical mind. *The American Psychologist*, *52*(1), 35–44. doi:10.1037/0003-066X.52.1.35 PMID:9017931

Hume, T., & Barry, J. (2015). Environmental Education and Education for Sustainable Development. In J. D. Wright (Ed.), *International Encyclopedia of the Social & Behavioral Sciences* (Vol. 7, pp. 733–739). Elsevier. doi:10.1016/B978-0-08-097086-8.91081-X

Hupy, J. P., Aldrich, S. P., Schaetzl, R. J., Varnakovida, P., Arima, E. Y., Bookout, J. R., Wiangwang, N., Campos, A. L., & McKnight, K. P. (2005). Mapping soils, vegetation, and landforms: An integrative physical geography field experience. *The Professional Geographer*, *57*(3), 438–451. doi:10.1111/j.0033-0124.2005.00489.x

Ingold, T. (2018). *L'anthropologie comme éducation*. Presses Universitaires de Rennes.

Intzidou, G., & Lambrinos, N. (2019). Investigation of the knowledge of the students of Elementary Schools in digital cartography. *Proceedings of 12th International Conference of Greek Geographical Company "INNOVATIVE GEOGRAPHIES: Mapping and Modeling our World"*, 1-4.

Intzidou, G., Lambrinos, N., Tourtouras, Ch., & Seroglou, F. (2021). Metadata: A pedagogical tool for the teaching of map projections in elementary schools. *European Journal of Geography*, *12*(3), 56–59. doi:10.48088/ejg.g.int.12.3.56.69

Israel, A. L. (2012). Putting geography education into place: What geography educators can learn from place-based education, and vice versa. *The Journal of Geography*, *111*(2), 76–81. doi:10.1080/00221341.2011.583264

Jacob, C. (2016). Por uma história cultural da cartografia. *Espaço e Cultura*, *39*(39), 221–236. doi:10.12957/espacoecultura.2016.31762

Jadallah, M., Hund, A. M., Thayn, J., Studebaker, J. G., Roman, Z. J., & Kirby, E. (2017). Integrating geospatial technologies in fifth-grade curriculum: Impact on spatial ability and map-analysis skills. *The Journal of Geography*, *116*(4), 139–151. doi:10.1080/00221341.2017.1285339

Jakab, I., Grezo, H., & Sevcik, M. (2016). Inquiry based and blended learning using geographical information system. In J. Novotna & A. Jancarik (Eds), *Proceedings of the 15th European Conference on e-Learning ECEL 2016* (pp. 287-295). Academic Press.

Jo, I., & Bednarz, S. W. (2009). Evaluating geography textbook questions from a spatial perspective: Using concepts of space, tools of representation, and cognitive processes to evaluate spatiality. *The Journal of Geography*, *108*(1), 1, 4–13. doi:10.1080/00221340902758401

Jo, I., Hong, J. E., & Verma, K. (2016). Facilitating spatial thinking in world geography using Web-based GIS. *Journal of Geography in Higher Education*, *40*(3), 442–459. doi:10.1080/03098265.2016.1150439

Jolivet, L., Dominguès, C., Mermet, E., & Seten, S. (2021). Lived and perceived space during lock-down in a sensitive map approach, *Proc. International Cartographic Association*, *4*, 50. doi:10.5194/ica-proc-4-50-2021

Jones, C. E., & Weber, P. (2012). Towards Usability Engineering for Online Editors of Volunteered Geographic Information: A Perspective on Learnability. *Transactions in GIS*, *16*(4), 523–544. doi:10.1111/j.1467-9671.2012.01319.x

Joublot Ferré, S. (2021), L'enquête paysagère, une méthode d'intelligibilité politique du monde? Étude de cas en formation d'enseignants. Projets de paysage, 24.

Justice, C., Warry, W., Cuneo, C., Inglis, S., Miller, S., Rice, J., & Sammon, S. (2002). *A grammar for inquiry: Linking goals and methods in a collaboratively taught social sciences inquiry course. In The Alan Blizzard Award Paper: The Award Winning Papers*. Society for Teaching and Learning in Higher Education and McGraw-Hill Ryerson.

Kahn, P. H. Jr. (2011). *Technological Nature. Adaptation and the Future of Human Life*. MIT Press. doi:10.7551/mitpress/7983.001.0001

Kahn, P. H., & Kellert, S. R. (Eds.). (2002). *Children and nature: Psychological, sociocultural, and evolutionary investigations*. MIT Press. doi:10.7551/mitpress/1807.001.0001

Kaipa, P. (2000). Knowledge architecture for the twenty-first century. *Behaviour & Information Technology*, *19*(3), 153–161. doi:10.1080/014492900406146

Kalifa, D. (2010). Enquête et « culture de l'enquête » au XIXe siècle. *Romantisme*, *149*(3), 3–23. doi:10.3917/rom.149.0003

Kapusta, A., Milewska, E., Sypniewski, J., & Śwital, P. (2020). *CO_2 się stanie w oceanie?* http://e-chemia.nazwa.pl/amgen/co2/

Karvánková, P., & Popjaková, D. (2018). How to link geography, cross-curricular approach and inquiry in science education at the primary schools. *International Journal of Science Education*, *40*(7), 707–722. doi:10.1080/09500693.2018.1442598

Keeling, A. (2008). "We are scholars": Using teamwork and problem-based learning in a Canadian regional geography course. *MountainRise*, *4*(3), 1–13.

Kessler, F. C., Battersby, S. E., Finn, M. P., & Clarke, K. C. (2017). Map Projections and the Internet. In M. Lapaine& E. Usery (Eds.), Choosing a Map Projection. Lecture Notes in Geoinformation and Cartography (pp. 117-202). Cham: Springer. doi:10.1007/978-3-319-51835-0_4

Keuning, J. (1955). The history of geographical map projections until 1600. *Imago Mundi, 12*(1), 1–24. doi:10.1080/03085695508592085

Khuri, S., & Reed, M. (2020). *Tips and tools for making your online meetings and workshops more interactive*. https://www.fasttrackimpact.com/post/tips-and-tools-for-making-your-online-meetings-and-workshops-more-interactive

Kilpatrick, W. H. (1918). The project method. *Teachers College Record, 19*, 319–335.

Kim, M., & Yoon, O. (2013). Development and application of a sense of place test instrument: A case study of gender differences of elementary students. *The Journal of The Korean Association of Geographic and Environmental Education, 21*(2), 17–28. doi:10.14697/jkase.2013.33.1.017

Klus-Stańska, D. (2000). *Konstruowanie wiedzy w szkole*. Wydawnictwo Uniwersytetu Warmińsko-Mazurskiego.

Knapp, C. (2005). The "I-thou" relationship, place-based education and Aldo Leopold. *The Journal of Environmental Education, 27*(3), 277–285.

Knoll, M. (1991). Europa-nicht America: Zumursprung der projekt methode in der pädagogik, 1702-1875. *Pädagogische Rundschau, 45*, 41–58.

Kolb, A. Y., Kolb, D. A., Passarelli, A., & Sharma, G. (2014). On Becoming an Experiential Educator: The Educator Role Profile. *Simulation & Gaming, 45*(2), 204–234. doi:10.1177/1046878114534383

Kowasch, M., & Lippe, D. F. (2019). Moral impasses in Sustainability Education? Empirical results from school geography in Austria and Germany. *Environmental Education Research, 25*(7), 1066–1082. doi:10.1080/13504622.2018.1557112

Kozel, S. (Org.). (2018). *Mapas Mentais. Dialogismo e Representações*. Appris.

Kozel, S. (2007). Mapas mentais - uma forma de linguagem: perspectivas metodológicas. In S. Kozel, J. C. Silva, & S. F. Gil Filho (Eds.), *Da percepção e cognição a representação: reconstrução teóricas da Geografia Cultural e Humanista* (pp. 114–138). Terceira Margem.

Krajcik, J. S., & Blumenfeld, P. C. (2005). Project-Based Learning. In R. K. Sawyer (Ed.), *The Cambridge handbook of the learning sciences* (pp. 317–333). Cambridge University Press. doi:10.1017/CBO9780511816833.020

Krajcik, J. S., & Shin, N. (2014). Project-based learning. In R. K. Sawyer (Ed.), *The Cambridge handbook ofthe learning sciences* (2nd ed., pp. 275–297). Cambridge University Press. doi:10.1017/CBO9781139519526.018

Krathwohl, D. R. (2002). A Revision of Bloom's Taxonomy: An Overview. *Theory into Practice, 41*(4), 212–218. doi:10.1207154304 21tip4104_2

Kriewaldt, J. (2006). The key role of metacognition in an inquiry-based geography curriculum. *Geographical Education, 19*, 24–30.

Kwon, Y., & Hwang, M. (2005). The significance of the sense of place in environmental education. [in Korean]. *Environmental Education, 18*(2), 55–65.

Kythreotis, A. P., Mercer, T. G., Howarth, C., Jonas, A., & Castree, N. (2019). *The 'New Civil Politics of Climate Change': Can schools, colleges and universities augment government policy action?* EAUC. https://www.eauc.org.uk/the_new_civil_politics

Kythreotis, A. P., & Mercer, T. G. (2020). Collaborative education as a new urban civil politics of climate change. In V. Castán Broto, E. Robin, & A. While (Eds.), *Climate urbanism: towards a critical research agenda* (pp. 195–210). doi:10.1007/978-3-030-53386-1_12

Lalli, M. (1992). Urban-related identity: Theory, measurement, and empirical findings. *Journal of Environmental Psychology, 12*(4), 285–303. doi:10.1016/S0272-4944(05)80078-7

Lambrinos, N. (2009). *Teaching about school geography.* GRAFIMA Press.

Lambrinos, N., & Asiklari, F. (2014). The introduction of GIS and GPS through local history teaching in primary school. *European Journal of Geography, 5*(1), 32–47.

Lapon, L., Ooms, K., DeMaeyer, P. (2020). The Influence of map projections on people's global-scale cognitive map: A worldwide study. *ISPRS International Journal of Geo-Information, 9*(4), 196. doi:10.3390/ijgi9040196

Lazzarotti, O. (2017). *Une place sur Terre ? Franz Schubert, de l'homme mort à l'habitant libre.* Auxerre: HDiffusion.

Lazzarotti, O. (2006). *Habiter : La condition géographique.* Paris: Belin.

Le Breton, D. (2000). *Éloge de la marche.* Éd. Métailié.

Le Moigne, J., Grubb, T. G., & Milner, B. C. (2013). IMAGESEER: NASA IMAGEs for Science, Education, Experimentation and Research. *IEEE Geoscience and Remote Sensing Magazine, 1*(1), 44–58. doi:10.1109/MGRS.2013.2244694

Lee, J. S., Blackwell, S., Drake, J., & Moran, K. A. (2014). Taking a Leap of Faith: Redefining Teaching and Learning in Higher Education Through Project-Based Learning. *The Interdisciplinary Journal of Problem-Based Learning, 8*(2). Advance online publication. doi:10.7771/1541-5015.1426

Lee, T. (1968). Urban neighbourhood as a socio-spatial schema. *Human Relations, 21*(3), 241–267. doi:10.1177/001872676802100303

Leinhardt, G., Stainton, C., & Bausmith, J. M. (1998). Constructing maps collaboratively. *The Journal of Geography, 97*(1), 19–30. doi:10.1080/00221349808978821

Lempert, R. J., & Groves, D. G. (2010). Identifying and evaluating robust adaptive policy responses to climate change for water management agencies in the American west. *Technological Forecasting and Social Change, 77*(6), 960–974. doi:10.1016/j.techfore.2010.04.007

Leopold, L. B., Clarke, F. E., Hanshaw, B. B., & Balsley, J. R. (1971). A procedure for evaluating environmental impact. Geological survey circular. doi:10.3133/cir645

Leroux, X., & Verherve, M. (2014). Ma petite géographie ou la fabrique de la représentation des lieux chez de jeunes élèves. *Mappemonde, 113*. http://mappemonde.mgm.fr/num41/articles/art14101.html

Levy, P., & Petrulis, R. (2012). How do first-year university students experience inquiry and research, and what are the implications for the practice of inquiry-based learning? *Studies in Higher Education, 37*(1), 85–101. doi:10.1080/03075079.2010.499166

Liceras, A. (2013). Didáctica del paisaje. Lo que es, lo que representa, cómo se vive. *Íber. Didáctica de Las Ciencias Sociales. Geografía e Historia, 74*, 85–93.

Likouri, A. A., Klonari, A., & Flouris, G. (2017). Relationship of Pupils' Spatial Perception and Ability with Their Performance in Geography. *Review of International Geographical Education Online, 7*(2), 154–170.

Lincolnshire County Council. (2021). Lincolnshire Climate Summit.

Linn, M. C., Davis, E. A., & Bell, P. (Eds.). (2004). *Internet environments for science education*. Lawrence Erlbaum Associates Publishers.

Lin, Y.-W. (2011). A qualitative enquiry into OpenStreetMap making. *New Review of Hypermedia and Multimedia, 17*(1), 53–71. doi:10.1080/13614568.2011.552647

Littlejohn, A., Gourlay, L., Kennedy, E., Logan, K., Neumann, T., Oliver, M., Potter, J., & Rode, J. A. (2021). Moving teaching online: Cultural barriers experienced by university teachers during covid-19. *Journal of Interactive Media in Education, 2021*(1), 1–15. doi:10.5334/jime.631

Loidl, M., Krampe, S., Zagel, B., & Pucher, G. (2014). *Aufbereitung von Open Street Map Daten für GIS-Modellierungen und Analysen*. Paper presented at the AGIT 2014, Salzburg. http://gispoint.de/fileadmin/user_upload/paper_gis_open/537543048.pdf

Lopes, A. R. C., & Richter, D. (2013). A construção de mapas mentais e o ensino de Geografia: articulações entre o cotidiano e os conteúdos escolares. *Revista Territorium Terram, 2*(3), 2-12. http://seer.ufsj.edu.br/index.php/territorium_terram/article/view/606

Louis, A. (2013). Ce que l'enquête fait aux études littéraires: à propos de l'interdisciplinarité. *Fabula / Les colloques*. https://www.fabula.org/colloques/document2097.php

Louv, R. (1990). *Childhood's future*. Houghton Mifflin.

Louv, R. (2011). *The nature principle. Human restoration and the end of nature-deficit disorder*. Algonquin Books of Chapel Hill.

Lussault, M. (2019). La spatialité d'abord. In Y. Calbérac, O. Lazzarotti, J. Lévy, & M. Lussault (Eds.), *Carte d'identités : L'espace au singulier* (pp. 7–24). Hermann. doi:10.3917/herm.povoa.2019.01.0007

Lynch, K. (1960). *The Image of the City*. The MIT Press.

Lyon, S. W., & Teutschbein, C. (2011). Problem-based learning and assessment in hydrology courses: Can non-traditional assessment better reflect intended learning outcomes? *Journal of Natural Resources and Life Sciences Education*, *40*(1), 199–205. doi:10.4195/jnrlse.2011.0016g

MacFarlane, R. (2013). *The Old Ways. A Journey on Foot*. Penguin Ed.

Maderuelo, J. (2005). La definición de paisaje. In *El paisaje. Génesis de un concepto* (pp. 15–40). Adaba Editores.

Mahabir, R., Stefanidis, A., Croitoru, A., Crooks, A. T., & Agouris, P. (2017). Authoritative and Volunteered Geographical Information in a Developing Country: A Comparative Case Study of Road Datasets in Nairobi, Kenya. *ISPRS International Journal of Geo-Information*, *6*(1), 25. doi:10.3390/ijgi6010024

Maitland, B. (1998). Problem-based learning for an architecture degree. In D. Boud & G. Feletti (Eds.), *The Challenge of Problem-Based Learning* (2nd ed., pp. 211–217). Kogan Page.

Maltby, E. (Ed.). (2009). *Functional Assessment of Wetlands. Towards Evaluation of Ecosystem Services*. CRC Press. doi:10.1201/9781439829141

Manduca, C. A., Iverson, E. R., Luxenberg, M., Heather Macdonald, R., McConnell, D. A., Mogk, D. W., & Tewksbury, B. J. (2017). Improving undergraduate STEM education: The efficacy of discipline-based professional development. *Science Advances*, *3*(2), e1600193. doi:10.1126ciadv.1600193 PMID:28246629

Manzanares, J. A., & Quintana, S. M. (2019). El uso de los itinerarios didácticos (SIG) en la educación ambiental. *Enseñanza de Las Ciencias: Revista de Investigación y Experinecias Didácticas*, *37*(2), 173–188.

Marín-Cepeda, S., & Fontal, O. (2020). Percepciones de estudiantes de Educación Secundaria Obligatoria en torno al Patrimonio. *Arte, Individuo y Sociedad*, *32*(4), 917–933. https://revistas.ucm.es/index.php/ARIS/article/view/64657/4564456554158

Markuszewska, I., Tanskanen, M., & Vila Subirós, J. (2018). New Ways to Learn Geography - Challenges of the 21st Century. *Quaestiones Geographicae*, *37*(1), 37–45. doi:10.2478/quageo-2018-0004

Marrón, M. J. (2012). Aproximación al concepto de escala en el primer ciclo de educación primaria. Presentación de un juego para su tratamiento desde la enseñanza active. *Didáctica Geográfica*, *13*, 93–112.

Martinich, J. A., Solarz, S. L., & Lyons, J. R. (2006). Preparing students for conservation careers through project-based learning. *Conservation Biology, 20*(6), 1579–1583. doi:10.1111/j.1523-1739.2006.00569.x PMID:17181792

Marx, R. W., Blumenfeld, P. C., Krajcik, J. S., & Soloway, E. (1997). Enacting project-based science: Challenges for practice and policy. *The Elementary School Journal, 97*(4), 341–358. doi:10.1086/461870

Mayring, P. (2000). Qualitative Content Analysis. *Forum Qualitative Social Research, 1*(2). Advance online publication. doi:10.17169/fqs-1.2.1089

Mayring, P. (2014). *Qualitative content analysis: theoretical foundation, basic procedures and software solution*. Beltz.

Meadows, M. E. (2020). Geography Education for Sustainable Development. *Geography and Sustainability, 1*(1), 88–92.

Meadows, M. E. (2020). Geography education for sustainable development. *Geography and Sustainability, 1*(1), 88–92. doi:10.1016/j.geosus.2020.02.001

Mekdjian, S., & Amilhat Szary, A.-L. (2015). *Cartographies traverses, des espaces où l'on ne finit jamais d'arriver*. Visioncarto.

Melgar, F., & Silvio, D. (2011). Salir del aula. Aprender de otros contextos: patrimonio natural, museos e Internet. *Revista Eureka Sobre Enseñanza y Divulgación de Las Ciencias, 8*(3), 323–333. http://revistas.uca.es/index.php/eureka/article/view/2720/2368

Mercer, T. G., Kythreotis, A. P., Robinson, Z., George, S., & Sands, D. (2019). Step out of your comfort zone, try discipline-based educational research. *Times Higher Education*. https://www.timeshighereducation.com/blog/step-out-your-comfort-zone-try-discipline-based-educational-research

Mercer, T. G., & Kythreotis, A. P. (2020). Towards Citizen Governance for Climate Change Education and Justice: A Science–Policy Perspective. In S. Myers, S. Hemstock, & E. Hanna (Eds.), *Science, Faith and the Climate Crisis* (pp. 79–92). Emerald Publishing Limited., doi:10.1108/978-1-83982-984-020201008

Mercer, T. G., Kythreotis, A. P., Robinson, Z. P., Stolte, T., George, S. M., & Haywood, S. K. (2017). The use of educational game design and play in higher education to influence sustainable behaviour. *International Journal of Sustainability in Higher Education, 18*(3), 359–384. Advance online publication. doi:10.1108/IJSHE-03-2015-0064

Mietzel, G. (2002). *Psychologia kształcenia*. Gdańskie Wydawnictwo Psychologiczne.

Mintz, S. (2020). *Top five strategies for integrating active learning into virtual classes*. THE Campus. https://www.timeshighereducation.com/campus/top-five-strategies-integrating-active-learning-virtual-classes

Mishra, P. (2019). Considering Contextual Knowledge: The TPACK Diagram Gets an Upgrade. *Journal of Digital Learning in Teacher Education, 35*(2), 76–78. doi:10.1080/21532974.2019.1588611

Mitchell, D. (2007). Muerte entre la abundancia: los paisajes como sistemas de reproducción social. In J. Nogué (Ed.), *La construcción social del paisaje* (pp. 85–110). Biblioteca Nueva.

Mohan, J. (1995). Thinking local: Service-learning, education for citizenship and geography. *Journal of Geography in Higher Education, 19*(2), 129–142. doi:10.1080/03098269508709297

Monmonier, M. S. (2004). *Rhumb Lines and Map Wars: A Social History of the Mercator Projection.* University of Chicago Press. doi:10.7208/chicago/9780226534329.001.0001

Monroe, M. C., Plate, R. R., Oxarart, A., Bowers, A., & Chaves, W. A. (2019). Identifying effective climate change education strategies: A systematic review of the research. *Environmental Education Research, 25*(6), 791–812. doi:10.1080/13504622.2017.1360842

Mooney, P., & Corcoran, P. (2012). The Annotation Process in OpenStreetMap. *Transactions in GIS, 16*(4), 561–579. doi:10.1111/j.1467-9671.2012.01306.x

Moore, J. L., Dickson-Deane, C., & Galyen, K. (2011). E-Learning, online learning, and distance learning environments: Are they the same? *Internet and Higher Education, 14*(2), 129–135. doi:10.1016/j.iheduc.2010.10.001

Morandi, F. (1997). *Modèles et méthodes en pédagogie.* Nathan.

Moreira, R. (2007). *Pensar e ser em geografia.* Contexto.

Morón, H., & Morón, M. C. (2012). Paisaje y patrimonio como medio para enseñanza y aprendizaje de la geografía. *I Congreso Internacional "El Patrimonio Cultural y Natural Como Motor de Desarrollo: Investigación e Innovación,"* 1670–1681. https://dialnet.unirioja.es/servlet/articulo?codigo=4160136

Moseley, C. (Ed.). (2010). *Atlas of the World's Language in Danger* (3rd ed.). UNESCO Publishing. http://www.unesco.org/languages-atlas/en/atlasmap.html

Muis, A.-S. (2016). Psychogéographie et carte des émotions, un apport à l'analyse du territoire? *Carnets de géographes, 9.* https://journals.openedition.org/cdg/713

Müller, A. M., Goh, C., Lim, L. Z., & Gao, X. (2021). COVID-19 Emergency eLearning and Beyond: Experiences and Perspectives of University Educators. *Education Sciences, 11*(1), 19. doi:10.3390/educsci11010019

Muñoz, E. M. (2010). *Propuesta metodológica para el reconocimiento de los sistemas fluviales. Una base para su aplicación en la toma de decisions* [Tesis de Doctorado]. Universidad de Castilla La Mancha. https://www.educacion.gob.es/teseo/imprimirFicheroTesis.do?idFichero=n%2BY%2FKH0JwYU%3D

Muñoz, E. M. (2016). Los cauces fluviales como ejes didácticos. In La investigación e innovación en la enseñanza de la Geografía (pp. 291-304). Universitat d'Alacant.

Muñoz, E. M., & Jerez, O. (2021). El itinerario didáctico como recurso para conocer los proyectos del programa Nosotros Proponemos. In Una forma diferente de educar a través de la ciudad. El Proyecto ¡Nosotros Proponemos! Ediciones de la Universidad de Castilla-La Mancha. doi:10.18239/atenea_2021.30.00

Muñoz, E. M., Español, I. M., & Florín, M. (2007). Fundamental Boundaries of the Fluvial Space. Application to environmental planning. Application to environmental planning. In *III International Symposium on Transboundary Water Management* (pp. 335-348). UCLM. UNESCO.

Muñoz, F. (2007). Paisajes aterritoriales, paisajes en huelga. In J. Nogué (Ed.), *La construcción social del paisaje* (pp. 293–323). Biblioteca Nueva.

Myers Kelson, A. C., & Distlehorst, L. H. (2000). Groups in Problem-Based Learning (PBL): Essential Elements in Theory and Practice. In D. H. Evensen & C. E. Hmelo (Eds.), *Problem-based learning: A research perspective on learning interactions* (pp. 166–183). Lawrence Erlbaum.

Myers, K., Doran, P., Cook, J., Kotcher, J., & Myers, T. (2021). Consensus revisited: Quantifying scientific agreement on climate change and climate expertise among Earth scientists 10 years later. *Environmental Research Letters, 16*(10), 104030. doi:10.1088/1748-9326/ac2774

Nam, H. (2005). The nature and approach of place-based learning in geography education. *Social Studies Education, 44*(3), 195-210.

National Research Council. (2006). *Learning to Think Spatially: GIS as a Support System in the K-12 Curriculum* (1st ed.). National Academies Press.

Neis, P., & Zielstra, D. (2014). Recent Developments and Future Trends in Volunteered Geographic Information Research: The Case of OpenStreetMap. *Future Internet, 6*(1), 76–106. Advance online publication. doi:10.3390/fi6010076

Nogueira, A. R. B. (1994). *Mapa mental: recurso didático no ensino de geografia no 1o grau* [Dissertação de Mestrado]. Universidade São Paulo.

Novak, J. D. (1990). Concept maps and vee diagrams: Two metacognitive tools for science and mathematics education. *Instructional Science, 19*, 29–52. doi:10.1007/BF00377984

Novak, J. D. (1991). Clarify with concept maps: A tool for students and teachers alike. *Science Teacher (Normal, Ill.), 58*, 45–49.

Novak, J. D. (2002). Meaningful learning: The essential factor for conceptual change in limited or appropriate propositional hierarchies (liphs) leading to empowerment of learners. *Science Education, 86*(4), 548–571. doi:10.1002ce.10032

O'Neill, P. A. (2000). The role of basic sciences in a problem-based learning clinical curriculum. *Medical Education, 34*(8), 608–613. doi:10.1046/j.1365-2923.2000.00629.x PMID:10964207

O'Steen, B., Perry, L., Cammock, P., Kingham, S., Pawson, E., Stowell, R., & Perry, D. (2010). Engaging teachers and learners through service-learning. In K.Weir & P. Coolbear (Eds.), *Good Practice Publication Grants*. Ako Aotearoa. https://ako.ac.nz/assets/Knowledge-centre/Engaging-teachers-and-learners-through-service-learning/Engaging-teachers-and-learners-through-service-learning.pdf

OECD. (2002). *Definition and selection of competences (DeSeCo): Theoretical and conceptual foundations. Strategy paper.* http://deseco.ch/bfs/deseco/en/index/02.parsys.34116.downloadList.87902

Okoń, W. (1987). *Wprowadzenie do dydaktyki ogólnej*. PWN.

Okoń, W. (2004). *Nowy słownik pedagogiczny*. PWN.

Olcina, J. (2013). *Gota fría*. https://www.levante-emv.com/opinion/2013/10/27/gota-fria-12856313.html

Oliveira, A. F., Carvalho, A. J. E., & Fernandes, C. C. P. (2020). Mapas mentais como suporte para interpretação de áreas imageadas por veículo aéreo não tripulado no ensino de cartografia. *Geomae*, *11*(2), 45–56.

Olmedo, E. (2017). L'expérimentretien comme méthode d'enquête. Cartographie sensible et terrains de recherche collaboratifs entre art et géographie. *Mappemonde*, *121*.

Olson, J. (2006). Map projections and the visual detective: How to tell if a map is equal-area, conformal, or neither. *The Journal of Geography*, *105*(1), 13–32. doi:10.1080/00221340608978655

Orte, D. (2007). *Territorio de abejas, paisaje de hombres*. Imanat Documentales.

Ortega, N. (2010). Paisaje e identidad en la cultura española moderna. In E. Martínez de Pisón & N. Ortega (Eds.), *El paisaje: valores e identidades* (pp. 47–68). Universidad Autónoma de Madrid - Fundación Duques de Soria. https://libros.uam.es/uam/catalog/view/5/6/25

Ott, A.-C. (2020). L'organisation de l'espace mondial chez les enfants: les contours d'une géopolitique enfantine. *L'Espace Politique*, *40*(1). https://journals.openedition.org/espacepolitique/7997

Ozdamli, F., & Asiksoy, G. (2016). Flipped classroom approach. *World Journal on Educational Technology: Current Issues*, *8*(2), 98–105.

Özerol, G., Dolman, N., Bormann, H., Bressers, H., Lulofs, K., & Böge, M. (2020). Urban water management and climate change adaptation: A self-assessment study by seven midsize cities in the North Sea Region. *Sustainable Cities and Society*, *5*, 55. doi:10.1016/j.scs.2020.102066

Palsky, G. (2013). Cartographie participative, cartographie indisciplinée. *L'Information Geographique*, *77*(4), 10–25. doi:10.3917/lig.774.0010

Pamuk, S. (2012). Understanding pre-service teachers' technology use through TPACK framework. *Journal of Computer Assisted Learning*, *28*(5), 425–439. doi:10.1111/j.1365-2729.2011.00447.x

Panciera, K., Halfaker, A., & Terveen, L. (2009). Wikipedians are born, not made: a study of power editors on Wikipedia. *Proceedings of the ACM 2009 international conference on Supporting group work*. 10.1145/1531674.1531682

Partoune, C. (1999). *L'apport du géographe dans un projet interdisciplinaire*. http://www.lmg.ulg.ac.be/competences/chantier/contenus/cont_interdis1.html#ancre555514

Pawson, E., Fournier, E., Haigh, M., Muniz, O., Trafford, J., & Vajoczki, S. (2006). Problem-based learning in geography: Towards a critical assessment of its purposes, benefits and risks. *Journal of Geography in Higher Education*, *30*(1), 103–116. doi:10.1080/03098260500499709

Pérez, M., & Morón, H. (2016). El itinerario didáctico como herramienta para la activación patrimonial: una experiencia docente desde el parque de María Luisa. *Campo Abierto, 35*(2), 83–94. https://mascvuex.unex.es/revistas/index.php/campoabierto/article/view/2750/pdf_29

Perkins, C. (2014). Plotting practices and politics: (im) mutable narratives in OpenStreetMap. *Transactions of the Institute of British Geographers*, *39*(2), 304–317. doi:10.1111/tran.12022

Perkins, N., Hazelton, E., Erickson, J., & Allan, W. (2010). Place-based education and geographic information systems: Enhancing the spatial awareness of middle school students in Maine. *The Journal of Geography*, *109*(5), 213–218. doi:10.1080/00221341.2010.501457

Perrenet, J. C., Bouhuijs, P. A. J., & Smits, J. G. M. M. (2000). The suitability of problem-based learning for engineering education: Theory and practice. *Teaching in Higher Education*, *5*(3), 345–358. doi:10.1080/713699144

Peterson, E. G., Kolvoord, B., Uttal, D. H., & Green, A. E. (2020). High school students' experiences with geographic information systems and factors predicting enrollment in the geospatial semester. *The Journal of Geography*, *119*(6), 238–247. doi:10.1080/00221341.2020.1824009

Pfeil, U., Zaphiris, P., & Ang, C. S. (2006). Cultural Differences in Collaborative Authoring of Wikipedia. *Journal of Computer-Mediated Communication*, *12*(1), 88–113. doi:10.1111/j.1083-6101.2006.00316.x

Phang, C. W., Sutanto, J., Kankanhalli, A., Li, Y., Tan, B. C. Y., & Teo, H. H. (2006). Senior citizens' acceptance of information systems: A study in the context of e-government services. *IEEE Transactions on Engineering Management*, *53*(4), 555–569. doi:10.1109/TEM.2006.883710

Piaget, J., & Inhelder, B. (1947). *La représentation de l'espace chez l'enfant* (Ed1972). Ed. PUF.

Picanco, P. L., & Delazari, L. S. (2016). Usability Assessment of VGI Systems Interfaces in Feature Insert Task. *Boletim de Ciências Geodésicas*, *22*(3), 492–510. doi:10.1590/S1982-21702016000300028

Piotrowska, I., Cichoń, M., Mrula, A., & Nowak-Pierszalska, K. (2011). *Scenariusze zajęć z geografii dostosowane do strategii kształcenia wyprzedzającego – Kolegium Śniadeckich*. Pracownia Dydaktyki Geografii i Edukacji Ekologicznej, Wydział Nauk Geograficznych i Geologicznych, Uniwersytet im. Adama Mickiewicza w Poznaniu.

Piotrowska, I., Cichoń, M., Abramowicz, D., & Sypniewski, J. (2019). Challenges in geography education - a review of research problems. *Quaestiones Geographicae*, *38*(1), 71–84. doi:10.2478/quageo-2019-0009

Pope, J., & Morrison-Saunders, A. (2018). Teaching impact assessment. In *Special Publication Series, 10.* International Association for Impact Assessment.

Popkiewicz, M., Kardaś, A., & Malinowski, Sz. (2019). *Nauka o klimacie. Wyd. Sonia Draga Sp. z o.o. oraz Wyd.* Nieoczywiste.

Pretorius, R., Lonmbard, A., & Khotoo, A. (2016). Adding value to education for sustainability in Africa with inquiry-based approaches in open and distance learning. *International Journal of Sustainability in Higher Education*, *17*(2), 167–187. doi:10.1108/IJSHE-07-2014-0110

Pruneau, D., Khattabi, A., & Demers, M. (2010). Challenges and possibilities in climate change education. *US-China Education Review*, *7*(9), 15-24.

Puig, J., & Casas, M. (2017). El impacto ambiental: Un despertar ético valioso para la educación. *Teoría de la Educación*, *29*(1), 101–128. doi:10.14201/teoredu291101128

Puig, J., & Villarroya, A. (2013). Ecological quality loss and damage compensation in estuaries: Clues from a lawsuit in the Basque Country, Spain. *Ocean and Coastal Management*, *71*, 46–51. doi:10.1016/j.ocecoaman.2012.09.005

Pulido-Bosch, A., Pulido-Leboeuf, P., Molina-Sánchez, L., Vallejos, A., & Martin-Rosales, W. (2000). Intensive agriculture, wetlands, quarries and water management. A case study (Campo de Dalias, SE Spain). *Environmental Geology*, *40*(1-2), 163–168. doi:10.1007002540000118

Purwanto, U., Utaya, S., Handoyo, B., & Bachri, S. (2020). Transformation of Geospatial Technology Knowledge in Pre-service and Experienced Geography Teachers as Pedagogical Tools in the Technological-Pedagogical-Content Knowledge Framework. *International Journal of Learning, Teaching and Educational Research*, *19*(9), 58–76. doi:10.26803/ijlter.19.9.4

QAA. (2020a). *Guidance - Building a Taxonomy for Digital Learning*. https://www.qaa.ac.uk/docs/qaa/guidance/building-a-taxonomy-for-digital-learning.pdf

QAA. (2020b). Guidance - Preserving Quality and Standards Through a Time of Rapid Change: UK. *Higher Education.*

QAA. (2020c). *Guidance - Questions to Inform a Toolkit for Enhancing Quality in a Digital Environment.* https://www.qaa.ac.uk/docs/qaa/guidance/questions-to-inform-a-toolkit-for-enhancing-quality-in-a-digital-environment.pdf

Ralph, R. A. (2015). Post secondary project-based learning in science, technology, engineering and mathematics. *Journal of Technology and Science Education*, *6*(1), 26–35.

Rebolloso, R. (2008). El conocimiento en tiempos de globalizacion. *Ingenierías, 11*(38), 24–32. http://ingenierias2.uanl.mx/38/38_elconocimiento.pdf

Richter, D. (2010). *Raciocínio geográfico e mapas mentais: a leitura espacial do cotidiano por alunos do Ensino Médio* [Tese de Doutorado]. Universidade Estadual Paulista.

Richter, D. (2011). *O mapa mental no ensino de Geografia: concepções e propostas para o trabalho docente*. Cultura Acadêmica.

Richter, D., & de la Vega, A. G. (2019). O mapa no ensino de geografia: Uma análise do trabalho docente em Madrid (Espanha) e em Goiânia (Brasil). *Ateliê Geográfico, 13*(3), 46–65. doi:10.5216/ag.v13i3.60860

Richter, D., & Moraes, L. B. (2020). A Cartografia escolar na BNCC de Geografia do Ensino Fundamental: uma análise do pensamento espacial e do raciocínio geográfico. In E. M. B. Morais & D. Richter (Eds.), *Formação de professores e ensino de Geografia: contextos e perspectivas* (pp. 141–168). Editora Alfa.

Rigollet, N., A., (2017). Adolescences "Em Sampa". Représentations urbaines de lycéens français expatriés à Sao Paulo. *Carnets de géographes, 10.*

Rivero, M. P., & Gil, J. (2011). Didáctica de las ciencias sociales para educación infantil. Mira.

Rizqihandari, N., & Indratmoko, S. (2016). Using OpenStreetMap Data for Population Distribution Model. In B. Handoyo (Ed.), *Proceedings of the Lst International Conference on Geography and Education* (Vol. 79, pp. 244-247). Paris: Atlantis Press.

Roberts, M. (2017). Geography education is powerful if … *Teaching Geography, 42*(1), 6-9.

Roberts, M. (2013). *Geography through Enquiry*. Geographical Association.

Rockström, J., Steffen, W., Noone, K., Persson, Å., Chapin, F. S., Lambin, E. E., Lenton, T.M., Scheffer, M., Folke, C., Schellnhuber, H., Nykvist, B., De Wit, C.A., Hughes, T., van der Leeuw, S., Rodhe, H., Sörlin, S., Snyder, P.K., Costanza, R., Svedin, U., Falkenmark, M., Karlberg, L., Corell, R.W., Fabry, V.J., Hansen, J., Walker, B., Liverman, D., Richardson, K., Crutzen, P., & Folley, J. (2009). Planetary boundaries: Exploring the safe operating space for humanity. *Ecology and Society, 14*(2), 32. https://www.ecologyandsociety.org/vol14/iss2/art32/

Rød, J., Larsen, W., & Nilsen, E. (2010). Learning geography with GIS: Integrating GIS into upper secondary school geography curricula. *NorskGeografiskTidsskrift–Norwegian Journal of Geography, 64*(1), 21–35. doi:10.1080/00291950903561250

Rousell, A., & Zipf, A. (2017). Towards a Landmark-Based Pedestrian Navigation Service Using OSM Data. *ISPRS International Journal of Geo-Information, 6*(3), 22. doi:10.3390/ijgi6030064

Sadoń-Osowiecka, T. (2009). *Konstruowanie wiedzy geograficznej w klasach gimnazjalnych. Możliwości i zaniedbania*. Oficyna Wydawnicza Impuls.

Sánchez, L. E., & Morrison-Saunders, A. (2010). Teaching impact assessment: Results of an international survey. *Impact Assessment and Project Appraisal, 28*(3), 245–250. doi:10.3152/146155110X12791029734641

Sands, D., Parker, M., Hedgeland, H., Jordan, S., & Galloway, R. (2018). *Using concept inventories to measure understanding.* doi:10.1080/23752696.2018.1433546

Santa-Cecilia, F. J., García de la Vega, A., & Martín-Moreno, R. (2021). La erupción del volcán Cumbre Vieja en la Isla de La Palma (2021). El enfoque educativo de un volcán urbano. *Didácticas Específicas*, (25), 7–31. doi:10.15366/didacticas2021.25.001

Savery, J. R., & Duffy, T. M. (1995). Problem Based Learning: An instructional model and its constructivist framework. *Educational Technology*, *35*, 31–38.

Savery, J. S. (2006). Overview of PBL: Definitions and distinctions. *The Interdisciplinary Journal of Problem-Based Learning*, *1*(1), 9–20. doi:10.7771/1541-5015.1002

Savin-Baden, M. (2000). *Problem-Based Learning in Higher Education: Untold Stories.* Society for Research in Higher Education and Open University Press.

Savin-Baden, M. (2004). Understanding the impact of assessment on students in problem-based learning. *Innovations in Education and Teaching International*, *41*(2), 223–233. doi:10.1080/1470329042000208729

Savin-Baden, M., & Wilkie, K. (2004). *Challenging Research in Problem-Based Learning.* Open University Press.

Šavrič, B., Patterson, T., & Jenny, B. (2019). The Equal Earth map projection. *International Journal of Geographical Information Science*, *33*(3), 454–465. doi:10.1080/13658816.2018.1504949

Schmidt, M., Klettner, S., & Steinmann, R. (2013). *Barriers for contributing to VGI projects.* Paper presented at the 26th International Cartographic Conference, Dresden.

Schmidt, H. G., & Moust, J. H. C. (2000). Processes that shape small-group tutorial learning: A Review of Research. In D. H. Evensen & C. E. Hmelo-Silver (Eds.), *Problem-based learning: A research perspective on learning interactions* (pp. 19–52). Lawrence Erlbaum.

Schmidt, H. G., Vermeulen, L., & van der Molen, H. T. (2006). Long-term effects of problem-based learning: A comparison of competencies acquired by graduates of a problem-based and a conventional medical school. *Medical Education*, *40*(6), 562–567. doi:10.1111/j.1365-2929.2006.02483.x PMID:16700772

Schoepfer, I. (2021). Éditorial. GeoAgenda, 4, 3.

Scholz, M. A., Huynh, N. T., Brysch, C. P., & Scholz, R. W. (2014). An evaluation of university world geography textbook questions for components of spatial thinking. *The Journal of Geography*, *113*(5), 208–219. doi:10.1080/00221341.2013.872692

Schommer, M. (2019). Spatial orientation - competence expectations and common misconceptions based on map projections. In M. Gröger, C. Prust, A. Flügel (Eds.), Cultural appropriation of spaces and things, 20 – 30 October 2019 (pp. 231 – 244). Siegen: Universi.

Sebastiá, R., & Tonda, E. M. (2015). El concepto y representación del espacio geográfico en la enseñanza de la Geografía en los niveles educativos no universitarios. Análisis bibliométrico. In J. de la Riva, P. Ibarra, R. Montorio, & M. Rodrigues (Eds.), *Análisis espacial y representación geográfica: innovación y aplicación* (pp. 1505–1514). Universidad de Zaragoza-AGE.

SEDA. (2020). *SEDA - Home*. https://www.seda.ac.uk/

Serrano-Estrada, L., Martin, T. J., & Marti, P. (2021). Understanding city dynamics: Using geolocated social media in a problem-based activity as an investigative tool to enhance student learning. *Journal of Geography in Higher Education*, 1–22. Advance online publication. doi:10.1080/03098265.2021.2004390

Shamai, S. (1991). Sense of place: An empirical measurement. *Geoforum*, *22*(3), 347–358. doi:10.1016/0016-7185(91)90017-K

Shulman, L. (2005). Signature pedagogies in the professions. *Daedalus*, *134*(3), 52–59. doi:10.1162/0011526054622015

Silva, P., & Roque Ascenção, V. O. (2018). Por uma construção do raciocínio geográfico para além do pensamento espacial (spatial thinking). In *Anais do 5° Colóquio Internacional da Rede Latino-Americana de Investigadores de Didática de Geografia* (pp. 73–94). UFG.

Smith, D., Hoersch, A. L., & Gordon, P. R. (1995). Problem-based learning in the undergraduate geology classroom. *Journal of Geological Education*, *43*(4), 149–152. doi:10.5408/0022-1368-43.4.385

Smith, G. A. (2002). Place-based education: Learning to be where we are. *Phi Delta Kappan*, *83*(8), 584–594. doi:10.1177/003172170208300806

Smith, R. M. (1993). The triple jump examination as an assessment tool in the problem-based medical curriculum at the University of Hawaii. *Academic Medicine*, *68*(5), 366–371. doi:10.1097/00001888-199305000-00020 PMID:8484850

Soares, P. A. P. (2017) *O pensamento espacial: um desafio no Ensino da Geografia Escolar* [Dissertação de Mestrado]. Universidade de Coimbra.

Sobel, D. (2005). *Place-based Education: Connecting Classrooms & Communities*. The Orion Society.

Solem, M. (2001). Using geographic Information Systems and the Internet to Support Problem-based Learning. *Planet*, *2*(1), 22–24. doi:10.11120/plan.2001.00040022

Solent University. (2020). *Transforming Solent's learning delivery during Covid-19*. https://www.solent.ac.uk/news/transforming-solents-learning-delivery-during-covid-19

Spronken-Smith, R. (2005). Implementing a Problem-Based Learning Approach to Teaching Research Methods in Geography. *Journal of Geography in Higher Education*, *29*(2), 203–221. doi:10.1080/03098260500130403

Spronken-Smith, R. (2013). Towards securing a future for geography graduates. *Journal of Higher Education in Geography, 37*(3), 315–326. doi:10.1080/03098265.2013.794334

Spronken-Smith, R. A., Bullard, J. E., Ray, W., Roberts, C., & Keiffer, A. (2008). Where might sand dunes be on Mars? Engaging students through inquiry-based learning in Geography. *Journal of Geography in Higher Education, 32*(1), 71–86. doi:10.1080/03098260701731520

Spronken-Smith, R. A., Walker, R., Batchelor, J., O'Steen, B., & Angelo, T. (2012). Evaluating student perceptions of learning processes and intended learning outcomes under inquiry approaches. *Assessment & Evaluation in Higher Education, 37*(1-2), 57–72. doi:10.1080/02602938.2010.496531

Spronken-Smith, R., & Harland, T. (2009). Learning to teach with problem-based learning. *Active Learning in Higher Education, 10*(2), 138–153. doi:10.1177/1469787409104787

Spronken-Smith, R., & Kingham, S. (2009). Strengthening teaching and research links: The case of a pollution exposure inquiry project. *Journal of Geography in Higher Education, 33*(2), 241–253. doi:10.1080/03098260802276813

Spronken-Smith, R., & Walker, R. (2010). Can inquiry-based learning strengthen the links between teaching and disciplinary research? *Studies in Higher Education, 35*(6), 723–740. doi:10.1080/03075070903315502

Stalans, L. J. (2012). Frames, Framing Effects, and Survey Responses. In L. Gideon (Ed.), *Handbook of Survey Methodology for the Social Sciences* (pp. 75–90). Springer New York. doi:10.1007/978-1-4614-3876-2_6

Stańczuk-Gałwiaczek, M., Sobolewska-Mikulska, K., Ritzema, H., & Van Loon-Steensma, M. J. (2018). Integration of water management and land consolidation in rural areas to adapt to climate change: Experiences from Poland and the Netherlands. *Land Use Policy, 77*, 498–511. doi:10.1016/j.landusepol.2018.06.005

Staszak, J.-F. (2012). La construcción del imaginario occidental del 'allá' y la fabricación de las 'exótica': El caso de los koi moko maoris. In *Geografia de los imaginerarios* (pp. 179–210). Anthropos/Universidad Autónoma Metropolitana Iztapalapa.

Staver, J. R., & Bay, M. (1987). Analysis of the project synthesis goal cluster orientation and inquiry emphasis of elementary science textbooks. *Journal of Research in Science Teaching, 24*(7), 629–643. doi:10.1002/tea.3660240704

Steffen, W., Broadgate, W., Deutsch, L., Gaffney, O., & Ludwig, C. (2015). The trajectory of the Anthropocene: The great acceleration. *The Anthropocene Review, 2*(1), 81–98. doi:10.1177/2053019614564785

Steinmann, R., Häusler, E., Klettner, S., Schmidt, M., & Lin, Y. (2013). *Gender Dimensions in UGC and VGI: A Desk-Based Study.* Paper presented at the GI_Forum 2013, Wien.

Stephens, M. (2013). Gender and the GeoWeb: Divisions in the production of user-generated cartographic information. *GeoJournal, 78*(6), 981–996. doi:10.100710708-013-9492-z

Stepien, W. J., Gallagher, S. A., & Workman, D. (1993). Problem-based learning for traditional and interdisciplinary classrooms. *Journal for the Education of the Gifted, 16*(4), 338–357. doi:10.1177/016235329301600402

Stevens, A., & Coupe, P. (1978). Distortions in Judged Spatial Relations. *Cognitive Psychology, 10*(4), 422–437. doi:10.1016/0010-0285(78)90006-3 PMID:699514

Stevenson, J. A. (1930). *Metoda projektu w nauczaniu.* Książnica - Atlas, Lwów - Warszawa.

Sun, Y., Chan, R. C. K., & Chen, H. (2016). Learning with geographical sensitivity: Place-based education and its praxis. *The Professional Geographer, 68*(4), 574–583. doi:10.1080/00330124.2015.1121835

Sypniewski, J. (2015). *Influence of the strategy of the Inquiry Based Science Education on the development of natural interests of students and acquiring natural-geographical com-petences in the course of extracurricular activities of natural science activity group* [MA thesis]. Adam Mickiewicz University.

Sypniewski, J. (2019). Where the geographical expanse ends – Space education in primary school. Implementation of inquiry based science education (IBSE) in geography lessons in Polish school. *Miscellanea Geographica, 23*(4), 256–266. doi:10.2478/mgrsd-2019-0024

Szanto, C., & Bergsjö, A. (2018). Représenter l'expérience spatiale. Description et analyse d'une expérimentation graphique. *Projets de paysage, 18.* https://journals.openedition.org/paysage/1119

Szymański, M. S. (2000). *O metodzie projektów.* Wydaw. Żak.

Talbert, R. (2017). *Flipped learning: a guide for higher education faculty. Sterling.* Stylus Publishing, LLC.

Tarapiah, S., & Atalla, S. (2015). Public Transportation Management System based on GPS/WiFi and Open Street Maps. *International Journal of Advanced Computer Science and Applications, 6*(1), 189-194.

Teixeira, S. K. (2001). *Das imagens às linguagens do geográfico: Curitiba, a "capital ecológica"* [Tese de Doutorado]. Universidade de São Paulo.

The Comission on Geographical Education. (2016). *International Charter on Geography Eduction.* International Geographical Union.

Thielmann, T., van der Velden, L., Fischer, F., & Vogler, R. (2012). Dwelling in the web. Towards a googlization of space. *HIIG Discussion Paper Series, 2012*(3). doi:10.2139/ssrn.2151949

Thomas, R. (2007). La marche en ville. Une histoire de sens. *L'Espace Geographique, 36*(1), 15–26. doi:10.3917/eg.361.0015

Tolman, E. C. (1948). Cognitive maps in rats and men. *Psychological Review, 55*(4), 189–208. doi:10.1037/h0061626 PMID:18870876

Tourón, J., & Santiago, R. (2015). El modelo Flipped Learning y el desarrollo del talento en la escuela. *Review of Education, 368*, 196–231. doi:0.4438/1988-592X-RE-2015-368-288

Treby, E., Hewitt, I., & Shah, A. (2006). Embedding 'disability and access' into the geography curriculum. *Teaching in Higher Education, 11*(4), 413–425. doi:10.1080/13562510600874169

Truong, Q. T., de Runz, C., & Touya, G. (2019). Analysis of collaboration networks in OpenStreetMap through weighted social multigraph mining. *International Journal of Geographical Information Science, 33*(8), 1651–1682. doi:10.1080/13658816.2018.1556395

Tseng, W. K. (2014). The shortest overall distance of two piecewise rhumb-lines. *10th International Conference on Natural Computation (ICNC)*, 1153-1157. 10.1109/ICNC.2014.6976004

Tuan, Y. F. (1977). *Space and Place: The perspective of experience*. University of Minnesota Press.

Turner, A. J. (2006). *Introduction to neogeography*. O'Reilly.

Tzotzis, I. (2007). *The practical importance of map projections in the teaching of maps in school* [Doctoral Thesis]. Faculty of Primary Education, A.U.T.H., Thessaloniki, Greece.

Tzotzis, I., & Lambrinos, N. (2004). The cognitive competence of first grade of Gymnasium in the subject of mathematical geography. *Proceedings of 7th Pan-Hellenic Geographical Conference*, 2, 399-406.

UN Development Programme. (2020). *Human Development Report 2020 - The next frontier: Human development and the Anthropocene*. https://hdr.undp.org/sites/default/files/hdr2020.pdf

UN. (1987). *Report of the World Commission on Environment and Development: Our Common Future*. http://www.un.org/en/ga/search/view_doc.asp?symbol=A/42/427

UN. (2015). *Transforming our world: the 2030 Agenda for Sustainable Development.* https://undocs.org/A/RES/70/1

UNESCO. (1972). *Convention concerning the protection of the World Cultural and Natural Heritage*. https://whc.unesco.org/archive/convention-en.pdf

UNESCO. (2006). *Framework for the United Nations Decade of Education for Sustainable Development. International Implementation Scheme*. https://unesdoc.unesco.org/images/0014/001486/148650E.pdf

UNESCO. (2021). *Operational Guidelines for the Implementation of the World Heritage Convention.* https://whc.unesco.org/document/190976

University of Kent. (2020). *CSHE Events 2020/21 - Centre for the Study of Higher Education - University of Kent*. https://www.kent.ac.uk/cshe/news-events.html#

University of Lincoln. (2020). *Teaching Remotely Essentials – Remote Teaching*. https://remoteteaching.blogs.lincoln.ac.uk/teaching-remotely-essentials/

Unwyn, T. (1992). *The place of Geography*. Longman Group Limited.

USACE. (1994). *Sharing the challenge: floodplain management into the 21st century.* Report of the Interagency Floodplain Management Review Committee to the Administration Floodplain Management Task Force. https://usace.contentdm.oclc.org/digital/collection/p16021coll11/id/2471/

Valero, J. R. (2016). La fuente oral en la enseñanz a y la investigación geográficas. In L. Alanís, J. Almuedo, G. De Oliveira, R. Iglesias, & B. Pedregal (Eds.), *Nativos digitales y Geografía en el siglo XXI: educación geográfica y estilos de aprendizaje* (pp. 65–79). Asociación de Geógrafos Españoles. Grupo de Didáctica de la Geografía - Universidad Pablo de Olavide - Universidad de Alicante.

Valovicová, L., Vallo, D., & Kramáreková, H. (2019). Mercator projection as a tool for development of interdisciplinary reasons among physics, mathematics and geography. In *21st International Conference DIDFYZ, 9 – 12 October 2019*. Terchova: AIP.

Van der Schee, J. (2014). Looking for an international strategy for geography education. *Journal of Research and Didactics in Geography*, *1*(3), 9–13.

Van Uum, M., Verhoeff, R., & Peeters, M. (2017). Inquiry-based science education: Scaffolding pupils' self-directed learning inopen inquiry. *International Journal of Science Education*, *39*(18), 1–21. doi:10.1080/09500693.2017.1388940

Vandecasteele, A., & Devillers, R. (2015). Improving Volunteered Geographic Information Quality Using a Tag Recommender System: The Case of OpenStreetMap. In J. J. Arsanjani, A. Zipf, P. Mooney, & M. Helbich (Eds.), *Openstreetmap in Giscience: Experiences, Research, and Applications* (pp. 59–80). Springer-Verlag Berlin. doi:10.1007/978-3-319-14280-7_4

Villarroya, A., & Larrañeta, M. (2019). La regeneración de un bosque en el campus de la Universidad de Navarra. In R. Española de Aprendizaje-Servicio (Ed.), XII Encuentro estatal de Aprendizaje-Servicio - I Jornada Internumiversitaria: "El ApS en las universidades navarras. Academic Press.

Vogler, R., & Hennig, S. (2013). Providing Geomedia Skills beyond (Post)Secondary Education. *GI_FORUM - Journal for Geographic Information Science, 2013*(1), 317-327. doi:10.1553/giscience2013s317

Vogler, R., Hennig, S., & Ferber, N. (2017). Redressing the Exclusiveness. Challenges which Prevent New Users from Contributing to OSM. *GI_FORUM - Journal for Geographic Information Science, 2017*(2), 294-299. doi:10.1553/giscience2017_01_s294

Vrbik, D. (2016). Crowdsourcing of Local Spatial and Historical Knowledge. *GI_Forum, 2*, 109-122. doi:10.1553/giscience2016_02_s109

Vygotsky, L. S. (1978). *Mind in Society. The Development of Higher Psychological Processes* (2nd ed.). Harvard University Press.

Vygotsky, L. S. (2000). *A construção do pensamento e da linguagem.* Martins Fontes.

Wakabayashi, Y., & Matsui, Y. (2013). Variation of geospatial thinking in answering geography questions based on topographic maps. *Proceedings of the 26th International Cartographic Conference.*

Walkington, H. (2015). *Students as researchers: Supporting undergraduate research in the disciplines in higher education.* Academic Press.

Wallenhorst, N. (2020). *La vérité sur l'Anthropocène.* Éditions Le Pommier.

Walshe, N. (2017). Developing trainee teacher practice with geographical information systems (GIS). *Journal of Geography in Higher Education, 41*(4), 608–628. doi:10.1080/03098265.2017.1331209

Wawrzyniak, K., & Banaszak, K. (2014). Metoda projektów – dydaktyczną syntezą współczesnej wiedzy i doświadczeń w edukacji. In S. Dylak, G. Barabasz, D. Hejwosz-Gromkowska (Eds.), Metoda projektów w edukacji ponadpodstawowej (na przykładzie przedmiotów przyrodniczych) (pp. 100-119). UAM, OFEK, Samorząd Woj. Wielkopolskiego.

Weiss, R. E. (2003). Designing problems to promote higher order thinking. In Problem-Based Learning in the Information Age. San Francisco, CA: Jossey-Bass. doi:10.1002/tl.109

Werlen, B. (1993). *Society action and space: an alternative human geography.* Routledge.

Wiegand, P. (2001). Geographical information systems (GIS) in education. *International Research in Geographical and Environmental Education, 10*(1), 68–71.

Wiegand, P. (2006). *Learning and Teaching with Maps.* Routledge. doi:10.4324/9780203477793

Wilcke, H., & Budke, A. (2019). Comparison as a method for geography education. *Education in Science, 9*(3), 225–240. doi:10.3390/educsci9030225

Wilkerson, L., & Gijselaers, H. (Eds.). (1996). Bringing Problem-Based Learning to Higher Education: Theory and Practice, New Directions for Teaching and Learning, No. 68. San Francisco, CA: Jossey-Bass.

Wood, L., & Hartshorne, M. (2017, March 8). Literacy: The role of communication skills. *The Voice for Secondary Education.* https://www.sec-ed.co.uk/best-practice/literacy-the-role-of-communication-skills

Wood, D. (1993). *The Power of Maps.* Routledge. doi:10.1038cientificamerican0593-88

Wood, D. (2010). *Rethinking the Power of Maps.* Guilford Press.

Wood, D., Bruner, J. S., & Ross, J. (1976). The role of tutoring in Problem Solving. *Journal Child Psychologist Psychiatrics, 17*(2), 89–100. doi:10.1111/j.1469-7610.1976.tb00381.x PMID:932126

Woodhouse, J., & Knapp, C. (2000). *Place-based curriculum and outdoor and environmental education approaches.* ERIC Digest.

Woods, D. R. (2006). *Preparing for PBL* (3rd ed.). McMaster University.

Woods, D. R., Hrymak, A. N., Marshall, R. R., Wood, P. E., Crowe, C. M., Hoffman, T. W., Wright, J. D., Taylor, P. A., Woodhouse, K. A., & Bouchard, C. G. K. (1997). Developing Problem Solving Skills: The McMaster Problem Solving Program. *Journal of Engineering Education*, *86*(2), 75–91. doi:10.1002/j.2168-9830.1997.tb00270.x

Yang, A. R., Fan, H. C., & Jing, N. (2016). Amateur or Professional: Assessing the Expertise of Major Contributors in OpenStreetMap Based on Contributing Behaviors. *ISPRS International Journal of Geo-Information*, *5*(2), 15. doi:10.3390/ijgi5020021

Yli-Panula, E., Jeronen, E., & Lemmetty, P. (2020). Teaching and learning methods in geography promoting sustainability. *Education Sciences*, *10*(1), 5. doi:10.3390/educsci10010005

Zeleňáková, M., Diaconu, D. C., & Haarstad, K. (2017). Urban Water Retention Measures. *Procedia Engineering*, *190*, 419–426. doi:10.1016/j.proeng.2017.05.358

Zheng, B., Zhang, Y., & Chen, J. (2011). Preference to home landscape: Wildness or neatness? *Landscape and Urban Planning*, *99*(1), 1–8. doi:10.1016/j.landurbplan.2010.08.006

About the Contributors

Alfonso García de la Vega is a senior scientist in Geography Teaching at Universidad Autónoma de Madrid (Spain). Coordinator and Main Researcher of "Landscape, Heritage and Education" (Research Group, UAM). Director of Master's Degree in Museums and Natural Parks Education (UAM, 2011-2020). Associate Dean for Research and Innovation, Faculty of Teacher's Training and Education UAM, 2012-2014). His research interests are Landscape Education, Geography Teaching and Environmental Education and Geography. Also, he has interested on Geography Curriculum, Geospatial Thinking, Problem based Learning, Fieltrip and Fieldwork in Geography. He has a award in Geography Degree and a Spanish-Grant for PhD. He has supervised five doctoral theses and has another seven ongoing.

***

Dawid Abramowicz is a PhD student in the Laboratory of Didactics of Geography and Educational Research at the Faculty of Geographical and Geological Sciences at Adam Mickiewicz University in Poznań. His research interests include participatory processes in geographical education, distance learning, cityscapes, fieldwork and geodiversity. He is a geography teacher at a high school for autistic students. He is interested in alternative pedagogical approaches, especially the pedagogy of Celestin Freinet.

María Casas is an Environmental Biologist and teaching assistant at the University of Navarra. She combines research activity with teaching in Secondary Education. Her main research lines address the relationships between human beings and their environment, namely: environmental education, education for sustainable development, environmental ethics, and environmental impact. Her Ph.D. (2018) explored the didactic potential of landscape to foster the social change that the global crisis demands.

Małgorzata Cichoń has been conducting research in the field of geography didactics. Except the issues related to geographic education, research also concerns: methodology of fieldwork and geographical educational trails. In recent years, she has undertaken research in terms of key competences at all educational levels and the use of the project method. She participates in the implementation of several projects financed by the European Union within the European Social Fund.

Luis Erneta is an "Associate Professor" and PhD "Assistant Professor", accredited by the National Agency for Quality Assessment and Accreditation Spanish (ANECA). Extensive researcher career path in Environmental Biology, Geography and Lanscape Planning. Expertise on processing and editing of spatial data as well as in cartographic analysis and development. Specialization on Geographic Information Systems with ArcGIS, ESRI. Teaching experience in preuniversity and university as well as in continuous training.

Sabine Hennig is senior scientist at the University of Salzburg, Austria. Her research interests are in the field of Digital Nature Conservation, the development off user-centered applications addressing the needs of special needs users and geo participation. In this context, she focuses on the further development of approaches and methods regard co-design or participative design to involve the population efficiently and effectively in the development of applications. Further does her work concentrate on success factors of participatory initiatives and how these can be addressed appropriately according to the target group. In recent years she has been working in various school projects and Citizen Science initiatives as well as in the development and implementation of e-learning materials.

Georgia Intzidou is a primary school teacher with more than 12 years experience in teaching. She teaches at Mandoulides Primary School, a private school in Thessaloniki, Greece since 2010. During her work she studied for her MSc in "Maths and Science Teaching and ICT in Education: An Interdisciplinary Approach" at the Aegean University (2014-2016) and finally she did her PhD in Geography Teaching at the Department of Primary Education at Aristotle University of Thessaloniki, Greece (2018-2022). She has been awarded as Microsoft Innovative Educated Expert (2016-2017). Her research interests include teaching STEM, making educational tools, experiential activities at school, etc.

Sylvie Joublot Ferré has a doctorate in geography (ENS Lyon, EVS Laboratory), is an associate professor and has a degree in political science (IEP Lyon). She is a member of the editorial board of the journal L'information géographique and currently teaches at the Haute École Pédagogique in Lausanne. Her work in cultural geography examines spatialities, spatial experience, inhabitation and spatial practices as ways of deciphering the relationship with the world. Her field of investigation is bi-national (France/Switzerland). Echoing her thesis methodology, she explores, with the students, different approaches and graphs to give an account of spatial experience in the urban space of proximity: the in situ survey, urban wanderings, walking in the city, observation of landscapes from cultural and symbolic details, sensitive cartography, photographic plates.

Hannah Kelly graduated from the University of Lincoln in 2021 with a First Class honours degree in BSc Geography and has a key interest in sustainability and encouraging pro-environmental behaviours.

Simon Kingham is a professor of geography at Te Whare Wānanga o Waitaha | University of Canterbury NZ, where he has been since 2000. His teaching and research mainly focuses on urban issues specifically related to transport and wellbeing. He is passionate about community engaged learning and in 2005 was awarded a University of Canterbury Teaching Award, and in 2016 a University of Canterbury Sustainability Award for his role in teaching with Community Service. He is currently seconded two days a week to the Ministry of Transport as their Chief Science Advisor.

Andrew P. Kythreotis is a Senior Lecturer in Social and Political Geography at the School of Geography, University of Lincoln, UK, a Senior Researcher at the Tyndall Centre for Climate Change Research, University of East Anglia, UK, an Honorary Research Fellow at the School of Psychology, Cardiff University and an ESRC Place-based Climate Action Network Associate. His research and teaching revolve around the broad themes of climate change and the environment and how its policy, politics, governance is constructed around socio-spatial ontologies. He has advised national governments on climate change adaptation issues, co-founded the Lincoln Climate Commission, is on the NERC Review Panel for the UKRI Climate Resilience Programme European Commission grant reviewer and was an external reviewer for the 2017 and 2022 UK Climate Change Risk Assessment Evidence Reports.

Nikolaos Lambrinos is a Professor of Geography Teaching, as well as the founder and President of the Hellenic digital earth Centre of Excellence (https://www.auth.gr/en/units/19861). Among other topics, his research interests include new methods in teaching Geography at schools, GIS and Remote Sensing as tools for teaching Geography, the future of school Geography in Europe and Web mapping. Prof. Dr. Lambrinos has participated in 27 national and international projects and has a strong publication record. He has organized and participated in 71 national seminars in Geography teaching, GIS and web mapping. His website (http://labrinos.webpages.auth.gr/digital_geography) has been extensively used by Greek primary school teachers. He is also an active member of the International Cartographic Association (ICA), the Open Source Geospatial Foundation's Committee for Educational outreach (OSGeo

GeoForAll Action – http://www.geoforall.org), as well as EUROGEO (European Association of Geographers - http://www.eurogeography.eu).

Theresa G. Mercer is a Senior Lecturer in Biogeography and Planetary Health at the School of Geography, University of Lincoln, UK. She has held several academic postings at the University of Hull, UK, Cranfield University, UK, Cardiff University, UK, Keele University, UK and the University of Queensland, Australia. She is an interdisciplinary environmental scientist with broad interests in environmental toxicology, environmental management and Education for Sustainable Development (ESD). More recently she has branched out into geography pedagogy, particularly digital solutions and co-creation approaches that seek to promote inclusivity.

Celia Laguna Mora is Graduate in Marine Sciences and Master in Oceanology from the University of Cadiz. She combines her job as water expert in the Spanish environmental attorney´s office and her doctorate in Territory, Infrastructures and Environment from the Civil Engineering School of University of Castilla-La Mancha. Her experience is focused on aquatic ecosystem research, with substantial field practice, in-situ quality monitoring and laboratory work. She has worked with Geographic Information Systems (ArcGIS and ESRI) on processing and editing spatial data. She has taken part in several national and regional projects of applied research and academic innovation with different research groups, and she have authored and/or co-authored 5 scientific articles, 16 papers in conferences and seminars and 7 book chapters. Besides her scientific career, she was owner and teacher of a science academy where she taught scientific subjects (mathematics, physics, chemistry, science) for Primary, Secondary and High School and scientific workshops for primary and secondary education.

Elena María Muñoz Espinosa is Graduate in Environmental Sciences from the Autonomous University de Madrid, Master in Spatial Planning, Environment, Landscape and Sustainability and PhD European Degree from the University of Castilla-La Mancha. Her professional and research experience focuses on environmental and landscape assessment, as well as the interweaving of projects in the territory. Her main line of research deals with the didactic transposition of environment and landscape for the training of future teachers. Currently, she works as Assistant Professor at the University Autónoma de Madrid in the Department of Specific Didactics. She is a member of the Landscape, Heritage, and Education Research Group.

Ralf Ohlemüller is an environmental geographer at Te Whare Wānanga o Otāgo, University of Otago. In his teaching and research, he deals with the responses of the living environment to anthropogenic and natural environmental change. His work typically involves some aspect of spatial modelling and aims to address an applied conservation problem. He likes maps.

Iwona Piotrowska is a geographer, geography didactician and academic teacher, employed at the Laboratory of Geography Didactics and Educational Research, Faculty of Geographical and Geological Sciences, Adam Mickiewicz University in Poznań. The research and scientific studies concern human activity in geographical environment, changes in relief and land use, geomorphological and landscape aspects, methods of geography and nature education, application of neurodidactic achievements in the process of education, Content and Language Integrated Learning (CLIL), formation of geographical notions in bilingual education, geoinformation technologies and GIS in teaching, new strategic and methodical solutions in geographical education, as well as the conception of core curricula for geographical education at various educational levels. The results of her activities also include the development of new strategies, methods and didactic materials such as programmes, textbooks, scripts and educational concepts. She has worked in the expert team on the development of core curricula for general geography education at all educational levels. She is co-organiser of many scientific and educational events such as conferences, seminars, workshops, summer schools. She also acts as an expert of the Ministry of Science and Education in the field of school geography textbooks. For many years she has participated in various scientific-educational projects financed by the European Union. Iwona Piotrowska's biography has been included in the New Biographical Dictionary of Polish Geography Didacticians (Piskorz S., Tracz M., ed., 2015, Scientific Publishing House of the Pedagogical Academy in Krakow).

Denis Richter is a PhD Professor at the Federal University of Goiás/Brazil in the Institute of Socio-Environmental Studies. Coordinator of the Laboratory of Teaching and Research in Geographic Education and the Group of Studies and Research in Cartography for Schoolchildren, member of the Center for Teaching and Research in Geographic Education and the Latin American Network of Investigators of Didactics of Geography.

David Sands is a retired academic physicist with a variety of professional interests in physics, including most recently research into the foundations of thermodynamics. However, his greatest passion is education, in particular promoting and developing a deep understanding of concepts among students, as well as the critical thinking necessary to be able to use that knowledge. He is a passionate advocate of active learning in all its forms.

Rachel Spronken-Smith is a professor in higher education and geography. After nine years as an academic in geography at the University of Canterbury NZ, she moved into higher education at the University of Otago NZ, working initially as an academic developer, then as a head of department, and for the last nine years, as dean of the Graduate Research School. Her teaching has received numerous awards including at the national level, and her research interests are in geography in higher education, graduate outcomes, inquiry-based learning and doctoral education.

Jakub Sypniewski is a Geography teacher at the Primary School No. 83 "Łejery" in Poznań. From 2019 methodological advisor to primary school geography teachers at the Teacher Training Center in Poznań. He runs a cooperation network of teachers 'Pog(eo)dajmy' - its name derives from the Poznań dialect, in which the word godoć (godać) means to talk / speak. He is a PhD student in the Didactics of Geography and Educational Research at the Faculty of Geographical and Geological Sciences at the Adam Mickiewicz University in Poznań. In his research, he pays the most attention to the Inquiry Based Science Education (IBSE). Author of lesson scenarios published, among others as part of the Amgen Teach project, School of Exercises in Suchy Las and Krotoszyn, Geography at school. Co-author of the MAC textbook for geography and materials of the Demart publishing house, the ambassador of the Learnetic company.

Ana Villarroya is a Biologist and professor at the University of Navarra. She focuses her research on the interaction between humans and their environment from different perspectives. Her Ph.D. and post-doctoral work studied environmental offsets as part of the Environmental Impact Assessment procedure, while her current line of research aligns with environmental humanities through projects that explore our relationship to other living beings and look for ways to restore humans' link to nature. She teaches Landscape Restoration and Environmental Education courses in the Environmental Sciences degree.

Robert Vogler is senior scientist at the University of Salzburg, Austria, Department of Sociology and Social Geography. He works in the field of geography education and is involved in teacher initial- and in-service training activities. His research interests and teaching expertise are focussed on digital geography education, geovisualization, geomedia and (geo)-communication in education environments, spatially enabled learning and teaching, social construction of space, social geographic theory, and meanings of space in (geography) education.

Index

A

Acute Issue 51-56, 72
Anthropocene 259-261, 263, 266, 274-275, 277

B

Biodiversity 24, 259
Blended Learning 79, 81, 90, 95-97, 100

C

Changes in Ice Sheets 26
Climate Change 23-24, 28, 31, 34-36, 38, 44-45, 106, 125, 230, 259
co-creation 102-105, 110-111
Community of Practice (CoP) 21
Constructivist Learning Environment 52, 57, 66, 72
Cultural Heritage 228, 238, 250

D

Didactic Resources 117, 119, 123, 142
Didactic Strategies 54, 142, 228, 254
Distortions 168-172, 180, 182, 190, 193, 195

E

Ecology 54, 56, 119, 132, 228, 239, 241-242, 244, 246, 254
Education for Sustainability 119, 142
Education for Sustainable Development 106, 118-119, 123, 135-136, 260
Educational Innovative Strategies 254
Educational Strategies and Methods 44
Environmental education 79, 81, 83, 85-87, 89, 97, 118
exclusion 147-149, 161

F

Field Trip 79-81, 86-87, 89-97, 100
Fieldwork 15, 52, 54, 60-61, 79, 87, 92, 94, 100, 265, 274, 277
Flipped Classroom 31, 124, 135
Fluvial Ecology 254
Fluvial Function 254
Fluvial Geology and Geomorphology 254
Fluvial Heritage 238, 254
Fluvial Hydrology 254
Fluvial Landscape 228, 254
Functional River Delineation 254
future-proofing 102

G

Geographic Information Systems (GIS) 126, 135, 201, 223, 250
Geographical Education 23-25, 31, 34, 45, 136, 275
Geographical Imagination 281
Geographical Scenario 51-52
Geography education 24, 80, 84-85, 172, 176, 204
Geography Teaching 3, 51-52, 54, 58, 102, 110, 210-211, 214, 217-218, 220
Geoscenario 51-56, 58, 61-70, 72-73
Global Challenges 23-25, 44
Globalization 82, 117-118, 143, 208

Great Circle 168, 170-174, 180, 185, 191, 201

H

Heritage 68, 117-124, 131, 134-136, 143, 228-229, 233, 238-239, 250, 254
Hydrology 228, 239-241, 246, 254

I

Inquiry-Based Learning (IBL) 2, 21

K

Knowledge and Skills 31, 60, 106, 119, 150, 267, 281

L

Land Uses 69-70, 228
Landscape Teaching 51-52, 67, 72

M

Map Projection 169-174, 180, 182, 192-193, 195, 201
Mental maps 203-204, 214-218, 220-222, 224, 269
Misconceptions 44, 168-169, 174, 193, 195

O

OSM 144-147, 149-153, 156, 158, 161-162
Outdoor Field Trip 81, 87, 89-90, 92-94, 97, 100

P

Peters Projection 171
Place-Based Education 79-86, 96-97, 100
Primary Education 178, 228-229, 231, 233-234, 245, 248, 258
Primary School 174
Problem-Based Learning (PBL) 2, 22, 51-52, 55

R

Reflective Drawings 268-269
remote learning 102-105, 107, 109-111
Research Methods 1, 7, 13, 92
Rhumb Line 168, 170-174, 180, 185, 191, 202
River Corridor 255
River Delineation 254
River Space 230
Robinson Projection 171, 194

S

School Cartography 206, 210-212
Sense of Place 80-82, 91, 100
Service Learning 13, 134, 143
Spatial Experience 259, 261-264, 268, 274-275, 277, 281
Spatial Thinking 57, 73, 80, 175, 193, 202, 204-209, 218
Spatiality 203-205, 208, 222, 259, 261-263, 269-270, 274-275, 281
Sustainable Development 106, 117-119, 123, 132, 134-136, 260

T

Teaching Geography 1-2, 32, 128, 135, 175, 194, 203-204, 208, 259, 261
Teaching Resources 265

U

user experience 144, 156-158, 161

V

VGI 144-145, 150, 152, 161
Video Content 79, 81, 86-97, 100
Virtual Field Trip 81, 86, 89-96, 100

W

Water Management and Retention 26

www.ingramcontent.com/pod-product-compliance
Lightning Source LLC
LaVergne TN
LVHW061550290926
849162LV00004B/962